Climate Change Impacts on Water Resources

Climate Change Impacts on Water Resources

Edited by Erick West

SYRAWOOD
PUBLISHING HOUSE
New York

Published by Syrawood Publishing House,
750 Third Avenue, 9th Floor,
New York, NY 10017, USA
www.syrawoodpublishinghouse.com

Climate Change Impacts on Water Resources
Edited by Erick West

International Standard Book Number: 978-1-64740-437-6 (Hardback)

Cataloging-in-publication Data

Climate change impacts on water resources / edited by Erick West.
p. cm.
Includes bibliographical references and index.
ISBN 978-1-64740-437-6
1. Water-supply. 2. Water-supply--Climatic factors. 3. Water-supply--Effect of global warming on.
4. Climatic changes. I. West, Erick.
TD353 .C55 2023
628.1--dc23

TABLE OF CONTENTS

PREFACE

This book has been a concerted effort by a group of academicians, researchers and scientists, who have contributed their research works for the realization of the book. This book has materialized in the wake of emerging advancements and innovations in this field. Therefore, the need of the hour was to compile all the required researches and disseminate the knowledge to a broad spectrum of people comprising of students, researchers and specialists of the field.

Water resources refer to the wide range of natural waters that exist on the Earth, irrespective of their form. The waters of the lakes, oceans and rivers are the most accessible sources of usable water. Deep subsurface waters, permanent snowfields, groundwater and glaciers are other sources of accessible water. Changing weather patterns can result in climate change, which in turn can lead to water scarcity, extreme weather events, contaminated water supplies, and uncertain water availability. Such effects of climate change have an impact on the water resources in terms of variability, form, quantity, intensity and timing of precipitation. Furthermore, water quality is also affected by climate change in both coastal and inland regions. This book unravels the recent studies with respect to climate change impacts on water resources. It will also provide interesting topics for research, which interested readers can take up. This book is meant for students who are looking for an elaborate reference text on this subject.

At the end of the preface, I would like to thank the authors for their brilliant chapters and the publisher for guiding us all-through the making of the book till its final stage. Also, I would like to thank my family for providing the support and encouragement throughout my academic career and research projects.

Editor

1

Influence of Climate and Land use Change on the Groundwater System of the Veluwe, The Netherlands: A Historical and Future Perspective

Marjolein H. J. Van Huijgevoort [1,*], Bernard R. Voortman [1,2], Sjoerd Rijpkema [3,†], Kelly H. S. Nijhuis [1,‡] and Jan-Philip M. Witte [1,4]

1 KWR Water Research Institute, 3430BB Nieuwegein, The Netherlands; bernard.voortman@moisture-matters.nl (B.R.V.); kelly3x3@live.nl (K.H.S.N.); flip.witte@ecohydrologie.nl (J.-P.M.W.)
2 Moisture Matters, 3533ED Utrecht, The Netherlands
3 Vitens, 6802CA Arnhem, The Netherlands; S.Rijpkema@waterbedrijfgroningen.nl
4 Flip Witte Ecohydrologie, 6862DC Oosterbeek, The Netherlands
* Correspondence: marjolein.van.huijgevoort@kwrwater.nl
† Current address: Waterbedrijf Groningen, 9723DV Groningen, The Netherlands.
‡ Current address: Waterschap Vallei en Veluwe, 7324AX Apeldoorn, The Netherlands.

Abstract: Changes in land use and climate have a large influence on groundwater recharge and levels. In The Netherlands, precipitation shifts from summer to winter are expected, combined with an increase in summer temperature leading to higher evaporation. These changes in climate could threaten the fresh water supply and increase the importance of large groundwater reservoirs. Sustainable management of these groundwater reservoirs, therefore, is crucial. Changes in land use could help mitigate the effects of climate change by decreasing the evaporation. In this study, we investigate the effect of changes in climate and land use on a large groundwater reservoir in The Netherlands, the Veluwe, for a historical period (1850–2016) and in the future (2036–2065). During the historical period, evaporation increased due to conversions from heather and drift sand to pine forest across the Veluwe. This change in land use had a larger effect on the groundwater recharge than change in climate over the historical period. In the future, an increase in winter precipitation will lead to higher groundwater levels in the elevated parts of the region. Surrounding areas are more vulnerable to an increase in dry periods in the summer. Groundwater reservoirs provide an opportunity to store water during wetter periods, which could alleviate drought impacts in surrounding regions during dry periods. Land use change, such as conversion from pine forest to other land use types, is a possible measure to increase water availability.

Keywords: groundwater; land use; climate change; evaporation

---

1. Introduction

Climate change will affect the availability of fresh water resources in the future (e.g., [1,2]) and will, therefore, have consequences for nature (e.g., [3]), agriculture (e.g., [4]), industry (e.g., [5]) and drinking water supply (e.g., [6]). In The Netherlands, precipitation shifts from summer to winter are expected, combined with increased evaporation demands in summer due to higher temperatures [7]. This can lead to longer and more intense dry spells occurring more often across The Netherlands [7].

Groundwater recharge and levels are impacted by these changes in climate [8–10], but also by changes in land use [11–14]. Actual evaporation (sum of interception, soil evaporation and transpiration) is partly determined by vegetation type. The vegetation, therefore, plays a critical role in groundwater recharge, where, for example, pine forest has a higher evaporation than heather [15,16]. Differences in evaporation between land use types are significant with an actual evaporation of 201 mm/year for drift sand [17], 430 mm/year for heather [18] and 630 mm/year for pine forest [15]. Decisions about land use could thus be important to mitigate climate change effects on groundwater reservoirs.

It is important to consider the impacts of climate change and land use change on strategic groundwater reservoirs, because sustainable management of these groundwater reservoirs is crucial—e.g., in [19]—to ensure freshwater availability for all sectors (e.g., drinking water, nature and agriculture). Groundwater will be affected by climate change, but projected impacts are uncertain [20]. The resilience of groundwater resources to hydrological extremes, compared to other fresh water resources, means that groundwater reservoirs will become even more important in the future [20]. These reservoirs, however, are already being depleted in many regions [21], emphasizing the need for sustainable management.

The Veluwe area is a primary strategic groundwater system in The Netherlands [22] and an important region for drinking water supply [23]. The effects of changes in land use and climate on the water balance of the Veluwe are investigated in this study in a historical and future perspective. The changes that occurred in the past (from 1850 onward) can provide an important lesson for sustainable management of the system in the future.

The aim of this study is to place the hydrological consequences of climate change for the Veluwe in a historical perspective and to compare them with the effects of changes in land use. For this purpose, the effect of climate and land use changes on the groundwater reservoir of the Veluwe are quantified for (i) historical (1850–2016), and (ii) future (2036–2065) conditions. The impact of the changes is simulated with a groundwater model based on historical data and future scenarios.

2. Site Description

The Veluwe area (ca. 1250 km^2) is located in the centre of The Netherlands (Figure 1) and consists of ice-pushed ridges and fluvio–glacial complexes (Figure 2) [22,24,25]. It is an elevated sandy area (Figure 1), which contains a large aquifer of fresh groundwater. On top of the deep aquifer (to 200 m deep) lies a large unsaturated zone up to 60 m thick [22,23]. The unsaturated zone consists mainly of sediments of sands [22]. The soils vary from fine sand to coarse sand [22]. In the ice-pushed ridges, tilted and discontinues clay layers exist that have a large impact on the groundwater table creating a spatially discontinuous pattern [22]. Due to the expanse of the area and its large unsaturated zone, the groundwater system responds slowly to changes in meteorological conditions. Brooks and streams are found at the fringe of the sandy area, where water exfiltrates. Upward seepage occurs in surrounding lower lying agricultural areas and seepage dependent nature areas with a species rich, rare and endangered vegetation (especially *Junco-Molinion* and *Calthion* grasslands and *Circaeo-Alnenion* woodlands). No drainage systems exist in the central part of the area. Most of the area is designated nature reserve and land use mainly consists of forests (predominantly pine), heather and drift sand [26].

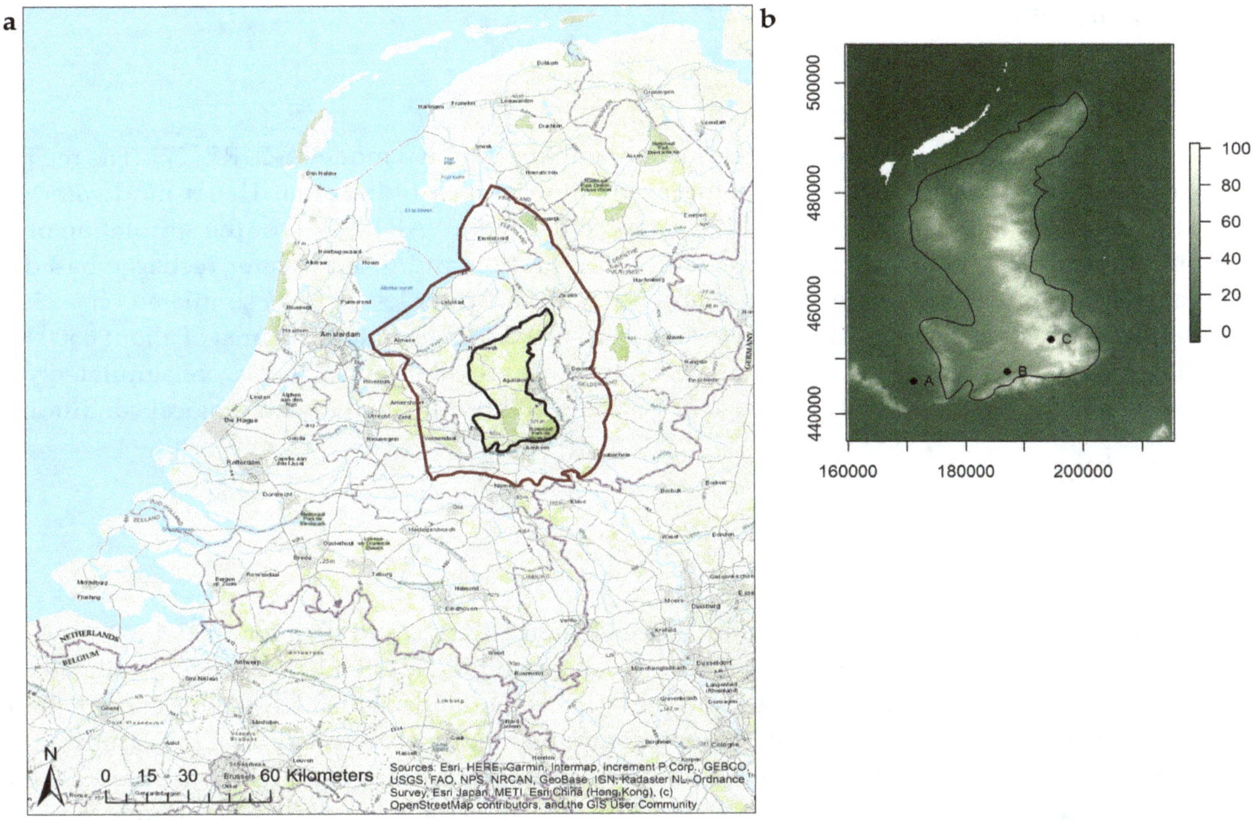

Figure 1. (**a**) Location of the Veluwe area in The Netherlands. The domain of the groundwater model is indicated in red, the Veluwe area in black. (**b**) Elevation (m a.s.l.) of the Veluwe area. Veluwe area is indicated in black and locations of the time series with letters (A: 171154,445864; B: 187208,447643; C: 194583,453461).

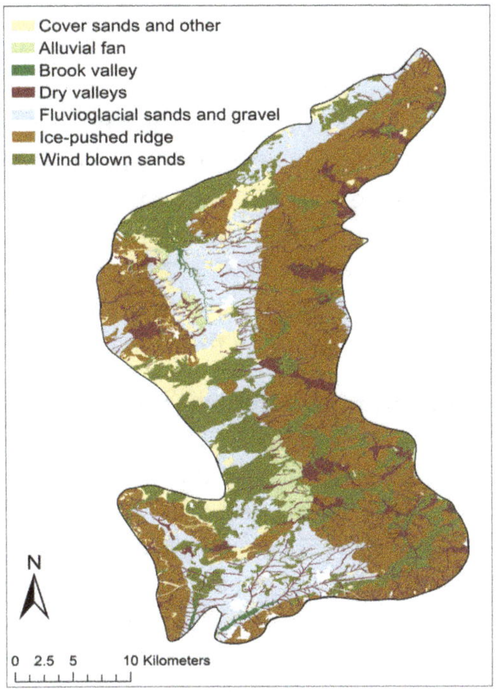

Figure 2. Main geomorphological features of the Veluwe based on Maas et al. [25].

3. Materials and Methods

3.1. Groundwater Model

Groundwater levels at the Veluwe were simulated with the regional model AZURE [27]. The regional model was chosen, because it includes the complex geomorphology of the area. The groundwater part of AZURE is based on MODFLOW [28] and it is coupled to MetaSWAP [29,30] for the simulation of the unsaturated zone. MetaSWAP simulates the actual evaporation and groundwater recharge based on meteorological conditions, soil properties and land use. Groundwater levels were simulated on a 250 by 250 m resolution on a daily basis for the Veluwe area and surroundings, in total an area of about 6500 km^2 (Figure 1). To study the effect of land use changes on the groundwater levels, they were simulated with AZURE for the period 1986–2016 with land use of 1850 and 2008 with equal meteorological conditions.

3.2. Data

3.2.1. Historical Conditions

Land use of the Veluwe area was determined from historical maps and data for 1850, 1900, 1960 and 2008 (Figure 4). For 1850, a digital map for the province Gelderland "Gelderland Histland" is available with historical land use. There are no earlier maps known for The Netherlands, which are reliable enough (based on level of detail and drawn to scale) to allow for a comparison with current land use [31]. Information for 1900 and 1960 is available from Historical Landuse Netherlands maps [32]. The available 1900 map is based on hand drawn historical maps with a scale of 1:50,000, which are digitised and automatically classified [32]. For 1960, topographical maps are available with a scale of 1:25,000. The current land use (2008) is taken from a database with maps of National Land Use Netherlands. This map is based on satellite data from 2007 and 2008 [33]. Information from all these maps was simplified in this study to the following land use types: grass, agriculture and bare soil, heather, broad leaf forest, pine forest, urban area, drift sand, water and other. In this study, land use was fixed for 30 year periods, so the development of vegetation was not taken into account; instead land use change was estimated as jumps between the different periods.

Besides changes in land use, there were also changes in groundwater abstractions during the historical conditions. The groundwater of the Veluwe is important for the drinking water supply in this part of The Netherlands and abstractions started early in the 20th century [23]. The exact amount of abstracted groundwater was not known for the entire period. Therefore, estimated amounts of total abstractions across the Veluwe for 4 specific years were used as reference, respectively, 0, 21, 130 and 130 Mm^3/year for 1895, 1930, 1995 and 2016. These reference numbers were interpolated linearly to create a time series for the entire period. To estimate the effect of the groundwater abstractions, the total was equally divided over the Veluwe area (1250 km^2) and subtracted from the recharge.

Meteorological data (temperature and precipitation) were available from a nearby weather station from the Royal Netherlands Meteorological Institute (KNMI) from 1851 to 2016. Reference evaporation (ET_{ref}) was derived from these data with Hargreaves and Allen [34] for the complete period. In the period 1957–2016, results from this method were calibrated with ET_{ref} values calculated with Makkink [35], because radiation data were available from 1957 onward. Evaporation from interception was calculated with the Rutter interception model [36]. The soil evaporation and transpiration were estimated with a Hydrus-1D model [37]. The Hydrus model was used to simulate different combinations of soil type and land use to a depth of 5 m with a free draining boundary condition, so in a groundwater independent way. The combinations of land use and soil type were based on soil maps and the described historical information of land use. Rooting depth was different for each vegetation type, ranging from 0.3 m for grass to 3 m for broadleaf and pine forest. Total actual evaporation (ET_{act}) is defined as the sum of evaporated interception (results from the interception model), and soil evaporation and transpiration (results from Hydrus model).

The groundwater recharge was calculated for three different scenarios: (1) land use change and groundwater abstractions, (2) land use change only, (3) no land use change or abstractions. Groundwater recharge in the first scenario is defined as precipitation minus evaporation from actual land use and groundwater abstractions. In the second scenario, recharge is calculated as the difference between precipitation and evaporation from actual land use. In the last scenario, recharge is determined as the difference between precipitation and evaporation with constant land use of 1850.

3.2.2. Future

To investigate the impact of climate change around 2050 (2036–2065) on the groundwater reservoir, climate scenarios specifically developed for The Netherlands by the KNMI [38] were used as input for the groundwater model. These climate scenarios are gridded datasets across The Netherlands [38]. The reference period for these scenarios is 1981–2010; input data for this period are based on observations and the existing climate change trend was removed. Four different climate scenarios are available, G_L, G_H, W_L and W_H. The G scenarios ("moderate") represent a low global mean temperature increase, while the W scenarios ("warm") correspond with a stronger temperature increase. The subscripts H and L relate to regional changes in air circulation patterns. The model projections represent a strong response to climate forcing (subscript H, "high value") with wetter winters and drier summers, and a relatively weak response (subscript L, "low value") with smaller changes in precipitation in both seasons. An overview of the changes in the most relevant meteorological variables in the four scenarios for the period 2036–2065, averaged over The Netherlands, is given in Table 1. For the Veluwe, all scenarios show an increase in winter precipitation, where the W_H and G_H scenarios show the largest increase (Figure 3a). In the summer months (June, July, August) the scenarios differ from each other. The W_H and G_H scenarios indicate a clear decrease in precipitation, whereas the G_L scenario predicts a slight increase in precipitation. The W_L scenario indicates an increase in June, precipitation similar to the reference period in July and a decrease in August. For ET_{ref}, all scenarios predict an increase in the period May–August (Figure 3b) with the W_H scenario predicting the largest increase. Changes in land use were not considered for this period.

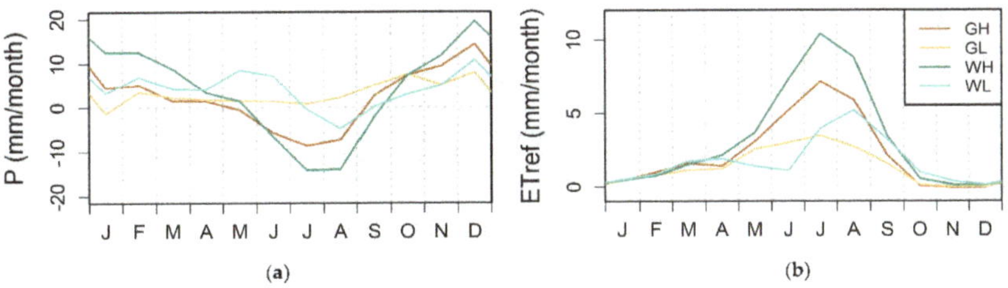

Figure 3. (**a**) Difference in monthly precipitation (P, mm) and (**b**) reference evaporation (ET_{ref}, mm) between the four scenarios and the reference period (1981–2010) averaged over 30 years for one representative grid cell at the centre of the Veluwe.

Table 1. Overview of the changes in the most relevant meteorological variables according to the climate change scenarios for the period 2036–2065 [38].

Variable	Indicator	Climate (1981–2010)	GL	GH	WL	WH
Temperature	Mean	10.1 °C	+1.0 °C	+1.4 °C	+2.0°C	+2.3 °C
Precipitation	Mean amount	851 mm	+4%	+2.5%	+5.5%	+5%
Solar radiation	Solar radiation	354 kJ/cm^2	+0.6%	+1.6%	−0.8%	+1.2%

4. Results

4.1. Analysis Historical Period

4.1.1. Historical Conditions

Considerable changes in land use occurred across the Veluwe between 1850 to 2008 (Figure 4). Heather decreased with 30% of the total area and drift sand with 26% of the total area. These land use types were replaced mainly with pine forest, which covered 44% of the total area in 2008. Due to the increase in population in the second half of the 19th century, available land had to be used in more productive ways [39]. In that period and in the beginning of the 20th century, heather areas were transformed to agricultural land where possible, but mainly into pine forests [39]. Especially in the drier areas of the Veluwe, the planting of pine forest took place on a large scale early in the 20th century [39].

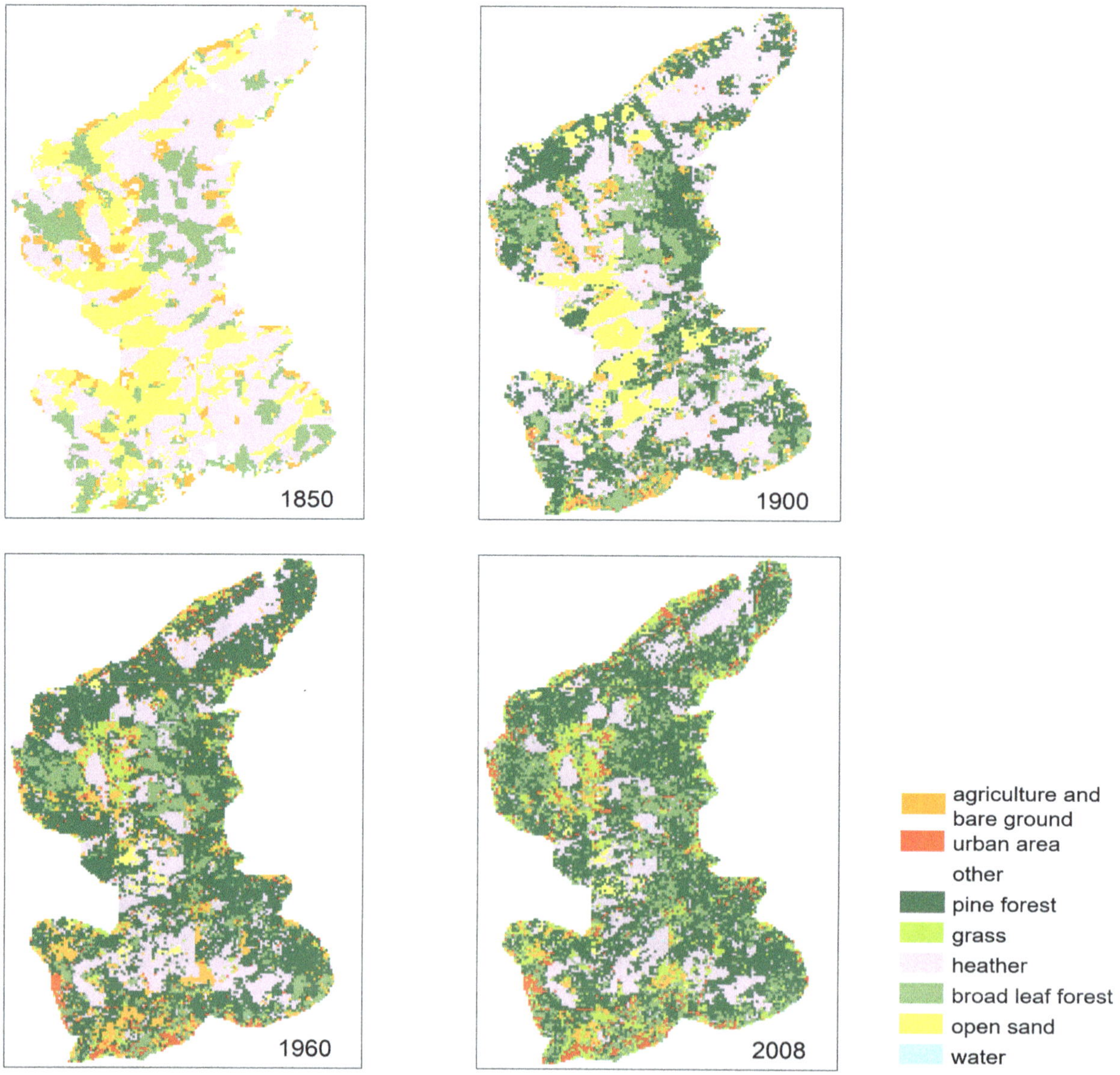

Figure 4. Historical land use across the Veluwe.

Besides changes in land use, climate changed as well. Precipitation and temperature observations were available from a nearby weather station from KNMI. Both precipitation and temperature increased from 1851 to 2016 (Figure 5). Mean yearly precipitation in the period 1851–1880 was 697 mm/year and increased to 845 mm/year in the period 1987–2016. In the period 1851–1880 mean yearly temperature was 9.6 °C and temperature increased to 10.3 °C in the period 1987–2016.

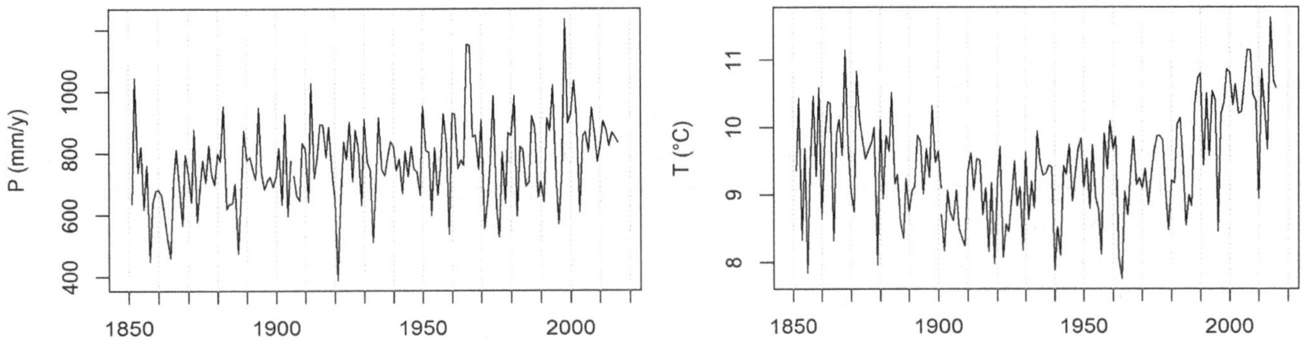

Figure 5. Yearly precipitation (P) and mean yearly temperature (T) for 1851–2016 based on observations.

4.1.2. Impact of Changes in Historical Period

ET_{act} was simulated for the period 1850–2016 for a scenario with land use change and without land use change (Figure 6). During the whole period, ET_{act} increased when land use change was taken into account. In the period 1987–2016, mean yearly ET_{act} for the entire Veluwe area was 153 mm/year (38.9%) higher than mean yearly ET_{act} in the period 1851–1880, due to the combination of changes in land use and climate (Figure 6). When land use changes are not taken into account, by keeping land use of the year 1850 constant over the whole period, ET_{act} increased with 35 mm/year (9.4%) between 1851–1880 and 1987–2016 (Figure 6). The increase in ET_{act} over the historical period was thus mainly caused by land use changes. The increase in temperature and precipitation over time (Figure 5) had a small influence on the ET_{act}. The change in pine forest over the Veluwe area in the historical period dominated the increase in ET_{act} (Figure 7). Heather and drift sand, with a lower ET_{act} compared to pine forest, decreased in area, hence total ET_{act} over the region increased.

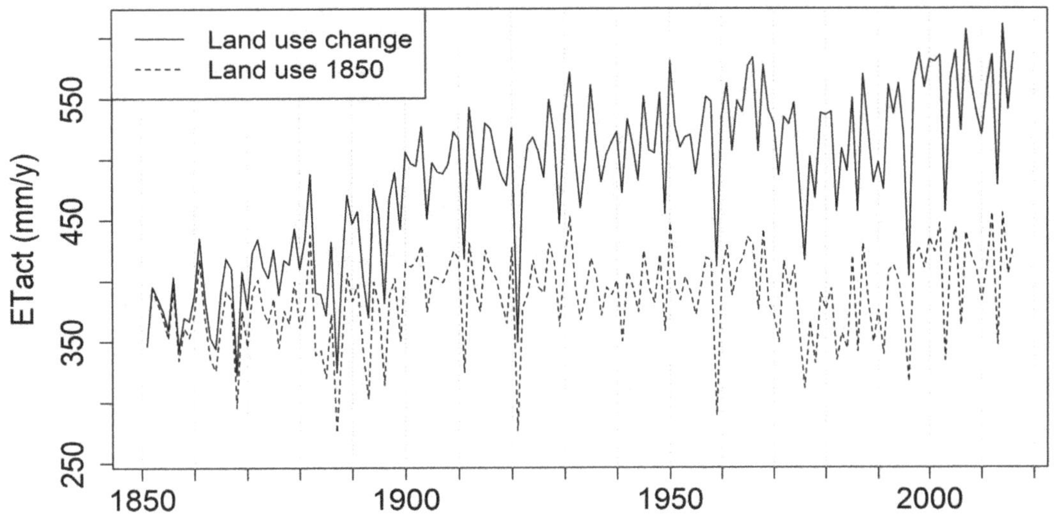

Figure 6. Mean yearly ET_{act} simulated with land use change and with a constant land use from 1850.

Groundwater recharge was affected by these changes in ET_{act} (Figure 8) as well as by groundwater abstractions in the region. Recharge was determined for three different scenarios: (1) land use change and groundwater abstractions, (2) land use change only, (3) no land use change or abstractions. The results will be discussed in this order. During the historical period, the estimated yearly groundwater recharge decreased by 101 mm/year (33%) for the scenario with actual land use, climate and groundwater abstractions between 1851–1880 and 1987–2016. When groundwater abstractions were excluded from the recharge estimate, the groundwater recharge decreased in the beginning of the 20th century with 37 mm/year (12%) in 1911–1940 compared to 1851–1880. After this period, recharge estimates increased again until 1987–2016, during which estimated average yearly recharge was equal to the recharge in the historical period 1851–1880, namely 306 mm/year. The increase in precipitation over the whole period (148 mm/year) was almost equal to the increase in ET_{act} (153 mm/year) due to changes in land use and climate. Without land use change, estimated groundwater recharge increased by 118 mm/year (36%) between 1851–1880 and 1987–2016. The estimated groundwater recharge in scenario 3 (no land use change) in 1987–2016 was 142 mm/year higher than in scenario 2 (land use change only). When including the groundwater abstractions (scenario 1), estimated recharge decreased by 101 mm/year compared to scenario 2 (land use change only). Land use change, therefore, caused the largest decrease in groundwater recharge over the whole period. After the period of large pine plantations in the beginning of the 20th century, however, groundwater abstractions had a larger influence than land use change.

This decrease in groundwater recharge over time due to land use change and water abstractions led to lower groundwater levels over the whole region (Figure 9). If land use would have been constant from 1851 onward, current mean groundwater levels (based on a 30 year period) would be higher across the Veluwe. There are, however, large differences due to the presence of clay and loam layers in the ice-pushed ridges across the region (Figure 2). In the highest parts of the Veluwe, mean groundwater levels would have been more than 10 m higher with a maximum difference of 13.5 m (Figure 9). At the fringes of the Veluwe, on the other hand, only small increases are found and even a decrease up to 0.7 m in a valley in the northern region of the Veluwe (Figure 9). In these regions, vegetation apparently has sufficient water by capillary flow from the shallow groundwater table to evaporate at a potential rate.

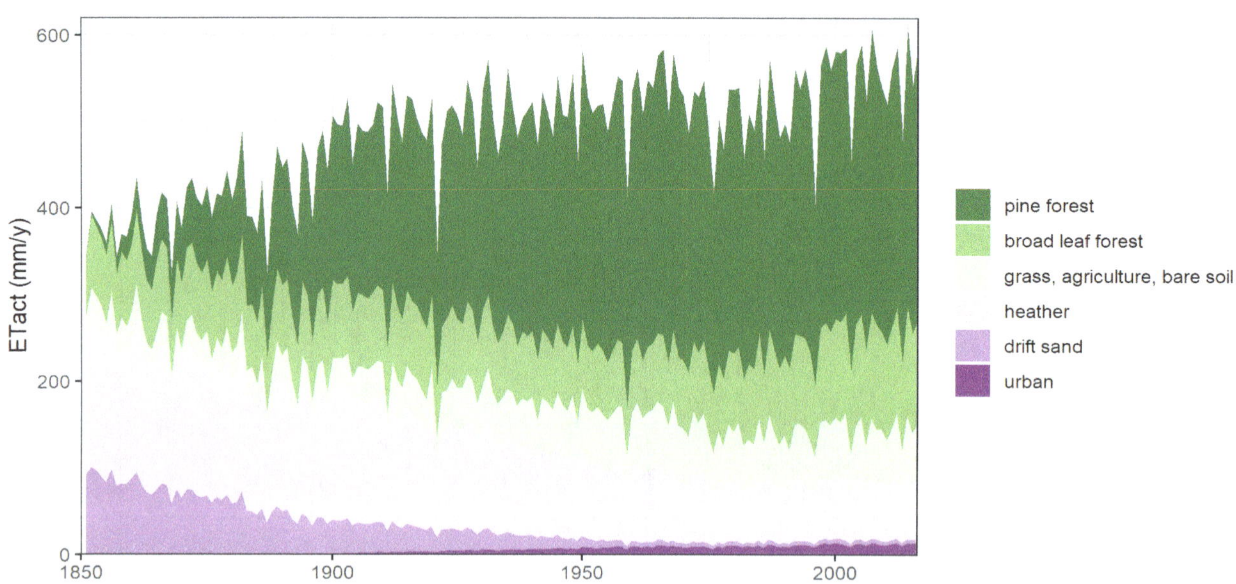

Figure 7. Mean yearly ET_{act} for the entire Veluwe area with the contribution of every land use type.

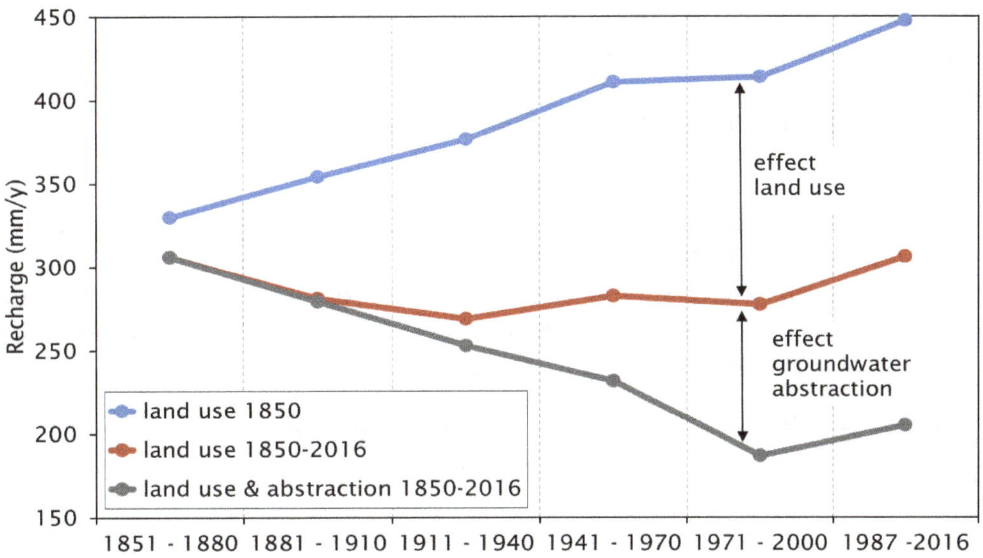

Figure 8. Groundwater recharge for the historical period based on three different scenarios: scenario with actual evaporation (ET_{act}) based on land use 1850, scenario with ET_{act} based on actual land use, scenario with ET_{act} with actual land use and including groundwater abstractions.

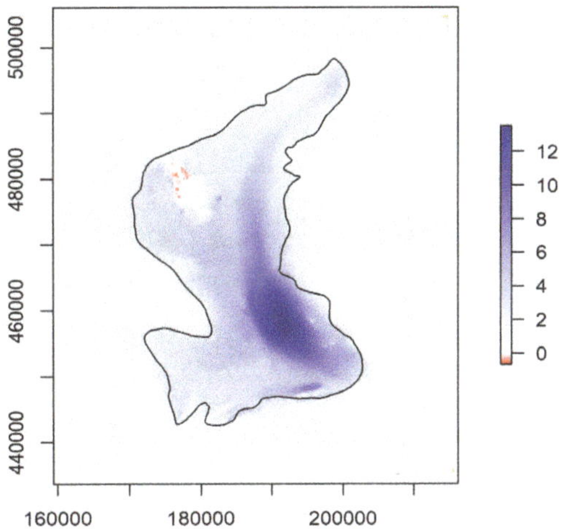

Figure 9. Difference in mean groundwater level (m) between historical land use (1850) and current land use (2008) simulated over 1986–2016. The boundary of the Veluwe area is indicated in black.

4.2. Analysis Future Period

Groundwater levels were simulated for the period 2036–2065 with four different KNMI scenarios (G_L, G_H, W_L and W_H). The differences in simulated mean groundwater levels between the scenarios and the reference period over the whole Veluwe area are given in Figure 10. Across the Veluwe, all scenarios indicate an increase in the mean groundwater levels, with a maximum increase of 4.8 m for the W_H scenario. The eastern ice-pushed ridge of the Veluwe showed the largest increase in the mean groundwater head, whereas lower lying areas showed small changes.

The change in mean groundwater levels across the Veluwe can be explained by the importance of winter precipitation in this region in combination with the uncertainty in potential and actual evaporation. The increase in winter precipitation in all scenarios leads to more recharge and higher groundwater levels, because no extra water is lost through evaporation. When soil moisture is limited, the soil will not be able to deliver enough water to meet the higher potential evaporation demands of the vegetation. This phenomenon especially occurs in coarse textured sandy soils with a groundwater level deeper than approximately 2 m below root zone depth, so capillary flow from the groundwater level to the root zone is absent.

The amount of precipitation in winter is thus very important for the recharge of the groundwater reservoir. This is also illustrated by the difference in reaction between the W_L and W_H scenario. The total increase in precipitation for these scenarios is similar (Table 1), yet the maximum increase in groundwater levels for the W_L scenarios is only 2.8 m versus 4.8 m in the W_H scenario. For the W_L scenario, precipitation increased more evenly over the year, not only in winter.

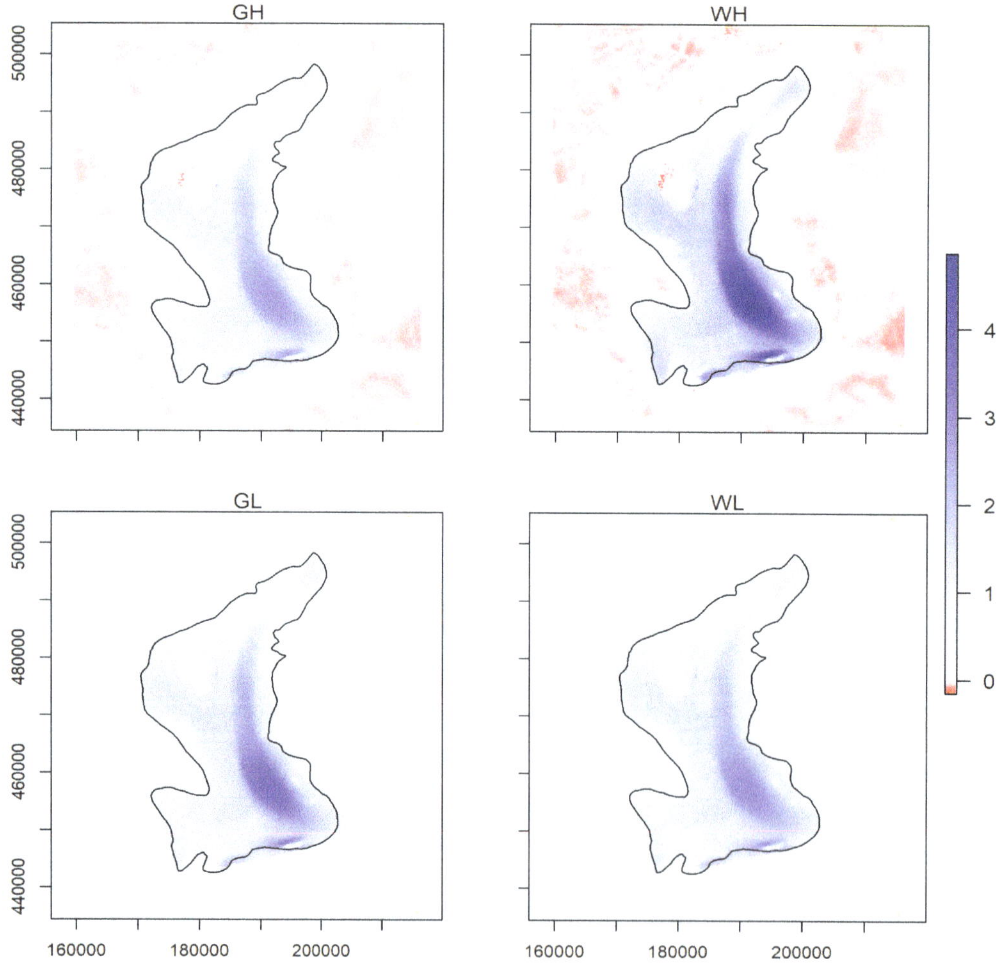

Figure 10. Changes in mean groundwater levels (m) between the climate change scenarios (2036–2065) and the reference period (1981–2010). The boundary of the Veluwe area is indicated in black.

The groundwater levels simulated with the G_H and W_H scenarios decreased in the surrounding agricultural areas of the Veluwe. In contrast to the elevated areas, these low-lying regions are thus prone to water shortages, especially in summer. These water shortages increase in the G_H and W_H scenarios (Figure 10). The different response to climate change of the low-lying regions is caused by the effect of shallow groundwater levels and the surface water levels in these regions. The shallow groundwater levels enable capillary supply of groundwater to the root zone in dry periods. On the other hand, the direct link to surface water levels leads to a fast response during periods with rain, because the extra water is directly discharged through the surface water. Differences in soil type (especially water holding capacity) also play an important role.

When focusing on changes over time (Figure 11) at selected locations across the southern part of the Veluwe (Figure 1), the different reaction to changes in climate between the Veluwe itself and the surrounding agricultural fields can be seen again. The elevated parts of the Veluwe show the slow response of the groundwater to interannual changes in recharge and climate change due to the large unsaturated zone and large drainage resistance in these regions (locations B and C). The response in groundwater levels is completely different in the low-lying regions, where groundwater levels are linked to surface levels and phreatic levels react quickly to precipitation (location A). At this location, the difference between the scenarios is mainly found during the summer, when scenarios G_H and W_H can lead to water shortages and thereby to decreases in crop yields.

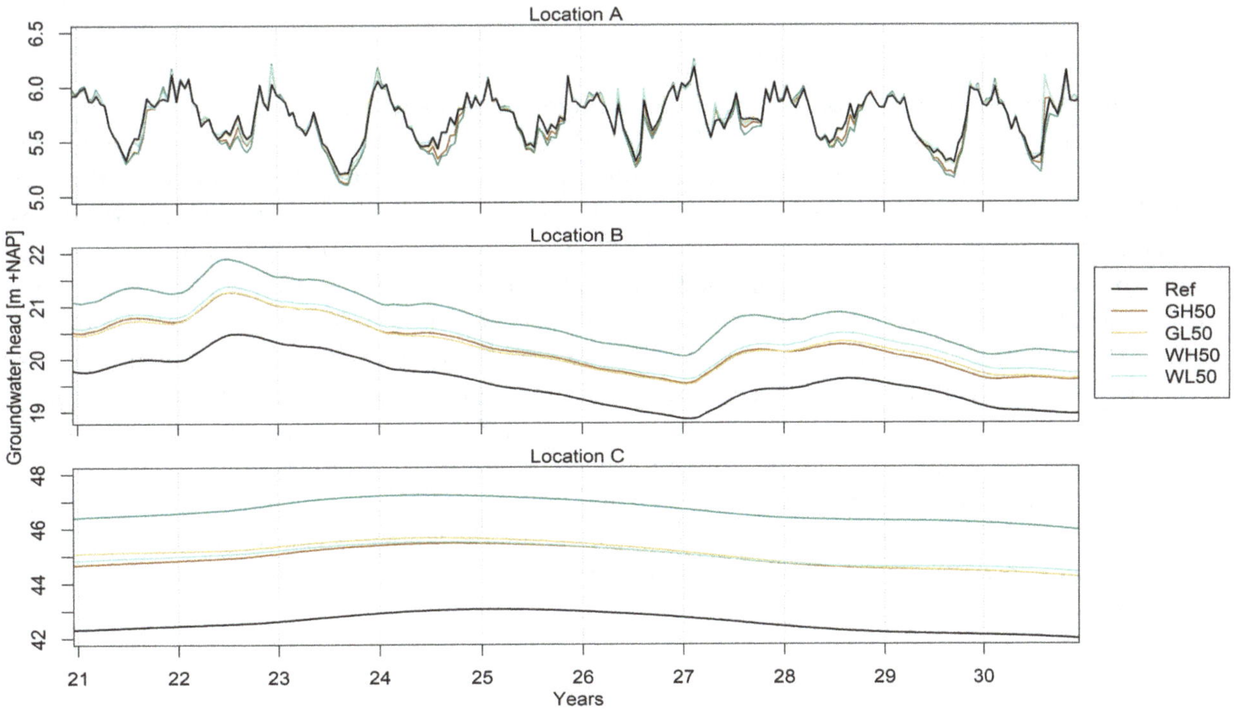

Figure 11. Time series of groundwater levels (m a.s.l) at three selected locations across the southern part of the Veluwe for the last 10 years of the reference period and the climate scenarios.

5. Discussion

5.1. Uncertainty Evaporation

The impact of changes in land use depends mainly on the ET_{act} values of the different vegetation types. It is difficult to measure ET_{act}, therefore, it is often estimated with a model. Here, Hydrus was used to estimate the ET_{act} values. The values for ET_{act} estimated with Hydrus for the different land use types corresponded well with values from literature. Average yearly actual evaporation for heather was slightly underestimated, with 361 mm/year compared to 430 mm/year in the literature [18], whereas evaporation from drift sands was overestimated by Hydrus with 380 mm/year compared to 201 mm/year in the literature [17]. Estimated actual evaporation for pine forest was equal to values in the literature with 622 mm/year compared to 630 mm/year [15]. The overestimation of evaporation from drift sands means change in groundwater recharge is underestimated, because the difference between evaporation from pine forest and drift sands is larger in reality than estimated in this study.

The changes in groundwater levels across the Veluwe area are simulated with AZURE and thus also influenced by the model choice. Recharge estimates depend on an accurate simulation of the ET_{act}, especially in dry conditions. Groundwater levels were too low to provide vegetation with enough capillary flow to maintain high ET values; therefore, ET_{act} did not increase during the summer months. The response of forest to dry conditions, however, is still uncertain. The ability of forest to regulate stomatal opening and the large rooting depth lead to the conservative character and persistence of evapotranspiration of forests during heatwaves [40]. Especially the persistence of evapotranspiration during dry conditions might be underestimated by the model. In order to determine the influence of climate change on the vegetation across the Veluwe in more detail, it is very important to measure the ET_{act} for different vegetation types during dry conditions, such as during the summer of 2018.

5.2. *Scenarios for the Future*

This study focused on the impact of land use change in the historical analysis and on the impact of climate change in the analysis for the future. The KNMI scenarios only include the changes in climate, so changes in land use and water use were not taken into account for the future. In recent years, the pressure on groundwater resources has already increased in The Netherlands. During the 2018 drought, water usage increased, both for drinking water and irrigation in The Netherlands [41]. Since these drought events are expected to occur more often in the future [42], it can be assumed that water use will increase. This also affects the groundwater levels, and projected increases in these levels based on climate change only will be too positive. Extensive changes in land use across the Veluwe are not expected because the region is a nature area (Natura 2000). This means certain regions and habitats are protected and land use may not be changed. Conversions from pine forest to broadleaf forest, however, are possible. This research focused on the impact of climate change only and therefore excluded these developments. For a more realistic estimate of groundwater levels in the future, other scenarios could be used in further research. For The Netherlands, Delta scenarios were developed that include changes in land use and socio-economic changes [43]. These scenarios include a range in possible developments—for example, expecting either much more or less water availability challenges.

5.3. *Relation to Previous Studies*

Several previous studies have taken into account changes in land use or climate across the Veluwe. The impact of land use changes on the groundwater levels was investigated by Gehrels [22]. In this study, hypothetical scenarios were simulated in which all coniferous forests were replaced with deciduous forest for the Veluwe or where forests were only converted in a small area close to a groundwater abstraction site. In both cases, the conversion had a significant effect on groundwater levels. Gehrels [22] showed that the impact of conversion in the smaller area was of comparable magnitude to the nearby abstraction. Land use changes that occurred during the historical period had even higher impacts, because changes in recharge are larger for the conversion between heather/drift sands to coniferous forest than for conversions of forests only.

Evaluating the effect of converting part of the pine forest of the Veluwe to heather and drift sands was also done by Aus der Beek et al. [44]. Again a significant increase in groundwater levels was found in the region that was converted. The converted region was hypothetical to estimate possibilities to increase recharge in the future as part of the sustainable management of groundwater. In this study, the historical, actual changes to land use and their impacts have been shown.

Climate change across the Veluwe was taken into account by Van Engelenburg et al. [23], who evaluate the scenario G_H. The changes in groundwater levels for this scenario from this study are similar to changes found by Van Engelenburg et al. [23]. They also indicated an increase in mean groundwater levels across the Veluwe. In this study, changes from all four climate scenarios were determined to include uncertainty in the predictions. The groundwater model in this study was updated since that in Van Engelenburg et al. [23] with a new parameterisation for heather [18]. Besides these differences, this study has shown the importance of land use changes compared to changes in climate and water abstractions.

6. Conclusions

The impacts of changes in land use and climate on a large strategic groundwater reservoir in The Netherlands, the Veluwe, were investigated for historical conditions (1851–2016) and future conditions (2036–2065). During the historical period, groundwater recharge and levels across the Veluwe decreased from 1850 onward. Over the whole period, land use change caused the largest decrease in groundwater recharge. In recent years, the groundwater abstraction contributed more to the decrease in groundwater recharge. Changes in climate would have resulted in higher groundwater recharge, but impacts of land use and abstractions were larger. Even though the impact of land use change depends on the actual evaporation, which is still not completely understood and depends on the chosen model, the change in recharge caused by the pine plantations was large enough to conclude that land use change had more impact than abstractions or climate for 1851–2016.

In future conditions, groundwater levels increased across the elevated parts of the Veluwe due to the predicted increase in winter precipitation. In the surrounding low-lying agricultural areas, drier conditions in summer led to lower groundwater levels in the G_H and W_H scenario. These regions could therefore be vulnerable to water shortages.

This study shows the importance of land use change compared to the effect of climate. The strategic groundwater reservoir of the Veluwe will be important in the future to alleviate drought impacts in surrounding areas. For sustainable management of the groundwater reservoir, land use change measures could be taken to increase the storage in the elevated parts and thereby mitigate the impact of climate change in low-lying regions even more.

Author Contributions: Conceptualization, M.H.J.V.H., B.R.V. and J.-P.M.W.; methodology, M.H.J.V.H., B.R.V., S.R. and J.-P.M.W.; software, S.R. and K.H.S.N.; formal analysis, M.H.J.V.H., B.R.V., S.R., K.H.S.N. and J.-P.M.W.; data curation, M.H.J.V.H., S.R. and K.H.S.N.; writing—original draft preparation, M.H.J.V.H.; writing—review and editing, M.H.J.V.H., B.R.V., S.R., K.H.S.N. and J.-P.M.W.; visualization, M.H.J.v.H and K.H.S.N. All authors have read and agreed to the published version of the manuscript.

Abbreviations

ET_{ref}	reference evaporation
ET_{act}	actual evaporation (sum of evaporated interception, soil evaporation and transpiration)
G_L	moderate climate scenario with weak response to changes in air circulation patterns
G_H	moderate climate scenario with strong response to changes in air circulation patterns
KNMI	Royal Netherlands Meteorological Institute
P	precipitation
W_L	warm climate scenario with weak response to changes in air circulation patterns
W_H	warm climate scenario with strong response to changes in air circulation patterns

References

1. Gosling, S.N.; Arnell, N.W. A global assessment of the impact of climate change on water scarcity. *Clim. Chang.* **2016**, *134*, 371–385. [CrossRef]
2. Schewe, J.; Heinke, J.; Gerten, D.; Haddeland, I.; Arnell, N.W.; Clark, D.B.; Dankers, R.; Eisner, S.; Fekete, B.M.; Colón-González, F.J.; et al. Multimodel assessment of water scarcity under climate change. *Proc. Natl. Acad. Sci. USA* **2014**, *111*, 3245–3250. [CrossRef] [PubMed]
3. Witte, J.P.M.; Runhaar, J.; Van Ek, R.; Van Der Hoek, D.C.J.; Bartholomeus, R.P.; Batelaan, O.; van Bodegom, P.M.; Wassen, M.J.; Van der Zee, S.E.A.T.M. An ecohydrological sketch of climate change impacts on water and natural ecosystems for The Netherlands: Bridging the gap between science and society. *Hydrol. Earth Syst. Sci.* **2012**, *16*, 3945–3957. [CrossRef]
4. Kummu, M.; Gerten, D.; Heinke, J.; Konzmann, M.; Varis, O. Climate-driven interannual variability of water scarcity in food production potential: A global analysis. *Hydrol. Earth Syst. Sci.* **2014**, *18*, 447–461. [CrossRef]
5. Van Vliet, M.T.H.; Yearsley, J.R.; Ludwig, F.; Vögele, S.; Lettenmaier, D.P.; Kabat, P. Vulnerability of US and European electricity supply to climate change. *Nat. Clim. Chang.* **2012**, *2*, 676–681. [CrossRef]
6. Delpla, I.; Jung, A.V.; Baures, E.; Clement, M.; Thomas, O. Impacts of climate change on surface water quality in relation to drinking water production. *Environ. Int.* **2009**, *35*, 1225–1233. [CrossRef]
7. Klein Tank, A.; Beersma, J.; Bessembinder, J.; Van den Hurk, B.; Lenderink, G. *KNMI'14 Climate Scenarios for The Netherlands*; Technical Report; KNMI: De Bilt, The Netherlands, 2014. Available online: http://projects.knmi.nl/publications/showAbstract.php?id=10756 (accessed on 5 October 2020).
8. Kløve, B.; Ala-Aho, P.; Bertrand, G.; Gurdak, J.J.; Kupfersberger, H.; Kværner, J.; Muotka, T.; Mykrä, H.; Preda, E.; Rossi, P.; et al. Climate change impacts on groundwater and dependent ecosystems. *J. Hydrol.* **2014**, *518*, 250–266. [CrossRef]
9. Kundzewicz, Z.W.; Döll, P. Will groundwater ease freshwater stress under climate change? *Hydrol. Sci. J.* **2009**, *54*, 665–675. [CrossRef]
10. Green, T.R.; Taniguchi, M.; Kooi, H.; Gurdak, J.J.; Allen, D.M.; Hiscock, K.M.; Treidel, H.; Aureli, A. Beneath the surface of global change: Impacts of climate change on groundwater. *J. Hydrol.* **2011**, *405*, 532–560. [CrossRef]
11. Allen, A.; Chapman, D. Impacts of afforestation on groundwater resources and quality. *Hydrogeol. J.* **2001**, *9*, 390–400. [CrossRef]
12. Scanlon, B.R.; Reedy, R.C.; Stonestrom, D.A.; Prudic, D.E.; Dennehy, K.F. Impact of land use and land cover change on groundwater recharge and quality in the southwestern US. *Glob. Chang. Biol.* **2005**, *11*, 1577–1593. [CrossRef]
13. Dams, J.; Woldeamlak, S.T.; Batelaan, O. Predicting land-use change and its impact on the groundwater system of the Kleine Nete catchment, Belgium. *Hydrol. Earth Syst. Sci.* **2008**, *12*, 1369–1385. [CrossRef]
14. Bosmans, J.H.C.; van Beek, L.P.H.; Sutanudjaja, E.H.; Bierkens, M.F.P. Hydrological impacts of global land cover change and human water use. *Hydrol. Earth Syst. Sci.* **2017**, *21*, 5603–5626. [CrossRef]
15. Dolman, H.; Moors, E.; Elbers, J.; Snijders, W.; Hamaker, P. *Het Waterverbruik Van Bossen in Nederland*; Technical Report; Alterra: Wageningen, The Netherlands, 2000. Available online: http://www.climatexchange.nl/projects/boshydrologie/Waterverbruik_Bossen.pdf (accessed on 5 October 2020).
16. Voortman, B.R.; Bartholomeus, R.P.; van der Zee, S.E.A.T.M.; Bierkens, M.F.P.; Witte, J.P.M. Quantifying energy and water fluxes in dry dune ecosystems of the Netherlands. *Hydrol. Earth Syst. Sci.* **2015**, *19*, 3787–3805. [CrossRef]
17. Stuyfzand, P.J. Hydrochemistry and Hydrology of the Coastal Dune Area of the Western Netherlands. Ph.D. Thesis, Vrije Universiteit Amsterdam, Amsterdam, The Netherlands, 1993.
18. Voortman, B.; van Huijgevoort, M.; Witte, J.P. *Verdamping Van Natuurterreinen Berekend Met AZURE. De Parametrisatie Van Heide op Basis Van Lysimetermetingen en een Vergelijking Met Literatuurcijfers*; Technical Report KWR2019.015; KWR: Nieuwegein, The Netherlands, 2019.
19. Gleeson, T.; Alley, W.M.; Allen, D.M.; Sophocleous, M.A.; Zhou, Y.; Taniguchi, M.; VanderSteen, J. Towards Sustainable Groundwater Use: Setting Long-Term Goals, Backcasting, and Managing Adaptively. *Groundwater* **2012**, *50*, 19–26. [CrossRef]
20. Taylor, R.G.; Scanlon, B.; Döll, P.; Rodell, M.; van Beek, R.; Wada, Y.; Longuevergne, L.; Leblanc, M.; Famiglietti, J.S.; Edmunds, M.; et al. Ground water and climate change. *Nat. Clim. Chang.* **2013**, *3*, 322–329. [CrossRef]

21. Bierkens, M.F.P.; Wada, Y. Non-renewable groundwater use and groundwater depletion: A review. *Environ. Res. Lett.* **2019**, *14*, 063002. [CrossRef]
22. Gehrels, J.C. Groundwater Level Fluctuations; Separation of Natural from Anthropogenic Influences and Determination of Groundwater Recharge in the Veluwe Area. Ph.D. Thesis, Vrije Universiteit Amsterdam, Amsterdam, The Netherlands, 1999.
23. Van Engelenburg, J.; Hueting, R.; Rijpkema, S.; Teuling, A.J.; Uijlenhoet, R.; Ludwig, F. Impact of Changes in Groundwater Extractions and Climate Change on Groundwater-Dependent Ecosystems in a Complex Hydrogeological Setting. *Water Resour. Manag.* **2018**, *32*, 259–272. [CrossRef]
24. van Alphen, H.J.; Alves, E.; aus der Beek, T.; Bruggeman, A.; Camera, C.; Fohrmann, R.; Fortunato, A.; Freire, P.; Iacovides, A.; Iacovides, I.; et al. *Characterization of the Catchments and the Water Systems*; Technical Report, BINGO Deliverable 3.1; 2016. Available online: http://www.projectbingo.eu/downloads/BINGO_Deliverable3.1_EC.pdf (accessed on 13 October 2020).
25. Maas, G.J.; van der Meij, W.M.; van Delft, S.P.J.; Heidema, A.H. *Toelichting bij de Legenda Geomorfologische Kaart Van Nederland 1:50,000 (2019)*; Wageningen Environmental Research: Wageningen, The Netherlands, 2019.
26. Nijhuis, K. Historisch Landgebruik Van de Veluwe in Hydrologisch Perspectief. Master's Thesis, Wageningen University: Wageningen, The Netherlands, 2017.
27. De Lange, W.; Borren, W. *Grondwatermodel AZURE versie 1.0. Actueel Instrumentarium voor de Zuiderzee Regio*; Technical Report; Deltares: Delft, The Netherlands, 2014.
28. Harbaugh, A.W. *MODFLOW-2005, the U.S. Geological Survey Modular Ground-Water Model—The Ground-Water Flow Process*; Technical Report; USGS. 2005. Available online: https://pubs.er.usgs.gov/publication/tm6A16 (accessed on 13 October 2020).
29. van Walsum, P.E.V.; Groenendijk, P. Quasi steady-state simulation of the unsaturated zone in groundwater modeling of lowland regions. *Vadose Zone J.* **2008**, *7*, 769–781. [CrossRef]
30. van Walsum, P.E.V.; Veldhuizen, A.A.; Groenendijk, P. *SIMGRO 7.2.0, Theory and Model Implementation*; Technical Report; Alterra: Wageningen, The Netherlands, 2011. Available online: http://nhi.nu/nl/files/8814/2662/2842/Handleiding_SIMGRO_7.2.0_Engels_Report_913_1_V7_2_0.pdf (accessed on 13 October 2020).
31. Koeman, C. *Geschiedenis Van de Kartografie Van Nederland: Zes Eeuwen Land-en Zeekarten en Stadsplattegronden*; Canaletto: Alphen aan den Rijn, The Netherlands, 1983.
32. Knol, W.C.; Kramer, H.; Gijsbertse, H. *Historisch Grondgebruik Nederland; een Landelijke Reconstructie van het Grondgebruik Rond 1900*; Alterra-Rapport 573; Alterra: Wageningen, The Netherlands, 2004. Available online: https://www.wur.nl/nl/Publicatie-details.htm?publicationId=publication-way-333238323739 (accessed on 13 October 2020).
33. Hazeu, G.W.; Schuiling, C.; Dorland, G.J.; Oldengarm, J.; Gijsbertse, H.A. *Landelijk Grondgebruiksbestand Nederland versie 6 (LGN6); Vervaardiging, Nauwkeurigheid en Gebruik*; Alterra-Rapport 2012; Alterra: Wageningen, The Netherlands, 2010. Available online: https://www.wur.nl/nl/Publicatie-details.htm?publicationId=publication-way-333931373630 (accessed on 13 October 2020).
34. Hargreaves, G.H.; Allen, R.G. History and evaluation of Hargreaves evapotranspiration equation. *J. Irrig. Drain. Eng.* **2003**, *129*, 53–63. [CrossRef]
35. Makkink, G.F. Testing the Penman formula by means of lysimeters. *J. Inst. Water Eng.* **1957**, *11*, 277–288.
36. Rutter, A.; Kershaw, K.; Robins, P.; Morton, A. A predictive model of rainfall interception in forests, 1. Derivation of the model from observations in a plantation of Corsican pine. *Agric. Meteorol.* **1971**, *9*, 367–384. [CrossRef]
37. Šimůnek, J.; Šejna, M.; Saito, H.; Sakai, M.; Van Genuchten, M.T. *The HYDRUS-1D Software Package for Simulating the Movement of Water, Heat, and Multiple Solutes in Variably Saturated Media*, Version 4.08; HYDRUS Software Series 3; Technical Report; Department of Environmental Sciences, University of California Riverside: Riverside, CA, USA, 2008.
38. KNMI. *KNMI'14: Climate Change Scenarios for the 21st Century—A Netherlands Perspective*; Technical Report; KNMI: De Bilt, The Netherlands, 2014.
39. Neefjes, J. *Landschapsbiografie Van de Veluwe: Historisch-Landschappelijke Karakteristieken en hun Ontstaan*; Rijksdienst voor het Cultureel Erfgoed and Staatsbosbeheer, Amersfoort, The Netherlands, 2018. Available online: https://library.wur.nl/WebQuery/groenekennis/2284674 (accessed on 13 October 2020).
40. Teuling, A.J.; Seneviratne, S.I.; Stöckli, R.; Reichstein, M.; Moors, E.; Ciais, P.; Luyssaert, S.; van den Hurk, B.; Ammann, C.; Bernhofer, C.; et al. Contrasting response of European forest and grassland energy exchange to heatwaves. *Nat. Geosci.* **2010**, *3*, 722–727. [CrossRef]
41. van den Eertwegh, G.; Bartholomeus, R.; de Louw, P.; Witte, F.; van Dam, J.; van Deijl, D.; Hoefsloot, P.; Clevers, S.;

Hendriks, D.; van Huijgevoort, M.; et al. *Droogte in Zandgebieden Van Zuid-, Midden-en Oost-Nederland: Rapportage Fase 1: Ontwikkeling Van Uniforme Werkwijze Voor Analyse Van Droogte en Tussentijdse Bevindingen*; Technical Report; 2019. Available online: https://edepot.wur.nl/511196 (accessed on 13 October 2020).

42. Philip, S.Y.; Kew, S.F.; van der Wiel, K.; Wanders, N.; van Oldenborgh, G.J. Regional differentiation in climate change induced drought trends in The Netherlands. *Environ. Res. Lett.* **2020**. [CrossRef]
43. Wolters, H.A.; van den Born, G.J.; Dammers, E.; Reinhard, S. *Deltascenario's Voor de 21e Eeuw, Actualisering 2017*; Technical Report; Deltares: Utrecht, The Netherlands, 2018. Available online: https://media.deltares.nl/deltascenarios/Deltascenarios_actualisering2017_hoofdrapport.pdf (accessed on 13 October 2020)
44. Aus der Beek, T.; Alves, E.; Becker, R.; Bruggeman, A.; Fortunato, A.B.; Freire, P.; Gragne, A.; van Huijgevoort, M.H.J.; Iacovides, A.; Iacovides, I.; et al. D3.4 Model Results for Water and Land Use Scenarios Completed and Analysed. Technical Report, BINGO Deliverable 3.4. 2018.

2

Climate Change Impact on Hydropower Resources in Gauged and Ungauged Lithuanian River Catchments

Darius Jakimavičius *, Gintaras Adžgauskas, Diana Šarauskienė and Jūratė Kriaučiūnienė

Laboratory of Hydrology, Lithuanian Energy Institute, Breslaujos st. 3, LT–444003 Kaunas, Lithuania; Gintaras.Adzgauskas@lei.lt (G.A.); Diana.Sarauskiene@lei.lt (D.Š.); jurate.kriauciuniene@lei.lt (J.K.)

* Correspondence: darius.jakimavicius@lei.lt

Abstract: Hydropower (potential and kinetic energy) is one of the most important renewable energy sources in the world. This energy is directly dependent on water resources and the hydrological cycle. Ongoing climate changes are likely to influence the availability/amount of this energy resource. The present study explores the relationship between climate changes and river runoff, projects future runoff in both gauged and ungauged river catchments, and then assesses how these alterations may affect the future hydropower resources in Lithuania. Runoff projections of the gauged rivers were evaluated applying Swedish Department of Climate hydrological model, and runoff of ungauged river catchments were estimated by created isoline maps of specific runoff. According to an ensemble of three climate models and two Representative Concentration Pathway scenarios, runoff and hydroelectric energy projections were evaluated for two future periods (2021–2040, 2081–2100). The results demonstrated a decrease in future river runoff. Especially significant changes are expected according to the most pessimistic RCP8.5 scenario at the end of the century. The projected changes are likely to bring a negative effect on hydropower production in the country. These findings could help understand what kind of benefits and challenges water resource managers may face in the future.

Keywords: HBV; Ungauged rivers; runoff projections; hydropower production

1. Introduction

Growth in the human population and relentless consumption is closely related to an increase in energy demands. However, the destabilized temperature equilibrium of the Earth forces the search for clean energy sources in order to reduce greenhouse gas emissions. Directive (EU) 2018/2001 [1] on the promotion of the use of energy from renewable sources obliges the EU Member States to achieve a 32% share of renewable energy in final energy consumption by 2030. According to Eurostat [2], in 2018, wind and hydropower each contributed around one-third of the total electricity from renewable sources, with wind power (36%) edging hydropower (33%) as the most important sources.

Hydropower (or hydroelectric energy) is the most important and world's largest source of renewable energy. It may harness the power of moving water (kinetic energy) or the water stored in dams (potential energy) and is often considered a very clean form of electricity generation. However, at present, most hydroelectric energy comes from hydropower plants that generate energy by damming of water. According to the International Energy Agency, hydropower accounts for about 17% of global electricity production. China is the largest producer of hydroelectricity, followed by Brazil, the United States, and Canada [3], while Africa has the highest percentage of untapped technical hydropower potential in the world [4].

Hydropower is a source of energy that changes over time because it is directly dependent on water resources and the hydrological cycle. In countries where the amount of electricity generated by hydropower plants is substantial, projections of hydropower potential are considered as a task of national importance. For example, in the United States, where hydropower is a key contributor to the renewable energy portfolio, the Department of Energy, directed by Congress, has conducted a second five-year assessment [5] examining the potential effects of, and risks from, global climate change associated with water supplies for federal hydroelectric power generation. This report indicated that the most critical climate change effect on hydrology is likely to be the trend toward earlier snowmelt and change in runoff seasonality. Under the projections of increasing winter–spring runoff and decreasing summer–fall runoff, water resource managers may need to consider different water use allocations. Rich in water resources, Canada, where national energy security is mostly dependent on the hydropower, is getting warmer and wetter with more contribution from rainfall than snow [6]. Consequently, under the continuation of climatic trends, production potential is expected to increase. However, the net gain/loss is subject to significant variations across different regions. In Norway, a country almost perfectly made for hydropower, a current concern about the future potential of this energy resource resulted in a report [7] that indicated the river flow increase due to the projected higher amounts of precipitation. The most significant increase in river flow occurs in winter, and at the same time, slightly smaller snowmelt floods are expected. By the end of the century (in 2071–2100), according to the RCP8.5 scenario, the total hydropower production in Norway is going to grow by 8% (compared to 1961–1990). In contrast, hydropower plants in Germany are affected negatively due to declining river discharge [8]. A special report from the German Advisory Council on the Environment [9] assumed that the hydropower potential for electricity generation in Germany is limited to about 28 TWh a^{-1} for orographic reasons. Thus, the additional development potential of hydropower is rated to be minimal. For the Alpine region, the average annual electricity generation of run-of-river plants in 2031–2050 (relative to 1961–1990) is estimated to decrease slightly for all climate scenarios considered (up to −8%) [10]. Whereas for Austria, two scenarios project a slight increase (not more than +5%), and according to the other two scenarios a slight decrease (not more than −5%) is expected. Most of the studies assess only the potential energy of water because a vast amount of hydropower is generated in large power plants having water storage reservoirs. However, the research of hydrokinetic energy resources remains limited. The potential of these resources on a national level has been previously assessed in Canada and the continental United States [11,12].

Hydroelectric energy potential varies significantly in time and space. Hydropower systems are dependent on water availability and can either increase competition or mitigate water scarcity [13]. In the long run, the success of hydropower generation and development will largely depend not only on the geographical region, economic opportunities, political approach, and environmental constrains, but also on future climate-induced river runoff changes. The National Energy Independence Strategy of Lithuania [14] stipulates to increase the share of electricity consumption from renewable energy sources up to 45% in 2030 and 80% in 2050 compared to the final electricity consumption. In this context, an assessment of the current and future potential of hydroelectric energy in Lithuania is very important. Jablonskis and Lasinskas [15] were the first to evaluate theoretical potential of hydropower in Lithuania. According to them, over 22,000 rivers (watercourses) with a total length about 77,000 km together with the slopes of the earth's surface produce a total of 688.8 MW of hydropower and about 6.0 billion kWh a^{-1}. However, the exploitation of hydropower resources is severely limited by environmental requirements; after assessing all nature protection restrictions, it is estimated that efficient technical energy will be 159.1 million kWh a^{-1}, which is only 2.6% of the total annual potential [16]. Currently, the total installed capacity of the hydropower plants is 128 MW (15.3% of all RES). This figure indicates the conventional hydropower resources: 101 MW of installed capacity at Kaunas Hydropower Plant (on the Nemunas—the largest Lithuanian river) and the rest at 97 small hydropower plants (on smaller rivers). River hydrokinetic resources, at the state level, have never been assessed. Gailiušis et al. [17] estimated that resources of hydrokinetic energy of Lithuanian small and medium-size rivers amount to 82.1 MW, but due to the specified exclusionary criteria (environmental constrains and weather conditions), the capacity of riverine hydrokinetic energy decreases to 13.6 MW; thus, only 0.7% of the total electric energy demand for the national economy can be met.

Up to now, no projections of future hydroelectric energy resources have been made in Lithuania. This study is designed to project the changes of river runoff and the total hydropower resources in both gauged and ungauged Lithuanian river catchments. The main objective is to assess the impact of climate change on the future of this renewable energy resource potential in Lithuania.

2. Materials and Methods

2.1. Study Area and Data

The object of this study were the following Lithuanian rivers: the Nemunas, Merkys, Šalčia, Ūla, Neris, Šventoji, Žeimena, Nevėžis, Šušvė, Dubysa, Šešupė, Jūra, Akmena, Šešuvis, Minija, Bartuva, Venta, Mūša, and Lėvuo (Table 1). Daily river discharge data from 25 water gauging stations (WGS) and daily air temperature and precipitation data from 15 meteorological stations (MS) were used from 1986 to 2005 (the period selected as a reference according to IPCC AR5 recommendations [18]) (Figure 1).

Table 1. Main characteristics of the studied gauging station catchments [19].

River-WGS	Characteristics				
	Average Discharge, m^3s^{-1}	Basin Area, km^2	Lakes, %	Wetlands, %	Woods, %
Nemunas-Druskininkai	200.2	37100	1.0	24.0	16.0
Merkys-Puvočiai	32.2	4220	0.9	10.0	46
Šalčia-Valkininkai	5.58	746	0.1	5.0	34.0
Ūla-Zervynos	4.77	679	0.3	11.0	84.0
Nemunas-Nemajūnai	246.0	42800	1.1	22.0	19.0
Neris-Jonava	168.0	24600	2.4	10.0	28.0
Šventoji-Ukmergė	44.1	5440	3.8	9.0	12.0
Žeimena-Pabradė	21.6	2580	7.0	10.0	37.0
Nevėžis-Dasiūnai	34.0	5530	0.2	5.0	18.0
Šušvė-Josvainiai	6.06	1100	0.1	14.0	21.0
Šušvė-Šiaulėnai	1.19	162	0.1	22.0	16.0
Dubysa-Padubysis	15.4	1900	0.6	11.0	14.0
Dubysa-Lyduvėnai	8.71	1130	0.7	13.0	14.0
Nemunas-Smalininkai	503.6	81200	1.5	15.0	21.0
Šešupė-Kudirkos Naumiestis	18.9	3210	2.0	11.0	17.0
Jūra-Tauragė	21.8	1690	0.2	6.0	20.0
Akmena-Paakmenis	4.61	314	0.8	11.0	5.0
Šešuvis-Skirgailai	15.5	1880	0.1	4.0	17.0
Minija-Kartena	17.3	1230	1.4	8.0	20.0
Bartuva-Skuodas	7.80	612	0.2	5.0	3.0
Venta-Leckava	30.6	4060	1.0	9.0	22.0
Venta-Papilė	10.4	1570	0.6	7.0	27.0
Mūša-Ustukiai	11.5	2280	0.8	3.0	14.0
Lėvuo-Pasvalys	5.43	1560	0.4	8.0	15.0
Lėvuo-Bernatoniai	3.73	1130	0.5	9.0	15.0

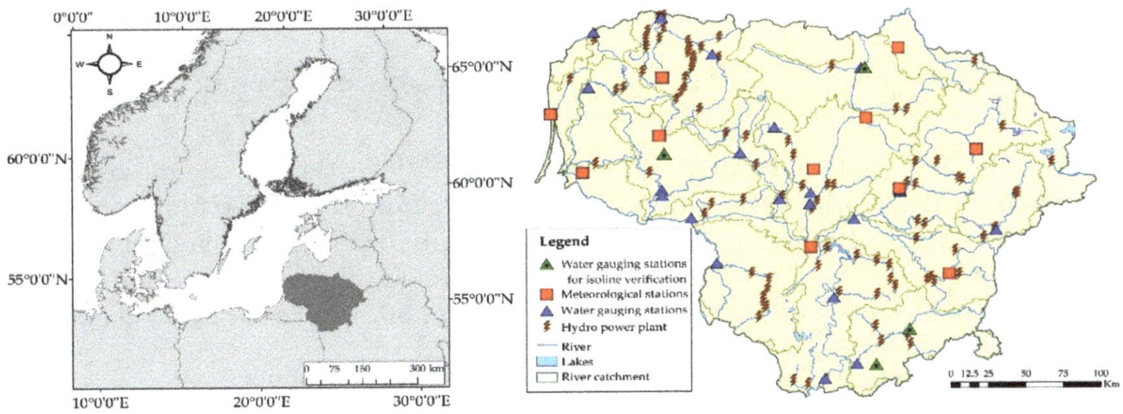

Figure 1. Study area and objects.

Kinetic hydropower resources were estimated in all main Lithuanian rivers; the boundaries of their catchments are indicated in Figure 1. Potential hydropower resources were assessed for 98 currently operating hydropower plants installed on 57 rivers. The near (2021–2040) and far future (2081–2100) projections of the kinetic and potential hydropower resources were made using daily step output data (average air temperature, °C; precipitation amount, mm) from ICHEC-EC-EARTH, MOHC-HadGEM2-ES, and MPI-M-MPI-ESM-LR regional climate models (a grid cell of 11 × 11 km). The data of regional climate models were derived from the EURO-CORDEX database (www.euro-cordex.net).

2.2. Methods

River flow is a key characteristic that is required for the assessment of river kinetic and potential resources. The main steps in the research were as follows (Figure 2): (i) adaptation of climate scenarios data for the Lithuanian territory according to three regional climate models and two RCP scenarios; (ii) evaluation of projections of gauged and ungauged rivers runoff under climate scenarios for two periods of the near (2021–2040) and far future (2081–2100); (iii) evaluation of projections of kinetic and potential riverine energy in the 21st century.

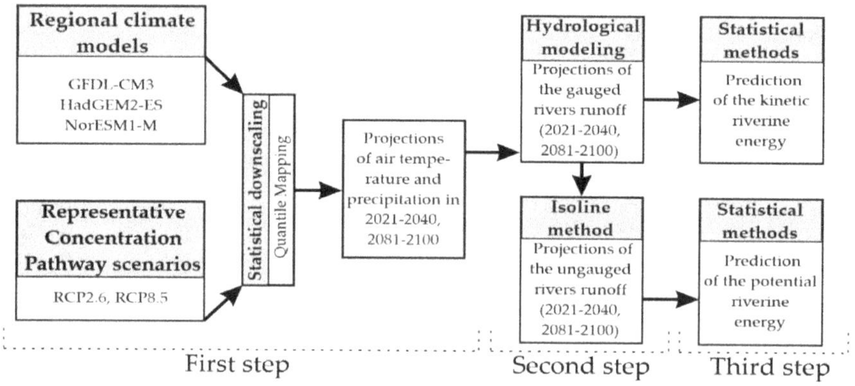

Figure 2. Principal scheme of the study.

2.2.1. Preparation of Regional Climate Models Data

Air temperature (T, °C) and precipitation (P, mm) data extracted from three regional climate models were used to simulate river runoff and to evaluate its projections in the future. The projections were prepared using two Representative Concentration Pathway scenarios: optimistic (RCP2.6) and pessimistic (RCP8.5) [18]. Adaptation of climate models data (air temperature and precipitation) for the Lithuanian territory was performed using the quantile mapping method. This method eliminates systematic errors in climate model data [20]. The nonparametric empirical quantile method is based on the concept that there is a transformation h, described by the following equation [21,22]:

$$St^{Obs} = h\left(St^{CM\ RP}\right) = ECDF^{OBS}\left(ECDF^{CM\ RP}\left(St^{CM\ Fut}\right)\right) \quad (1)$$

where St^{OBS} is the observed meteorological parameter, $St^{CM\ RP}$ is the climate model output for the reference period, $ECDF^{OBS}$ is the empirical cumulative distribution function for the observed period, $ECDF^{CM\ RP}$ is the empirical cumulative distribution function for the climate model reference period, and $St^{CM\ Fut}$ is a meteorological parameter, which is modeled by the climate model for the future period. All estimated results were compared with the values of the reference period (1986–2005).

2.2.2. Runoff Projections of the Gauged and Ungauged Rivers Catchments

The HBV model was used to simulate the runoff of gauged Lithuanian rivers. HBV model calculations were performed in three steps, estimating (1) precipitation amount entering the ground,

(2) slope runoff, and (3) runoff in the watercourse and runoff transformation. This model has previously been successfully applied in Lithuanian conditions [23–25]. HBV is based on a water balance equation [26]:

$$P - E - Q = \frac{d}{dt}[SP + SM + UZ + LZ + V] \tag{2}$$

where P is precipitation, E is evaporation, Q is discharge, SM is soil moisture, SP is snow pack, UZ is upper groundwater zone, LZ is lower groundwater zone, and V is lake or dam volume.

The daily values of Q in the Lithuanian river catchments from 25 water gauging stations (WGS) and of T and P from 15 meteorological stations (MS) (Figure 1) were used to create hydrological models. Information about the modeled catchment area, presence of lakes and forests, mean elevation (above sea level) of the area, and hydrometeorological data from WGS and MS were required as well. Main output data of the HBV model were time series of daily water discharge in the modeled river catchments.

The period from 1986 to 1995 was selected for the model calibration, whereas the period of 1996–2005 was used for validation. The calibration procedure consisted of changing model calibration parameters and comparing calculated discharge values with the measured ones. There are two ways for the evaluation of the calibration results [26]: (1) by visual comparison of hydrographs of computed and measured discharge values and (2) by calculation of correlation (R) between the observed and simulated discharge as well as by the Nash-Sutcliffe efficiency criterion (NSE) and the relative volume error (RE, %). Ideally, the correlation coefficient R should approach 1, but when it reaches 0.7, the model is considered to be properly calibrated [26]. The hydrological model can be regarded as calibrated when the NSE is not less than 0.5 [27].

When assessing the potential energy resources of ungauged Lithuanian rivers, there is a problem of a lack of hydrological data. Therefore, it was not possible to develop hydrological models for the rivers in which observations were carried out only briefly or not at all. For this reason, a method of specific runoff (q, $l\ s^{-1}\ km^{-2}$) mapping was chosen to project the runoff of ungauged rivers. Using this method, isoline maps of specific runoff of Lithuanian rivers were created in the following order:

(1) Based on the results of hydrological modeling, monthly, seasonal, and average annual runoffs (Q, $m^3\ s^{-1}$) were calculated at the studied WGSs for the periods of 1986–2005, 2021–2040, and 2081–2100 under RCP2.6 and RCP8.5 scenarios;

(2) The discharge of the gauged rivers Q ($m^3\ s^{-1}$) was converted to specific runoff q ($l\ s^{-1}\ km^{-2}$) according to [19]:

$$q = \frac{1000 \times Q}{A} \tag{3}$$

where q is the specific discharge, $l\ s^{-1}\ km^{-2}$, Q is the discharge, $m^3\ s^{-1}$, and A is the river catchment area in the conversion place, km^2;

(3) Using the Spline method of interpolation (in ArcGIS), isoline maps of monthly, seasonal, and annual q were created for the whole territory of Lithuania for the past (1985–2005) and future (2021–2040 and 2081–2100) periods;

(4) In the ungauged river catchments, specific runoff q was calculated from the coordinates of hydropower plant places in the created isoline maps;

(5) For ungauged rivers where water resources were assessed, q was converted to Q [19]:

$$Q = \frac{q \times A}{1000} \tag{4}$$

In order to assess the reliability of the created map of specific runoff isolines, four rivers (the Akmena, Lėvuo, Šalčia, and Ūla), having discharge observation data of 1986–2005 and which are not used for the mapping, were selected. These rivers belong to different hydrological regions of Lithuania (western, central, and south-eastern), which are characterized by specific conditions of runoff formation (different precipitation amount, slope, soils, etc.) [28]. Therefore, when verifying the isoline map, it was necessary to test whether the calculated values of monthly, seasonal, and annual specific runoff of the selected rivers for 1986–2005 corresponded to the runoff values extracted from the created isoline map of the specific runoff.

2.2.3. Assessment of Hydropower Resources of Lithuanian Rivers

Assessment of Riverine Hydrokinetic Resources

The authors of this article, together with their colleagues, developed a methodology for the assessment of Lithuanian river hydrokinetic resources and assessed these resources based on historical data [17,29]. However, the work performed did not provide a projection of hydrokinetic resources for the 21st century.

The assessment of hydrokinetic resources was accomplished for three periods: the reference period (1986–2005), against which the results of the projections were compared, and future periods (2021–2040 and 2081–2100). Kinetic resources were assessed in the major Lithuanian rivers (the Nemunas and Neris) and separately in the smaller ones. The assessment of technically available resources (the flow technical capacity obtained from a hydrokinetic device) was made using the following equation [17]:

$$P_{TK} = \eta \frac{\rho}{2} v^3 A_D N \tag{5}$$

where P_{TK} is the technical hydrokinetic capacity, W, η is the device power coefficient (average value is 0.30 according to [30]), ρ is the fluid density, kg m^{-3}, v is the flow velocity, m s^{-1}, A_D is the device swept area, m^2, and N is the number of devices.

The technical hydrokinetic capacity of a given river segment depends on the number of devices (N) used for energy production. The number of turbines in a given river segment of length L (m) was calculated according to [17]:

$$N = \frac{B}{2D}\frac{L}{10D} = \frac{BL}{20h^2} \tag{6}$$

$$E_{TK} = P_{TK}\, t_{TK} \tag{7}$$

where N is the number of the devices, L is the segment length, m, B is a part of the river channel width, where the depth is greater than the turbine diameter, m, h is the bed depth equal to the device turbine diameter, m, E_{TK} is the hydrokinetic generation, Wh, and t_{TK} refers to the hydrokinetic devices' working time, hours.

The next stage of this assessment was the estimation of river segments in the investigated rivers in which the production of hydrokinetic energy is possible. River bed depth (h, m), width (B, m), and flow velocity (v, m s^{-1}) are the main characteristics that determine the suitability of a particular segment for hydrokinetic energy production. The assessment of hydrokinetic energy resources was performed in the river segments with a minimum depth of 0.5 m and a minimum flow velocity of 0.4 m s^{-1}.

In order to evaluate bed depth (h), width (B), and flow velocity (v) in a selected river segment, hydromorphological relationships combining the main flow parameters with their forming factors were proposed. Such relationships were created for Lithuanian small and medium-sized rivers using 1540 discharge measurements of 86 rivers [31]:

$$H = 0.29\,\overline{Q}^{0.45} k^{0.39} I^{-0.2} \tag{8}$$

$$B = 8\overline{Q}^{0.30} k^{0.08} I^{-0.2} \tag{9}$$

$$v = 0.43\overline{Q}^{0.25} k^{0.53} I^{0.4} \tag{10}$$

where h is the average bed depth, m, B is the average bed width, m, v is the average flow velocity in the cross-section, m/s, $\overline{Q}$ is the average annual discharge, m^3 s^{-1}, I is the bed slope, k is the discharge modular coefficient, and Q_i is the discharge.

Assessment of Potential Hydropower Resources

Potential hydropower resources were assessed at existing hydropower plants, without planning for their future growth. Currently, there are 98 operating hydroelectric power plants (HPP) (Figure 1), which have been built on 57 Lithuanian rivers. The projection of potential hydropower resources was performed using discharge rates derived from the specific runoff isoline maps generated for different future periods.

The potential hydropower resources of the selected HPP were evaluated using the formula [32]:

$$P_P = 7\,\Delta H\,Q \tag{11}$$

$$E_P = P_P\,t_P \tag{12}$$

where P_P is the hydropotential capacity, kW, E_P is the potential generation, kWh, ΔH is the height of the HPP dam head, m, Q is the monthly discharge ($m^3\ s^{-1}$), and t_P is the HPP working time, hours (in Lithuania, on average, 4000 h a^{-1}, according to [32]).

3. Results

3.1. Calibration and Validation of Hydrological Models of Gauged and Ungauged Rivers

Hydrometeorological information of the period of 1986–2005 (the reference period) was used to create the model. The 1986–1995 was selected for the model calibration, whereas the period of 1996–2005 was used for validation. The calibration process has to be performed until the correlation coefficient R is the greatest, and the total deviation is the least. The R of the selected river catchments ranged from 0.75 to 0.89 during the calibration period and from 0.60 to 0.83 during the validation period (Table 2). NSE and RE (%) are also very important indicators describing the accuracy of calibration and validation. During the calibration period, the NSE varied from 0.60 to 0.80, and during the validation period it varied from 0.50 to 0.75. Meanwhile, the difference between the measured and modeled discharge (RE, %) was mostly positive during the calibration period and negative by a similar amount during the validation period (Table 2). Taking into account these results, it could be stated that the hydrological models of the selected rivers were properly calibrated and could be applied to project Lithuanian river runoff in the near and far future periods according to different climate scenarios.

Table 2. Results of calibration and validation of hydrological models.

River-WGS	Calibration			Validation		
	R	NSE	RE, %	R	NSE	RE, %
Nemunas-Druskininkai	0.85	0.72	−5.2	0.68	0.54	7.3
Merkys-Puvočiai	0.77	0.60	−4.3	0.65	0.52	6.3
Nemunas-Nemajūnai	0.85	0.72	−6.6	0.68	0.53	7.0
Neris-Jonava	0.87	0.76	−9.2	0.60	0.59	3.9
Šventoji-Ukmergė	0.75	0.64	2.6	0.68	0.64	12.9
Žeimena-Pabradė	0.81	0.66	−5.5	0.73	0.54	7.8
Nevėžis-Dasiūnai	0.86	0.73	−11.0	0.77	0.64	16.1
Šušvė-Josvainiai	0.83	0.69	3.8	0.63	0.53	5.6
Dubysa-Lyduvėnai	0.87	0.76	−2.9	0.73	0.61	3.1
Nemunas-Smalininkai	0.89	0.80	−7.3	0.65	0.50	7.8
Šešupė-Žiotys	0.83	0.70	−9.8	0.66	0.57	12.6
Jūra-Žiotys	0.82	0.67	−3.6	0.75	0.60	3.3
Minija-Kartena	0.88	0.77	3.8	0.83	0.70	−1.1
Nemunas-Žiotys	0.89	0.79	−6.2	0.75	0.62	6.8
Bartuva-Skuodas	0.77	0.60	−7.3	0.75	0.54	8.7
Venta-Leckava	0.88	0.77	−2.6	0.81	0.75	3.5
Mūša-Ustukiai	0.79	0.72	−4.3	0.77	0.68	7.3
Lėvuo-Pasvalys	0.83	0.68	−6.8	0.69	0.67	14.0

The isoline mapping verification method was described in more detail in the methodology (Section 2.2.2.). It compares q, which was determined in two ways: (1) exported from a runoff isoline map using ArcGIS; or (2) calculated using real observation data at a particular WGS. Using the isoline method, the average annual q differed only by −7.7%–+3.8% from the calculated q based on observational data (Table 3). The difference in Q varied from −11.5 to +11.5% in individual seasons. Considering that in practice, the discharge is determined with an error of 40%, the result obtained was reliable and suitable for use in further studies [33–36].

Table 3. Mapping verification.

River–WGS	Winter	Spring	Summer	Autumn	Annual
Akmena-Paakmenis	−7.7	−10.0	−9.6	−4.7	−7.7
Lėvuo-Pasvalys	7.0	−6.1	3.0	−11.5	−1.7
Šalčia-Valkininkai	−2.2	4.8	11.5	3.00	3.8
Ūla-Zervynos	−5.3	−4.9	2.7	0.5	−2.4

3.2. Projections of River Runoff

3.2.1. Projections of Runoff in Gauged River Catchments

Runoff projections of the gauged rivers were evaluated using the hydrological modeling method. The reference period's runoff and its projections in the near and far future were modeled according to an ensemble of three climate models and two RCP scenarios. Changes in the average annual discharge in the future were presented as deviations (%) from the discharge of the reference period (Table 4).

Table 4. Results of the modeling.

River-WMS	Q (1986–2005), $m^3 s^{-1}$	Change (in %) Compared to 1986–2005			
		2021–2040		2081–2100	
		2.6	8.5	2.6	8.5
Hydrological region of western Lithuania					
Bartuva-Skuodas	7.83	0.1	0.4	−5.2	−8.2
Minija-Kartena	15.8	−4.8	−0.6	−4.5	2.3
Jūra-Žiotys	39.2	−7.5	−4.0	−7.9	−10.4
Nemunas-Žiotys	677	−6.2	−5.0	−7.2	−6.8
Hydrological region of central Lithuania					
Nevėžis-Dasiūnai	32.2	−7.2	−4.0	−6.5	−9.9
Šušvė-Josvainiai	5.79	−7.2	−4.0	−6.6	−10.9
Dubysa-Lyduvėnai	8.60	−3.4	1.3	−4.0	1.3
Nemunas-Smalininkai	487	−8.3	−4.8	−5.5	−18.9
Venta-Leckava	30.0	−7.2	−4.0	−7.2	−5.0
Mūša-Ustukiai	10.9	−10.6	−6.7	−11.9	−22.7
Lėvuo-Bernatoniai	3.61	−10.8	−5.5	−9.3	−18.1
Šešupė-Žiotys	35.5	−15.2	−15.2	−14.9	−31.2
Hydrological region of south-eastern Lithuania					
Nemunas-Druskininkai	196	−9.7	−7.6	−7.2	−24.6
Merkys-Puvočiai	32.2	−5.9	−2.2	−1.3	−16.9
Nemunas-Nemajūnai	239	−8.9	−6.6	−6.3	−22.5
Neris-Jonava	159	−6.5	−0.2	−3.2	−12.8
Šventoji-Ukmergė	42.9	−5.2	0.8	−2.9	−7.7
Žeimena-Pabradė	22.8	−3.2	4.1	2.7	−9.9

In Lithuania, rivers are divided into three hydrological regions according to their runoff regime: western Lithuania, central Lithuania, and south-eastern Lithuania. Figure 3 shows the rivers' hydrographs representing these regions: the Jūra from western Lithuania, the Mūša from central Lithuania, and the Merkys from the hydrological region of south-eastern Lithuania.

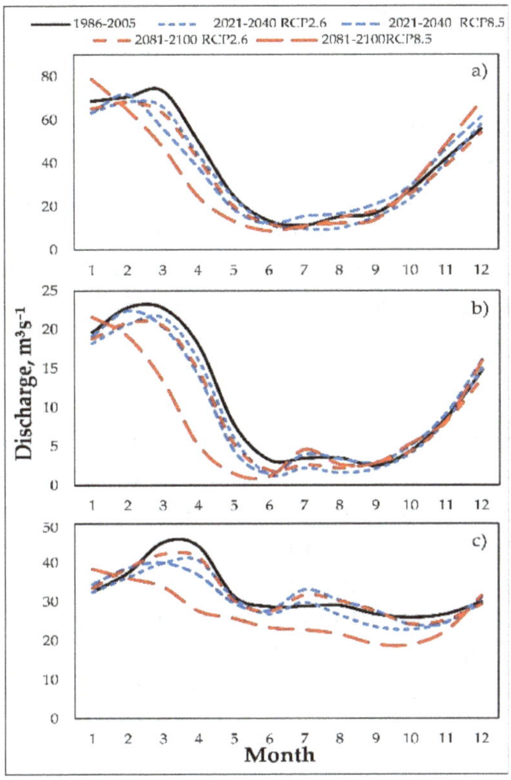

Figure 3. Projected discharge variation in the investigated rivers in the 21st century: (**a**) Jūra River, (**b**) Mūša River, and (**c**) Merkys River.

The hydrological region of western Lithuania receives the highest amount of precipitation in Lithuania (about 850 mm); half of this amount flows into rivers. It rains heavily in autumn and winter; floods in winter are of similar size and, in some cases, even larger than spring floods. The characteristic features of this hydrological region are reflected in the hydrograph of the Jūra River (Figure 3a). As can be seen, this indicates an increased runoff in late autumn and high runoff during the winter season. It also lacks explicit spring floods, while the summer runoff makes up a third of the average annual. The projections showed that, in the near future, the Jūra runoff will decrease from 4.0 to 7.5% as compared to the reference period. Even greater changes are projected in the far future when, according to the RCP2.6 scenario, the Jūra discharge is going to decline by 7.9% and under the RCP8.5 scenario by 10.4%. The largest runoff redistribution is expected in the far future during the spring season, when the expected discharge values may decrease from 15.5% (RCP2.6) to 42.5% (RCP8.5). Significantly smaller changes are likely to occur in the summer season—from 6.2 to 19.3% under the same scenarios. However, in the winter and autumn seasons, the RCP8.5 predicts 8.8% and 5.6% growth, respectively, relative to the reference period.

The hydrological region of central Lithuania is dominated by impermeable clay soils; thus, in winter, almost all the snow melt water runs off into rivers, without being absorbed into the soil, causing severe spring floods. In summer, there is little rainfall, and since the rivers are not fed by groundwater, they are dwindling. All these features are characteristic of the Mūša runoff. In the near future, according to the applied emission scenarios, the Mūša runoff will decrease by 6.6–10.7%, and in the far future, from 11.9% under the RCP2.6 scenario and to 22.7% under the RCP8.5 scenario (Figure 3b). The largest redistribution of the runoff between seasons is likely in the distant future. During the spring season, according to RCP8.5, the discharge is expected to decrease by 59.0%, relative to the reference period. According to the same scenario, the discharge will change slightly (−0.9%) during the winter season, decrease by 17.2% in the summer season, and increase by 6.2% in autumn.

In the hydrological region of south-eastern Lithuania, there are many lakes and forests that naturally regulate river runoff. In addition, the predominant coniferous forests evaporate little water. Rivers flood the least in spring, as water is easily absorbed into the sandy soil. These rivers are abundantly fed by groundwater, so they do not dry up, even in dry summers. Due to the physical-geographical conditions of this hydrological region and the changing climate in the near future, the runoff in the Merkys will be lower from 2.2 to 5.8% compared to the reference period. However, at the end of the century, it will change more significantly and will lower from 1.2 (RCP2.6) to 16.9% (RCP8.5) (Figure 3c). During the same period, significant seasonal runoff changes are expected in the Merkys under the RCP8.5 scenario. In the winter season, the runoff will increase to 6.2%, while in spring, summer, and autumn, it will decrease by 28.1%, 21.7%, and 23.8%, respectively, relative to the reference period.

The regularities of the runoff change in the three studied rivers (Figure 3) are typical for other rivers located in the same hydrological region. According to the available data, the least river runoff will change in western Lithuania: in the far future, according to the most pessimistic scenario, the average river discharge may decrease by 10.4%, relative to the reference period. The most unfavorable situation is possible in the hydrological region of central Lithuania, where the decline of the runoff is projected to be up to 31.2%. Meanwhile, in the hydrological region of south-eastern Lithuania, the most significant decrease of the runoff can be 24.6%. It should be emphasized that this region includes two large rivers, the Nemunas and the Neris, the major parts of which are located in Belarus. Thus, the change in their runoff is influenced by the larger basin area and more diverse physical-geographical conditions.

In runoff projections, uncertainty analysis plays a critical role. The uncertainty of Lithuanian river runoff projections was assessed in our previous study [37]. The evaluation of uncertainty due to different sources of origin, such as climate scenarios (RCPs), global climate models (GCMs), and statistical downscaling (SD) methods, was accomplished for the selected Lithuanian rivers from different hydrological regions (western, central, and south-eastern). The near (2021–2040) and far future (2081–2100) projections were developed according to three RCPs, three GCMs, and three SD methods. In the western hydrological region, GCMs were identified as the most dominant uncertainty source (41.0–44.5%) in the runoff projections. Moreover, the uncertainty of runoff projections from central and south-eastern regions of Lithuania was related to SD methods, and the range of uncertainties fluctuated from 39.4% to 60.9%. The rivers selected for the current study belong to the same hydrological regions. The runoff projections for 2021–2040 and 2081–2100 were created using to the same RCP scenarios and two GCMs as in [37]. Therefore, the tendencies of uncertainty assessment in the projections generated in [37] and the current study projections should be similar.

3.2.2. Projections of Runoff in Ungauged River Catchments

The method of specific runoff (q) mapping was used to project the runoff of ungauged rivers. The average values of q, according to the ensemble of three models in the historic (reference) period (1986–2005), are presented in Figure 4. The distribution of q in the studied area was quite diverse; it varied from 3.0 to 14.0 l s^{-1} km^{-2} (Figure 4). The observed differences were determined mainly by physical-geographical conditions. In the hydrological region of western Lithuania, there was more precipitation, high river slopes, and better conditions for fast water drainage; therefore, here, q varied from 8.0 to 14.0 l s^{-1} km^{-2}. The specific runoff gradually decreased, shifting to the hydrological region of central Lithuania. The slopes of the rivers in this region are small and impermeable soils are widespread, which facilitated more evaporation, so in this region, the range of q ranged from 3.0 to 8.0 l s^{-1} km^{-2}. Going east (to the south-eastern region), since elevation gets higher and the slopes of the area rise, q gradually increased and varied from 6.0 to 10.0 l s^{-1} km^{-2}. In this region, the greater values of q were also determined by the higher amount of precipitation (than in the hydrological region of central Lithuania) and abundant underground feeding. Using the q maps, which are given in $l\ s^{-1}\ km^{-2}$ and Equation (4), we could calculate Q in a specific river of interest.

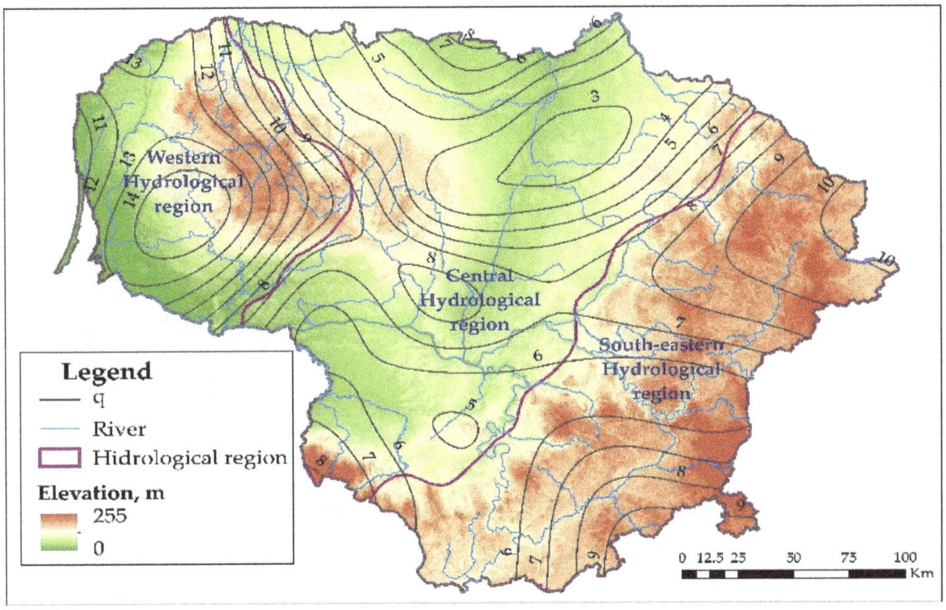

Figure 4. Spatial distribution of specific runoff (q, $l\ s^{-1}\ km^{-2}$) in the territory of Lithuania according to the ensemble of three models in the reference period (1986–2005).

Four maps in Figure 5 show the change (%) in the projections of the average annual specific runoff relative to the reference (1986–2005) period in the three hydrological regions. The smallest changes in q are predicted in the hydrological region of western Lithuania. In the near future, depending on the applied RCP scenario, the average q of the whole region will be 2.2% to 4.5% lower than in the reference period. Small changes in q are likely in the far future as well. Depending on the RCP scenario, the average values of q of the hydrological region will get smaller by 5.5–5.9% according to RCP2.6 and RCP8.5 scenarios, respectively, compared to the reference period. More considerable changes are possible in the hydrological region of south-eastern Lithuania; in this region, in the near future, the average region q will decrease from 0.7% to 5.5% depending on the RCP scenario, and in the far future from 1.8% (under RCP2.6) to 13.3% (under RCP8.5). The biggest changes are projected in the hydrological region of central Lithuania. In the near future, the average q is expected to decline by at least 5.8% to 9.1%, and by the end of the century, from 8.7 (RCP2.6) to 16.6% (RCP8.5) compared to the reference period. Values of the specific runoff (q) differ among rivers in the same hydrological region. This is due not only to local physical and geographical conditions, which are heterogeneous in individual hydrological region, but also to microclimatic features.

3.3. Projections of River Kinetic Energy

The assessment of kinetic energy resources of Lithuanian rivers was performed by applying technical restrictions, according to which the depth of the river bed must be greater than 0.5 m, and the flow velocity must exceed 0.4 $m\ s^{-1}$. These two minimum conditions must be met for hydrokinetic turbines (devices that convert the kinetic energy of flowing water into electricity) to be installed in the studied river segments. During this study, the kinetic energy resources of the main Lithuanian rivers (excluding the major rivers Nemunas and Neris) were estimated, which amounted to 242 GWh a^{-1} in the reference period (Table 5).

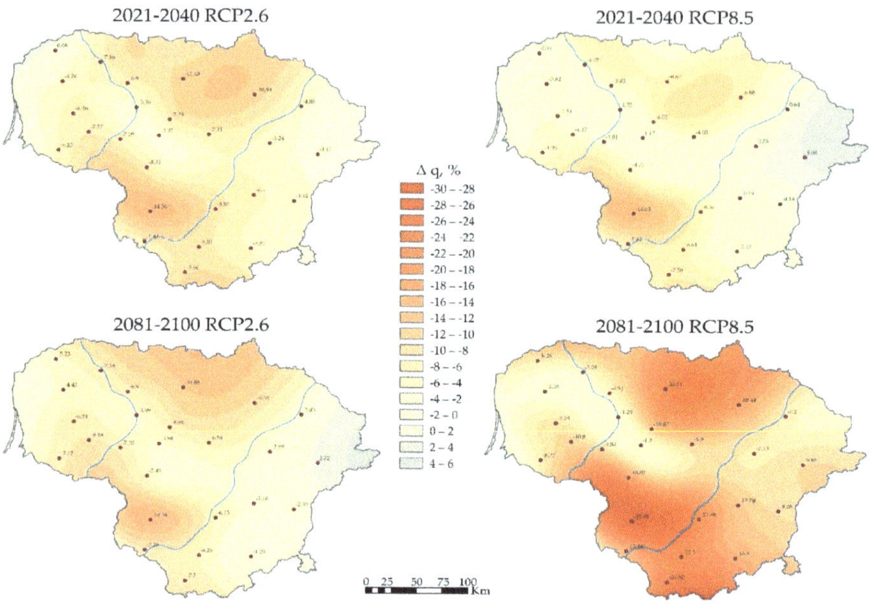

Figure 5. Changes (in %) of specific runoff in Lithuanian rivers in the 21st century relative to 1986–2005.

Table 5. Kinetic resources of the main Lithuanian rivers (E_{KT}, GWh a^{-1}) and projection of their changes (in %) in the near and far future relative to 1986–2005.

River	E (1986–2005), GWh a^{-1}	Change (in %) Compared to 1986–2005			
		2021–2040		2081–2100	
		2.6	8.5	2.6	8.5
Hydrological region of western Lithuania					
Bartuva	2.9	1.0	1.1	−6.2	−11.1
Venta	24.2	−8.0	−4.5	−8.3	−5.7
Jūra	24.6	−9.0	−4.1	−8.8	−12.4
Minija	26.2	−6.3	−0.6	−5.9	−2.3
Hydrological region of central Lithuania					
Lėvuo	2.7	−11.4	−11.5	−13.7	−25.9
Šušvė	7.4	−9.7	−9.4	−11.4	−17.4
Nevėžis	8.1	−9.1	−6.1	−10.1	−18.4
Mūša	10.1	−12.7	−8.0	−14.9	−27.4
Dubysa	17.7	−3.6	1.3	−4.8	−1.8
Šešupė	27.8	−17.5	−18.8	−18.2	−36.2
Hydrological region of south-eastern Lithuania					
Žeimena	14.6	−4.8	4.3	2.7	−12.3
Merkys	33.1	−7.2	−2.7	−1.7	−19.4
Šventoji	42.5	−4.2	1.5	−2.5	−6.1
Major rivers					
Neris	336.3	−6.7	−0.8	−3.2	−12.8
Nemunas	1087	−7.8	−6.3	−7.6	−15.7
Summary					
Without Nemunas and Neris	241.9	−7.9	−4.4	−8.0	−14.5
Nemunas and Neris	1108.6	−7.0	−3.2	−5.1	−14.1
All rivers	1350.5	−7.8	−4.3	−7.6	−14.4

In the hydrological region of western Lithuania, the hydrokinetic energy resources are projected to decrease the least (Table 5). In the near future, the Bartuva River resources will hardly change, while in other rivers of this region, they are expected to decline to 9%. In the long run, hydrokinetic resources could decrease by up to 12%.

In central Lithuania, the riverine kinetic resources will shrink the most. Smaller changes in hydropower are projected in the near future (19%); meanwhile, at the end of the century, the resources of the Lėvuo, Mūša, and Šešupė rivers may decrease very significantly—in the case of the RCP8.5 scenario, by 25.9%, 27.4%, and 36.2%, respectively, compared to the reference period.

In the near future, both positive and negative changes in the projections have been identified in the hydrological region of south-eastern Lithuania—from −7.2 to 1.5%. However, in the distant future, in the case of the RCP8.5 scenario, a decrease in this energy from 6.1% (in the Šventoji) to 19.4% (in the Merkys) compared to the reference period is possible.

The assessment of the projections of the total kinetic energy resources of Lithuanian rivers (excluding the Nemunas and Neris) indicated a decrease of 3.8–7.9% in the near future and 6.9–12.8% in the far future under the RCP2.6 and RCP8.5 scenarios, respectively, compared to the reference period (Figure 6a). Not only annual but also seasonal changes in kinetic resources are projected.

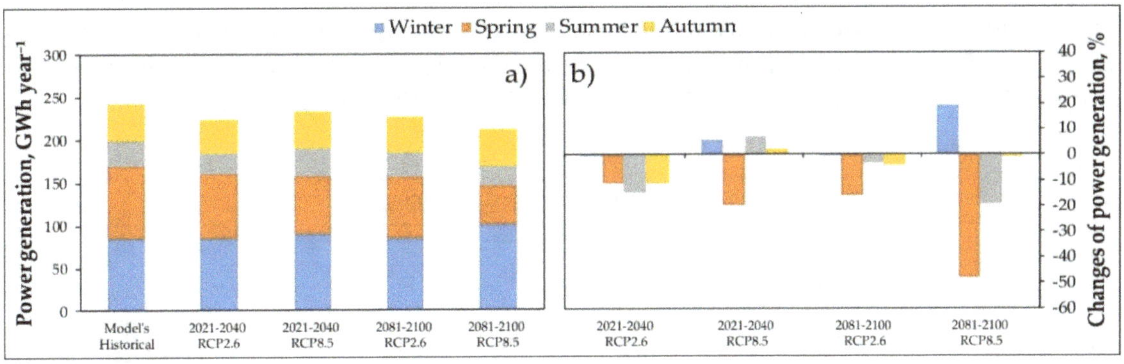

Figure 6. Projections of kinetic resources of the main Lithuanian rivers (except the Nemunas and Neris) in the near and far future in different seasons (**a**) and resource changes in comparison with 1986–2005 (**b**).

In both future periods, under both RCP scenarios, hydrokinetic resources will likely decrease (by 11.3–47.9%) during the spring season, but will increase during the winter season (up to 19.1%). The decline of the resources is also expected during the summer season due to reduced river flow rates and depth. These trends in the change of hydrokinetic resources are determined by the projected redistribution of the river runoff between seasons in the future.

The two largest rivers in Lithuania, the Nemunas and Neris, have favorable hydrokinetic energy generation conditions due to a high flow rate and depth. If hydrokinetic turbines were installed along the entire length of these rivers, it would be possible to produce 1109 GWh a^{-1} (Table 5). In the near future, the Nemunas and Neris rivers' total kinetic resources are projected to decrease from 5.0 to 7.5%, depending on the chosen RCP scenario. In the long term, the changes will be even more significant from 6.6% (RCP2.6) to 15.1% (RCP8.5) compared to the reference period (Figure 7a). The seasonal variation trends of the Nemunas and Neris rivers' kinetic resources were similar to those found for smaller rivers (Figure 7b).

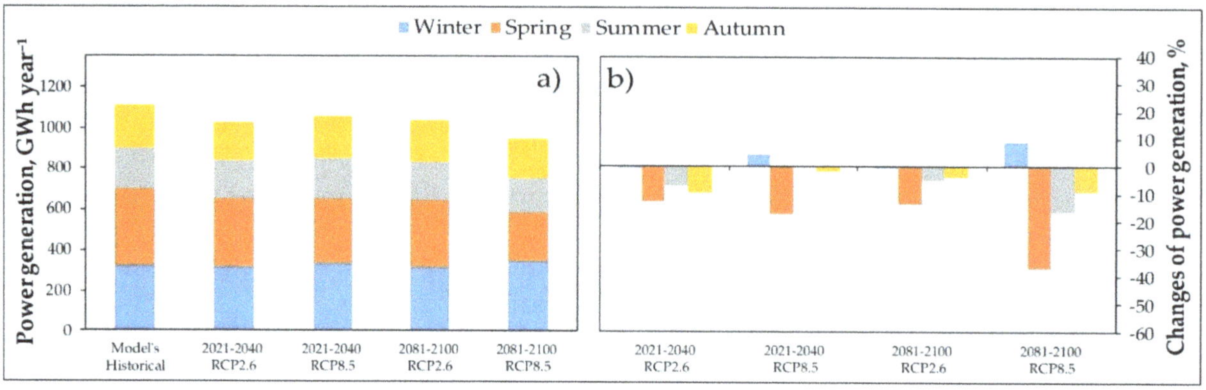

Figure 7. Projection of seasonal kinetic resources of the Nemunas and Neris revers in the near and far future (**a**) and their change compared to 1986–2005 (**b**).

3.4. Projections of River Potential Energy

During the reference period, 97 small HPPs generated 75 GWh a^{-1} of electricity. In 2021–2040, depending on the RCP scenario, the hydropower production of small hydropower plants is projected to decrease from 4.5% to 8.6% compared to the reference period. In 2081–2100, an even larger decrease in hydropower is expected: from 7.1% (RCP2.6) to 13.7% (RCP8.5) (Figure 8a).

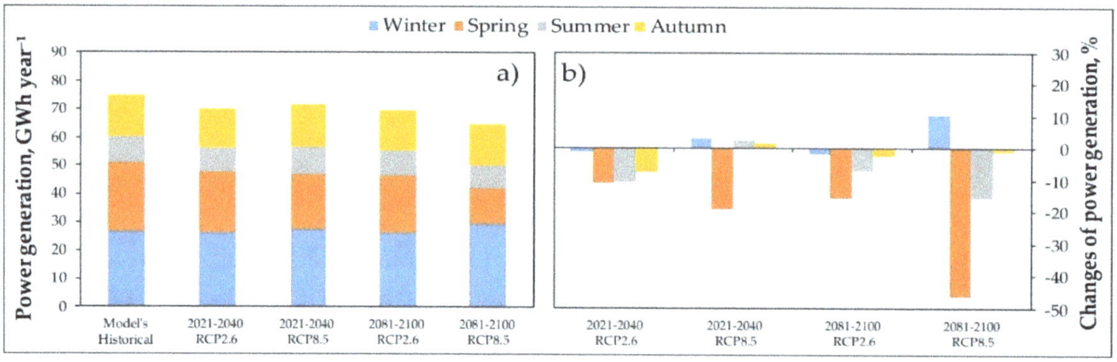

Figure 8. Projection of seasonal energy production in small Lithuanian hydropower plants in the near and far future (**a**) and its change compared to 1986–2005 (**b**).

An analysis of the seasonal energy production revealed the greatest changes in spring and summer (Figure 8b). According to the applied scenarios, the production will decrease to 19.3% in the spring season and to 10.8% in the summer season in the near future. In the far future, compared to the reference period, the largest changes are possible under the RCP8.5 scenario, which projects an increase of 10.4% during the winter season, a decrease of 46.4% and 15.4% in the spring and summer seasons, respectively, and an only slight change in autumn.

The largest HPP of Lithuania is built on the Nemunas River in Kaunas. It is significant for the Lithuanian energy system because of the amount of electricity produced and as a provider of the system services and contributor, ensuring the system's stability. This is the only power plant that can restore the power system in an accident (black start). In 1986–2005, Kaunas HPP produced 386 GWh a^{-1} of electricity. The projections indicated that the hydropower production of Kaunas HPP would decrease from 6.9% (RCP2.6) to 11.4% (RCP8.5) in the near future (Figure 9) and decline from 7.4% to 30.6% under the selected scenario in the far future. As shown in Figure 9, electricity production will gradually decrease during spring (up to 44%) and increase slightly during the winter season (8%). These changes are expected to occur due to the altered runoff in the Nemunas River.

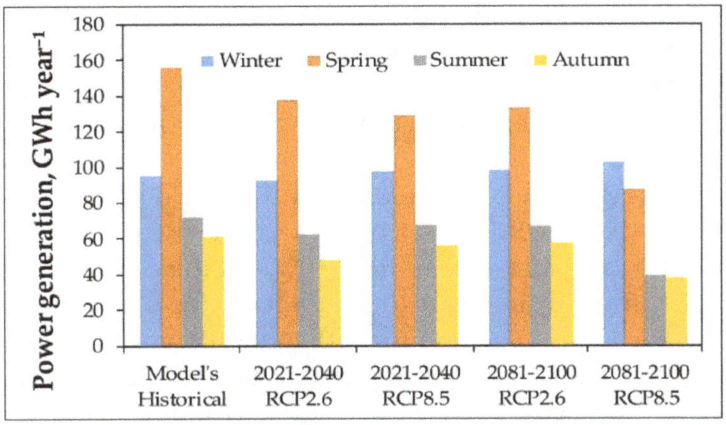

Figure 9. Projection of the seasonal energy production in Kaunas HPP in the near and far future and its change compared to 1986–2005.

4. Discussion and Conclusions

This study set out to assess the future potential of hydropower resources in the Lithuanian river catchments. Two challenges accompanied the investigation: projection of ungauged river runoff and assessing the total potential of hydroelectric energy according to the new climate scenarios.

The projection of hydropower resources is a complex process. When managing water resources, water specialists often face the problem of insufficient data or a complete lack of data from hydrological observations. For the present assessment of the total potential of hydroelectric energy, it was necessary to include the runoff data of ungauged rivers. Information transfer from gauged to ungauged Lithuanian river catchments was performed using isoline maps created by interpolating specific runoff derived from the hydrological modeling. Such spatial proximity approach is one of the earliest and most widely used regionalization methods [38–40]. This method enabled us to get the data of ungauged catchments necessary for the projection described in the paper. The average annual specific runoff differed only by −7.7%–+3.8% from the calculated one, based on observational data using the isoline method. That confirms that the selected regionalization scheme was successful.

The effects of climate change on the hydrological regime of rivers are well studied. Due to changing precipitation and air temperature patterns in the Baltic Sea basin, alterations in the annual and seasonal runoff distribution are observed and they are projected to continue in the future [41–45]. The aforementioned major drivers result in higher winter runoff and lower runoff during other seasons. However, in different regions of the planet, runoff projections may vary due to dissimilarities in natural (primarily climatic) conditions [46–50]. The present study results demonstrate that even among different hydrological regions of Lithuania, river runoff projections may differ. The most significant and intimidating changes are likely to occur under the most pessimistic scenario (RCP8.5) at the end of the 21st century. In the rivers of the hydrological region of western Lithuania, the runoff will decrease from 4.5 (RCP2.6) to 10.4% (RCP8.5), in central Lithuania from 4.0 (RCP2.6) to 31.2% (RCP8.5), and in south-eastern Lithuania from 1.3 (RCP2.6) to 24.6% (RCP8.5) relative to the reference period. The central Lithuanian hydrological region will experience the greatest decline of runoff according to all projections and in both the near and far future periods. This finding corresponds with a previous study [51] that highlighted vulnerability and a high risk of an increasing number of rivers drying up in response to climate change in this particular hydrological region.

Changing climate and runoff patterns provide additional uncertainty for hydropower generation. IPCC Special Report on Renewable Energy Sources and Climate Change Mitigation [13] states that the impacts of climate change on hydropower generation are likely to be small on a global basis. However, still significant regional changes in river flow volumes and timing may pose challenges for planning. Hamududu and Killingtveit [52] assessed that, globally, hydroelectric energy generation is predicted to change very little (by 0.46 TWh, i.e., less than 1% of the 2005 generation level) by the year 2050.

According to Turner et al. [53], by the end of the century, depending on the used general circulation model, a change in net global hydropower production of between −8% and +5% under RCP8.5 is projected. Lehner et al. [54] estimated, for the whole of Europe, a decline of the gross hydropower potential by about 6% by the 2070s according to the moderate climate and global change scenario. The present results demonstrate a clear shrink of hydropower potential in Lithuania due to the already discussed decline of river runoff. In 2081–2100, small hydropower plants' energy production is expected to lower from 7.1% (under RCP2.6) to 13.7% (under RCP8.5), whereas the reduced river runoff in spring will result in a decrease of almost 50%. In the only large Kaunas HPP, a decline from 7.4% to 30.6%, depending on the scenario in the far future, is projected. Such a decrease in Kaunas HPP electricity production is 1.6 times higher than the annual electricity production of all small hydropower plants. Thus, the loss of such energy would be significant for Lithuanian hydropower production. This estimation is also crucial concerning the possibilities of future development of small hydropower. In neighboring Latvia, it has been assumed that a projected decrease of the mean annual river runoff [41] will bring negative changes to hydropower supply as well, whereas in north Estonia, the climate change impact on hydropower potential is likely to be positive [55], likewise in Norway [7].

An object of the vast majority of studies on hydropower resources are conventional power plants that harness the potential energy of dammed water. Hydrokinetic energy is one among promising renewable energy resources; however, it has yet to be proven commercially viable [56]. Although some hydrokinetic power potential assessments have been accomplished and are available publicly, the authors of the present paper could not find any study dedicated to projections of this riverine hydroelectric energy resource. The current findings revealed that future climate alterations would bring runoff modifications and, consequently, changes of hydrokinetic power potential. In the near future, the total kinetic resources of two major Lithuanian rivers are projected to decrease from 5.0 to 7.5%, and in the far future period, from 6.6% to 15.1% depending on the chosen RCP scenario. The hydrokinetic resources of the main Lithuanian rivers (excluding the Nemunas and Neris) will decrease from 4.3 (RCP2.6) to 7.8% (RCP8.5) in the near future, while it will decrease twice as much compared to the reference period in the distant future.

As hydroelectric power plants' development using river potential (head) energy is limited in Lithuania due to environmental constraints, installing kinetic (non-head) hydroelectric power plants could be a great alternative in the future. In general, the Nemunas and Neris might have considerable potential as they have favorable hydrokinetic energy generation conditions due to a high flow rate and depth. Small conventional hydropower generation in smaller rivers should not be developed. In the projected conditions of limited water availability, finding the balance between human (energy production) and environment (suitable habitat for aquatic species) needs might be very complicated and hardly achievable. The provided projections could help understand what kind of benefits and challenges water resource managers may face in the future and how to transform the country's electricity sector development into something more sustainable.

Author Contributions: D.J., J.K. and G.A. designed the study; G.A., D.J. and D.Š. contributed to data collection and analysis; D.J. and G.A. did modelling of processes; all authors contributed to the writing and editing of the manuscript. All authors have read and agreed to the published version of the manuscript.

Abbreviations

HBV	Hydrological model developed by the Swedish Department of Climate;
RCP	Representative Concentration Pathway is a greenhouse gas concentration trajectory proposed by the Intergovernmental Panel on Climate Change;
RCP2.6, RCP8.5	Representative Concentration Pathway scenarios;
EU	European Union;
RES	Renewable energy sources;

WGS	Water gauging stations;
MS	Meteorological stations;
IPCC AR5	The Fifth Assessment Report of the United Nations Intergovernmental Panel on Climate Change;
ICHEC-EC-EARTH	Regional climate in the Great Lakes Water Balance; U.S. Gmodel developer by the Irish Centre for High-End Computing;
MOHC-HadGEM2-ES	Regional climate model developer by the Met Office Hadley Centre for Climate Science and Services, United Kingdom;
MPI-M-MPI-ESM-LR	Regional climate model developer by the Max Planck Institute for Meteorology, Germany;
EURO-CORDEX	The European branch of the CORDEX initiative and will produce data of the climate simulations in the 21st century;
ArcGIS	A geographic information system software developed by the Environmental Systems Research Institute, United States of America;
HPP	Hydropower plant;
R	Correlation coefficient;
NSE	Nash–Sutcliffe model efficiency coefficient;
RE	Difference between the measured and modeled discharge, %.

References

1. The European Parliament and Council of the European Union. Directive (EU) 2018/2001 on the promotion of the use of energy from renewable sources. *Off. J. Eur. Union* **2018**, *L328*, 82–209.
2. Eurostat. *Wind and Water Provide Most Renewable Electricity*; 2020. Available online: https://ec.europa.eu/eurostat/en/web/products-eurostat-news/-/DDN-20200129-1 (accessed on 20 October 2020).
3. IEA. *Key World Energy Statistics 2020*; OECD Publishing: Paris, France, 2020. [CrossRef]
4. IHA. *Hydropower Status Report. Trends and Insights*; International Hydropower Association: London, UK, 2019.
5. United States Department of Energy. *Effects of Climate Change on Federal Hydropower*; The Second Report to Congress; U.S. Department of Energy: Washington, DC, USA, 2017; p. 35.
6. Jabbari, A.A.; Nazemi, A. Alterations in Canadian hydropower production potential due to continuation of historical trends in climate variables. *Resources* **2019**, *8*, 163. [CrossRef]
7. Koestler, V.; Østenby, A.; Arnesen, F.; Birkeland, C.; Haddeland, I. *Climate Change Impact on Norwegian Hydro Power*; Report 50-2019; Norwegian Directorate for Water Resources and Energy: Oslo, Norway, 2019; p. 32. (In Norwegian)
8. Koch, H.; Vögele, S.; Hattermann, F.F.; Huang, S. The impact of climate change and variability on the generation of electrical power. *Meteorol. Z.* **2015**, *24*, 173–188. [CrossRef]
9. German Advisory Council on the Environment. *Pathways towards a 100% Renewable Electricity System*; Geschäftsstelle Des Sachverständigenrates Für Umweltfragen (SRU): Berlin, Germany, 2011. Available online: http://www.umweltrat.de/SharedDocs/Downloads/EN/02_Special_Reports/2011_10_Special_Report_Pathways_renewables.pdf?__blob=publicationFile (accessed on 19 October 2020).
10. Wagner, T.; Themeßl, M.; Schüppel, A.; Gobiet, A.; Stigler, H.; Birk, S. Impacts of climate change on stream flow and hydro power generation in the Alpine region. *Environ. Earth. Sci.* **2017**, *76*, 4. [CrossRef]
11. Canadian Hydraulics Centre at the NRC. *Assessment of Canada's Hydrokinetic Power Potential: Phase I Report—Methodology and Data Review*; Canadian Hydraulics Centre at the NRC: Ottawa, ON, Canada, 2010; p. 72.
12. Electric Power Research Institute. *Assessment and Mapping of the Riverine Hydrokinetic Energy Resource in the Continental United States, Technical Report*; EPRI: Palo Alto, CA, USA, 2012; p. 80.
13. Kumar, A.; Schei, T.; Ahenkorah, A.; Rodriguez, R.C.; Devernay, J.M.; Freitas, M.; Hall, D.; Killingtveit, Å.; Liu, Z. Hydropower. In *IPCC Special Report on Renewable Energy Sources and Climate Change Mitigation*; Edenhofer, O., Pichs-Madruga, R., Sokona, Y., Seyboth, K., Matschoss, P., Kadner, S., Zwickel, T., Eickemeier, P., Hansen, G., Schlömer, S., et al., Eds.; Cambridge University Press: Cambridge, UK; New York, NY, USA, 2011; pp. 437–496. [CrossRef]

14. National Energy Independence Strategy of Lithuania, State News, 2012-07-10, No 80-4149. Available online: https://e-seimas.lrs.lt/portal/legalAct/lt/TAD/TAIS.429490/asr (accessed on 20 October 2020).
15. Jablonskis, J.; Lasinskas, M. Lithuanian river cadaster (discharges, slopes, capacities). *Vilnius Lith.* **1962**, *3*, 640. (In Lithuanian)
16. Jablonskis, J.; Jurgelėnaitė, A.; Tomkevičienė, A. Hydropower in environment protection context. *Power Eng.* **2007**, *3*, 48–56. (In Lithuanian)
17. Gailiušis, B.; Jakimavičius, D.; Šarauskienė, D.; Jurgelėnaitė, A. Assessment of hydrokinetic resources of small and medium-size rivers: The Lithuanian case. *Baltica* **2017**, *30*, 23–30. [CrossRef]
18. IPCC. *Climate Change 2013: The Physical Science Basis. Contribution of Working Group I to the Fifth Assessment Report of the Intergovernmental Panel on Climate Change*; Stocker, T.F., Qin, D., Plattner, G.-K., Tignor, M., Allen, S.K., Boschung, J., Nauels, A., Xia, Y., Bex, V., Midgley, P.M., Eds.; Cambridge University Press: Cambridge, UK; New York, NY, USA, 2013; p. 1535. [CrossRef]
19. Gailiušis, B.; Jablonskis, J.; Kovalenkovienė, M. The Lithuanian rivers. Hydrography and runoff. *Lith. Engl. Summ.* **2001**, 792. (In Lithuanian)
20. Beyer, R.; Krapp, M.; Manica, A. An empirical evaluation of bias correction methods for palaeoclimate simulations. *Clim. Past* **2020**, *16*, 1493–1508. [CrossRef]
21. Gudmundsson, L.; Bremnes, J.B.; Haugen, J.E.; Engen–Skaugen, T. Technical Note: Downscaling RCM precipitation to the station scale using statistical transformations—A comparison of methods. *Hydrol. Earth Syst. Sci. Discuss.* **2012**, *16*, 3383–3390. [CrossRef]
22. Sunyer, M.A.; Hundecha, Y.; Lawrence, D.; Madsen, H.; Willems, P.; Martinkova, M.; Vormoor, K.; Burger, G.; Hanel, M.; Kriaučiūnienė, J.; et al. Inter–comparison of statistical downscaling methods for projection of extreme precipitation in Europe. *Hydrol. Earth Syst. Sci.* **2015**, *19*, 1827–1847. [CrossRef]
23. Kriauciuniene, J.; Meilutyte-Barauskiene, D.; Rimkus, E.; Kazys, J.; Vincevicius, A. Climate change impact on hydrological processes in Lithuanian Nemunas river basin. *Baltica* **2008**, *21*, 51–61.
24. Jakimavičius, D.; Kriaučiūnienė, J.; Šarauskienė, D. Impact of climate change on the Curonian Lagoon water balance components, salinity and water temperature in the 21st century. *Oceanologia* **2018**, *60*, 378–389. [CrossRef]
25. Akstinas, V.; Meilutytė-Lukauskienė, D.; Kriaučiūnienė, J. Consequence of meteorological factors on flood formation in selected river catchments of Lithuania. *Meteorol. Appl.* **2019**, *26*, 232–244. [CrossRef]
26. SMHI. *Integrated Hydrological Modelling System, Manual, Version 5.8*; Swedish Meteorological and Hydrological Institute: Norrköping, Sweden, 2005; p. 115.
27. Ritter, A.; Muñoz-Carpena, R. Performance evaluation of hydrological models: Statistical significance for reducing subjectivity in goodness-of-fit assessments. *J. Hydrol.* **2013**, *480*, 33–45. [CrossRef]
28. Kriaučiūnienė, J.; Meilutytė-Barauskienė, D.; Reihan, A.; Koltsova, T.; Lizuma, L.; Šarauskienė, D. Variability in temperature, precipitation and river discharge in Baltic States. *Boreal Environ. Res.* **2012**, *17*, 150–162.
29. Jakimavičius, D.; Gailiušis, B.; Šarauskienė, D.; Jurgelėnaitė, A.; Meilutytė-Lukauskienė, D. Assessment of the riverine hydrokinetic energy resources in Lithuania. *Baltica* **2014**, *27*, 141–150. [CrossRef]
30. Lalander, E. Modelling Hydrokinetic Energy Resources for In-Stream Energy Converters. Ph.D. Thesis, Department of Engineering Sciences, Uppsala University, Uppsala, Sweden, 2010; p. 57.
31. Gailiušis, B. River run-off management in the Lithuanian SSR. *Vilnius Lith.* **1986**, 156. (In Russian)
32. Punys, P.; Kvaraciejus, A.; Dumbrauskas, A.; Šilinis, L.; Popa, B. An assessment of micro-hydropower potential at historic watermill, weir, and non-powered dam sites in selected EU countries. *Renew. Eng.* **2019**, *133*, 1108–1123. [CrossRef]
33. Winter, T.C. Uncertainties in estimating the water balance of lakes. *J. Am. Water Resour. Assoc.* **1981**, *17*, 82–115. [CrossRef]
34. Sattary, J.; Judeh, W.A.; Warren, S. The impact of measurement uncertainty on the water balance. *Water Environ. J.* **2002**, *16*, 218–222. [CrossRef]
35. Neff, B.P.; Nicholas, J.R. *Uncertainty in the Great Lakes Water Balance*; U.S. Geological Survey Scientific Investigations Report 2004–5100; U.S. Geological Survey: Reston, VA, USA, 2005; p. 42.
36. McMillan, H.K.; Westerberg, I.K.; Krueger, T. Hydrological data uncertainty and its implications. *Wiley Interdiscip. Rev. Water* **2018** *5*, e1319. [CrossRef]
37. Akstinas, V.; Jakimavičius, D.; Meilutyte-Lukauskiene, D.; Kriaučiuniene, J.; Šarauskiene, D. Uncertainty of annual runoff projections in Lithuanian rivers under a future climate. *Hydrol. Res.* **2020**, *51*, 257–271. [CrossRef]

38. Oudin, L.; Andre´assian, V.; Perrin, C:; Michel, C.; Le Moine, N. Spatial proximity, physical similarity, regression and ungaged catchments: A comparison of regionalization approaches based on 913 French catchments. *Water Resour. Res.* **2008**, *44*, W03413. [CrossRef]
39. Razavi, T.; Coulibaly, P. Streamflow prediction in ungauged basins: Review of regionalization methods. *J. Hydrol. Eng.* **2013**, *18*, 958–975. [CrossRef]
40. Hong, M.; Zhang, R.; Wang, D.; Qian, L.; Hu, Z. Spatial interpolation of annual runoff in ungauged basins based on the improved information diffusion model using a genetic algorithm. *Discrete Dyn. Nat.* **2017**, *2017*, 4293731. [CrossRef]
41. Apsite, E.; Bakute, A.; Elferts, D.; Kurpniece, L.; Pallo, I. Climate change impacts on river runoff in Latvia. *Clim. Res.* **2011**, *48*, 57–71. [CrossRef]
42. Kayhko, J.; Apsite, E.; Bolek, A.; Filatov, N.; Kondratyev, S.; Korhonen, J.; Kriaučiūnienė, J.; Lindstrom, G.; Nazarova, L.; Pyrh, A.; et al. *Recent Change—River Run-off and Ice Cover. Second Assessment of Climate Change for the Baltic Sea Basin*; Springer: Heidelberg, Germany, 2015; pp. 99–116.
43. Stonevičius, E.; Rimkus, E.; Štaras, A.; Kažys, J.; Valiuškevičius, G. Climate change impact on the Nemunas River basin hydrology in the 21st century. *Boreal Environ. Res.* **2017**, *22*, 49–65.
44. Piniewski, M.; Szcześniak, M.; Huang, S.; Kundzewicz, Z.W. Projections of runoff in the Vistula and the Odra river basins with the help of the SWAT model. *Hydrol. Res.* **2018**, *49*, 303–317. [CrossRef]
45. Šarauskienė, D.; Akstinas, V.; Kriaučiūnienė, J.; Jakimavičius, D.; Bukantis, A.; Kažys, J.; Povilaitis, A.; Ložys, L.; Kesminas, V.; Virbickas, T.; et al. Projection of Lithuanian river runoff, temperature and their extremes under climate change. *Hydrol. Res.* **2018**, *49*, 344–362. [CrossRef]
46. Villarini, G. On the seasonality of flooding across the continental United States. *Adv. Water Resour.* **2016**, *87*, 80–91. [CrossRef]
47. Štefunková, Z.; Hlavčová, K.; Lapin, M. Runoff change scenarios based on regional climate change projections in mountainous basins in Slovakia. *Contrib. Geophys. Geod.* **2013**, *43*, 327–350. [CrossRef]
48. Devkota, L.P.; Gyawali, D.R. Impacts of climate change on hydrological regime and water resources management of the Koshi River Basin. *Nepal. J. Hydrol. Reg. Stud.* **2015**, *4*, 502–515. [CrossRef]
49. Gusev, Y.M.; Nasonova, O.N.; Kovalev, E.E.; Ayzel, G.V. Impact of possible climate changes on river runoff under different natural conditions. *Proc. Int. Assoc. Hydrol. Sci.* **2018**, *379*, 293–300. [CrossRef]
50. Champagne, O.; Arain, M.A.; Leduc, M.; Coulibaly, P.; McKenzie, S. Future shift in winter streamflow modulated by the internal variability of climate in southern Ontario. *Hydrol. Earth Syst. Sci.* **2020**, *24*, 3077–3096. [CrossRef]
51. Šarauskienė, D.; Akstinas, V.; Nazarenko, S.; Kriaučiūnienė, J.; Jurgelėnaitė, A. Impact of physico-geographical factors and climate variability on flow intermittency in the rivers of water surplus zone. *Hydrol. Process.* **2020**, 1–13. [CrossRef]
52. Hamududu, B.; Killingtveit, A. Assessing climate change impacts on global hydropower. *Energies* **2012**, *5*, 305–322. [CrossRef]
53. Turner, S.; Hejazi, M.; Kim, S.; Clarke, L.; Edmonds, J. Climate impacts on hydropower and consequences for global electricity supply investment needs. *Energy* **2017**, *141*, 2081–2090. [CrossRef]
54. Lehner, B.; Czisch, G.; Vassolo, S. The impact of global change on the hydropower potential of Europe: A model-based analysis. *Eng. Policy* **2005**, *33*, 839–855. [CrossRef]
55. Tamm, O.; Luhamaa, A.; Tamm, T. Modeling future changes in the North-Estonian hydropower production by using SWAT. *Hydrol. Res.* **2016**, *47*, 835–846. [CrossRef]
56. Wellinghoff, J.; Pederson, J.; Morenoff, D.L. Facilitating hydrokinetic energy development through regulatory innovation. *Energy Law J.* **2008**, *29*, 397–420.

3

Evaluation of the Impact of Climate Change on Runoff Generation in an Andean Glacier Watershed

Rossana Escanilla-Minchel [1,*], Hernán Alcayaga [1], Marco Soto-Alvarez [1], Christophe Kinnard [2] and Roberto Urrutia [3]

1 Escuela de Ingeniería en Obras Civiles, Universidad Diego Portales, Santiago 8370109, Chile; hernan.alcayaga@udp.cl (H.A.); marco.sotoal@gmail.com (M.S.-A.)
2 Department of Environmental Sciences, University of Québec at Trois-Rivières, Trois-Rivières, QC G8Z 4M3, Canada; christophe.kinnard@uqtr.ca
3 Facultad de Ciencias Ambientales/Centro EULA-Chile de la Universidad de Concepción, Concepción 409541, Chile; rurrutia@udec.cl
* Correspondence: rescanillam@gmail.com

Abstract: Excluding Antarctica and Greenland, 3.8% of the world's glacier area is concentrated in Chile. The country has been strongly affected by the mega drought, which affects the south-central area and has produced an increase in dependence on water resources from snow and glacier melting in dry periods. Recent climate change has led to an elevation of the zero-degree isotherm, a decrease in solid-state precipitation amounts and an accelerated loss of glacier and snow storage in the Chilean Andes. This situation calls for a better understanding of future water discharge in Andean headwater catchments in order to improve water resources management in glacier-fed populated areas. The present study uses hydrological modeling to characterize the hydrological processes occurring in a glacio-nival watershed of the central Andes and to examine the impact of different climate change scenarios on discharge. The study site is the upper sub-watershed of the Tinguiririca River (area: 141 km^2), of which nearly 20% is covered by Universidad Glacier. The semi-distributed Snowmelt Runoff Model + Glacier (SRM+G) was forced with local meteorological data to simulate catchment runoff. The model was calibrated on even years and validated on odd years during the 2008–2014 period and found to correctly reproduce daily runoff. The model was then forced with downscaled ensemble projected precipitation and temperature series under the RCP 4.5 and RCP 8.5 scenarios, and the glacier adjusted using a volume-area scaling relationship. The results obtained for 2050 indicate a decrease in mean annual discharge (MAD) of 18.1% for the lowest emission scenario and 43.3% for the most pessimistic emission scenario, while for 2100 the MAD decreases by 31.4 and 54.2%, respectively, for each emission scenario. Results show that decreasing precipitation lead to reduced rainfall and snowmelt contributions to discharge. Glacier melt thus partly buffers the drying climate trend, but our results show that the peak water occurs near 2040, after which glacier depletion leads to reducing discharge, threatening the long-term water resource availability in this region.

Keywords: glacier; climate change; hydrological modelling; projections of climate impacts

---

1. Introduction

Water from glacier melt plays a vital role as a water resource for population located downstream in high mountain watersheds [1]. This natural water storage is being affected by climate change, as the reduced volume of glacier causes net losses in water storage, leading to significant but ephemeral increases in runoff during the ablation period [2]. Worldwide glacier retreat is threatening the sustainability of water resources in mountain regions, demonstrating the need to quantify the effects of climate change in these areas [3].

The central Andes are the main source of freshwater in Chile. The Andean cryosphere acts as a natural reservoir, storing large volumes of water in solid state. As such, high mountain watersheds exert a significant control on the production of runoff during the summer ablation period [4–6]. This, along with the constant population increase and urban expansion towards higher elevations areas, has led to an increase in water consumption [7]. Therefore, pressure on water resources from different users and consumption sector (sanitary, agricultural, mining, hydroelectrical, tourism, etc.), along with climatic disturbances, has significantly increased the uncertainty regarding the availability of future water and its management [8].

Climate change has become one of the greatest challenges facing society [9], especially due to the uncertainty caused by impacts on water resources. Worldwide projections indicate an increase in average temperature, while average precipitations are projected to decrease in the dry regions of mid, subtropical latitudes [10]. These climate projections should lead to an increase in the elevation of the zero-degree isotherm, which will cause a decrease in solid precipitation [11], generating significant glacier retreat accompanied by calving and melting of important glacier masses. Therefore, watersheds with significant glacier and snow cover will be highly exposed to variations caused by climate change [12]. From a hydrological perspective, warming temperatures in nivo-glacial watersheds can cause a decrease in solid-state precipitation, which reduces snow accumulation and translates into variations of maximum discharges and changes in overall water balance [13]. In the context of the central Andes, a general drying trend has been observed over the last century especially over the last decade [14], along with warming temperatures [15] and rising snowlines [16]. Glaciers in the northern and central Andes have been retreating over the 20th century [17–19], and shifted from a slightly positive (2000–2009) to a strongly negative mass balance in 2009–2018 [20], in line with the mega drought occurring in Chile since 2010 [14]. However, the response of river discharge to ongoing and future climate and cryosphere changes can be highly catchment specific and requires detailed modeling at the catchment scale [21].

A central concept on the consequences of climate change on runoff from glacierized basins is that of the peak water: warming will typically cause a transient increase in runoff from increased glacial melt, until a point where the depletion of glacier mass causes the runoff to decrease [3]. The timing of the peak water depends on the climate trajectory, so that it is essential to determine the peak water timing for different emission scenarios in order to establish mitigation measures against the long-term decrease in glacial runoff and its consequences on ecosystems and access to fresh water [22].

In this context, analyzing the consequences of climate change becomes crucial in areas where a glacial and snow regime predominates [23]. Hydrological models are a key tool for this task, as they are capable of simulating the response of hydrological processes to climate, allowing possible changes in runoff generation to be projected and assessed [24]. There is a wide range of modelling strategies and approaches, the most common model classification being based on the level of spatial simplification (spatially distributed, semi-distributed and lumped) and model complexity (physically based and conceptual), among others [24]. The level of spatial simplification depends on the amount of data available in the study, while model complexity reflects the detail of process representation within the model. Complex models require more input data and require specifying many unknown parameters, leading to equifinality uncertainties that add up to structural uncertainties and climate scenarios uncertainties [25–27].

The main objective of this study was to evaluate the impact of climate change on surface runoff generation in an Andean glacial watershed. To this end, the semi-distributed Snowmelt Runoff Model [28] was coupled with the Glacier module proposed by Ismail and Bogacki [29], calibrated and validated against runoff observations. Subsequently, outputs from seventy ensemble GCM climate scenarios under two emission scenarios, RCP 4.5 and RCP 8.5, were used to generate the precipitation and temperature time series through statistical downscaling. These time series where used to force the hydrological model, generating an ensemble of seventy discharge projections, providing a range of variations from present to the end of the century and allowing to estimate the impact of future climate conditions on the hydrological regime of the watershed.

2. Study Area

The study area is located in the central Chilean Andes in the O'Higgins Region (Figure 1a). It is a headwater sub-watershed of the Tinguiririca River watershed, which contains Universidad Glacier, the largest glacier in Chile outside Patagonia [30]. The Tinguiririca River (14,177 km^2) is characterized by a mixed rain-snow hydrological regime, with two maximum discharge peaks each year. The first maximum occurs during winter (April to October) in response to increased rainfall at lower elevations, while the second peak occurs during the ablation period (November to March) in response to the melting of snow and glacier ice. The Universidad Glacier contributes between 20 and 25% of the total river runoff, based on available runoff observations [31].

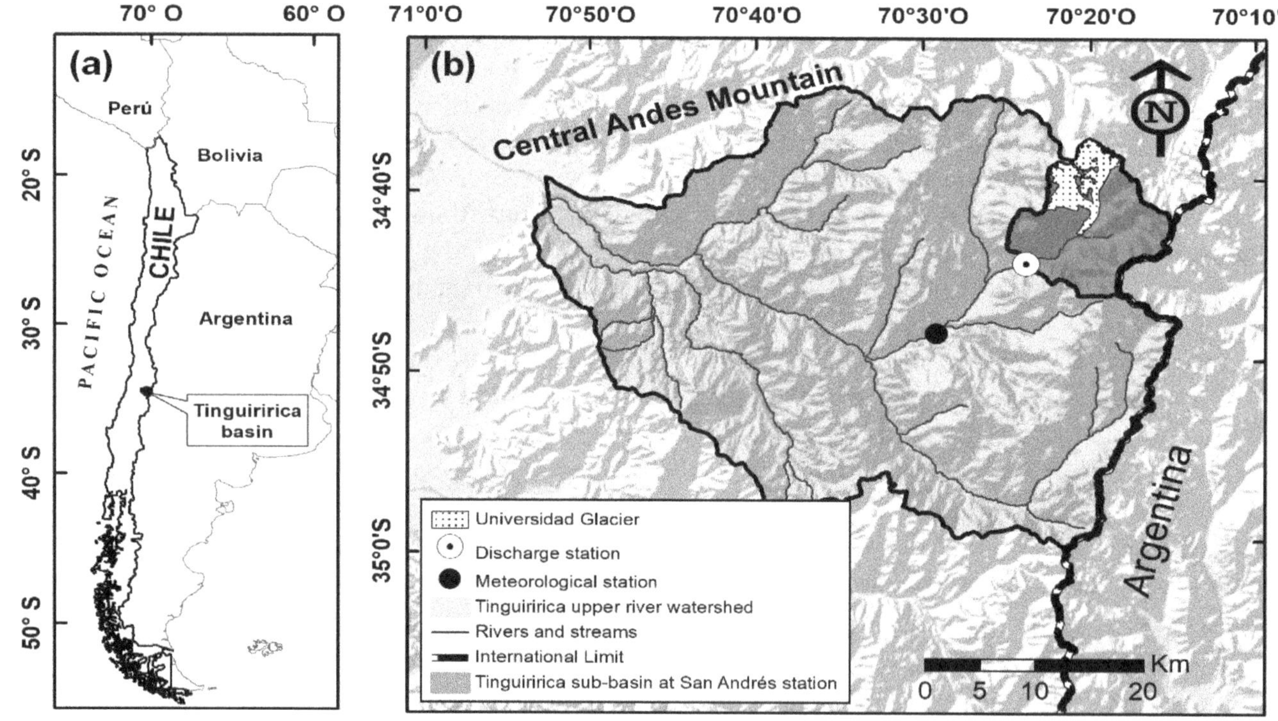

Figure 1. Location of study watershed, Universidad Glacier: (**a**) Geographic location of the Tinguiririca River watershed and international borders; (**b**) location of the upper Tinguiririca River watershed and studied sub-basin.

The sub-watershed was delimited based on the location of the San Andrés discharge monitoring station, which belongs to the hydroelectric plant of the same name, approximately 1500 m downstream of the glacier front at an altitude of 1724 m.a.s.l. (Figure 1b). This sub-watershed has an area of 140.73 km^2, 19.6% of which is glacier cover.

The Universidad Glacier is classified as a valley glacier and has a total area of 27.44 km^2, with an altitude range of 2427 to 4944 m.a.s.l. It has two main accumulation zones (Figure 2), which join in the ablation zone, located below 2790 m.a.s.l. Due to the enormous precipitation deficit resulting from dry summers and the ongoing drought in Central Chile [14], the runoff contribution of the glacier is vital for the region [32], whose consumptive water use is mainly for agricultural irrigation.

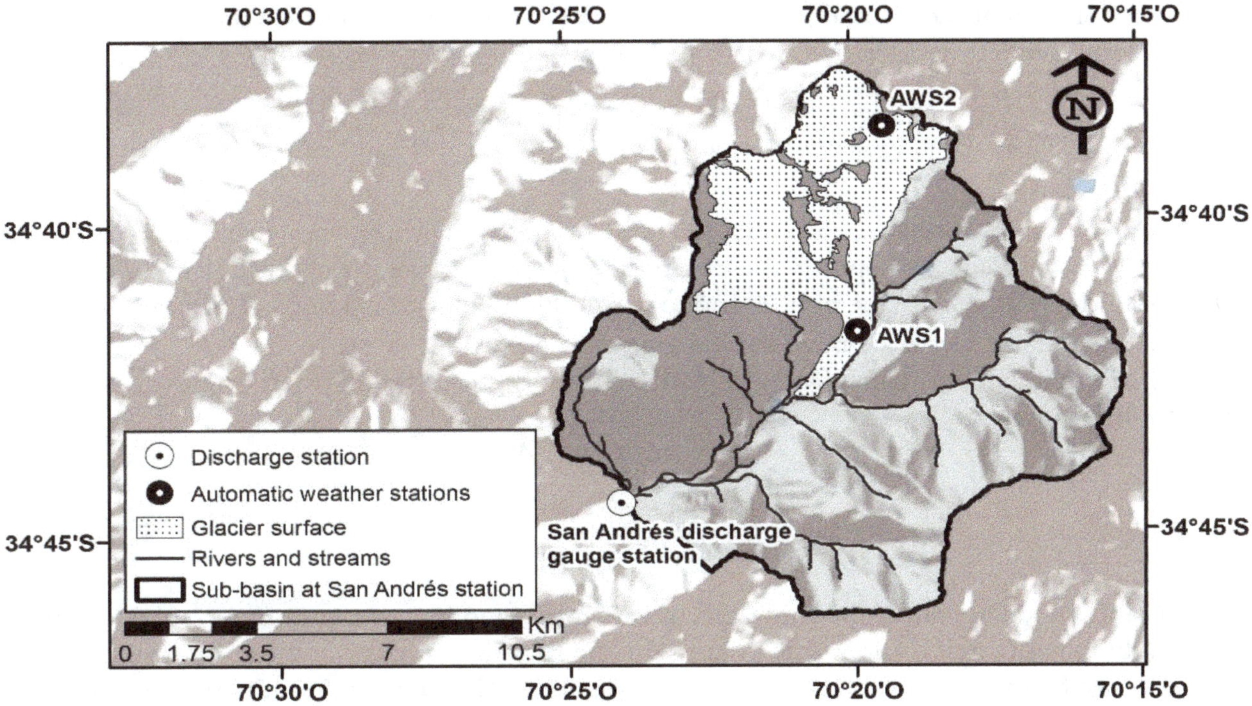

Figure 2. Delimitation of the sub-watershed upstream of the San Andrés discharge gauging station, as well the surface of Universidad Glacier, the water network and on-glacier automatic weather stations (AWS1 and AWS2).

In 2012, the Environmental Science Center EULA-Chile installed two automatic weather stations on Universidad Glacier (Figure 2) [30]. The first station (AWS1) is located in the ablation zone at 2790 m.a.s.l., while the second (AWS2) is located in the accumulation zone at 3629 m.a.s.l. [33]. These stations collected hourly data from November 2012 to November 2014. The closest permanent weather station is the Río Tinguiririca aguas abajo junta Río Azufre, hereafter denoted 'Río Tinguirirca station', located 18 km downstream of the San Andrés g auging station (Figure 1b) and in operation since 1980. EULA-Chile also collected glaciological information on Universidad Glacier. Mass balance was calculated for 2012–2014 using the glaciological method at a network of stakes and snow pits along the glacier centerline of the two glacier subbasins, and displacement was monitored at the stakes [30]. Mass-balance observations were used as a reference to project the future area of the glacier through a volume–area relationship.

3. Methods

3.1. Hydrological Model

3.1.1. Model Structure

The Snowmelt Runoff Model + Glacier is a semi-distributed hydrological model for simulating runoff on a daily scale in watersheds where snow and glacier melt dominate the runoff generation process. It has been successfully applied in over 100 snow-affected watersheds around the world, with areas up to 900,000 km^2 and elevations up to 8840 m.a.s.l. [28]. The standard version has two runoff components, i.e., snow and rain; therefore, for this study a third module was used to add the glacier component [29]. The model has been used in other studies to simulate the hydrological response of watersheds in Chile. The effectiveness of the model has been notably verified in the Tinguiririca River at Bajo Briones, downstream of the studied sub-watershed, where a correlation coefficient of 0.88 was found between observed and simulated streamflow [34].

The model estimates the total quantity of water produced by glacier melt and snow, along with liquid precipitation in the watershed, with a daily time step through Equation (1):

$$Q_{n+1} = \sum_{i=1}^{m}\left\{[M_{n,i} + R_{n,i} + G_{n,i}]\cdot\frac{10,000}{86,400}\right\}\cdot(1 - k_{n+1}) + Q_n k_{n+1} \quad (1)$$

where Q is the average daily discharge ($m^3\ s^{-1}$); M, R and G are average daily snowmelt, liquid precipitation, and glacier melt (cm day^{-1}), respectively; k is the recession coefficient (dimensionless); n is the index of the current day; i and m are the indices and total number of elevation bands. The expression $\left[\frac{10,000}{86,400}\right]$ represents a conversion factor to obtain the daily discharge units ($m^3\ s^{-1}$).

Daily runoff due to snowmelt (M), glacier melt (G) and precipitation (R) in cm day^{-1} is calculated using Equations (2)–(4):

$$M_{n,i} = c_S \cdot ddf \cdot T_{n,i} \cdot A_{S_{n,i}} \quad (2)$$

$$G_{n,i} = c_G \cdot a_G \cdot T_{n,i} \cdot A_{G_{n,i}} \quad (3)$$

$$R_{n,i} = c_R \cdot P_{n,i} \cdot A_{R_{n,i}} \quad (4)$$

where S, G and R indices refer to the snow, glacier and rain components of runoff; A represents the areas dominated by the different components in km^2; c is the runoff coefficient in these areas; ddf represents the degree-day factor for snow and a_G for glacier ice (cm $°C^{-1}$ day^{-1}); T is the number of positive degree days of air temperature (°C day) and P is liquid precipitation (cm). The SRM+G calculates snow and glacier melt based on positive temperatures; therefore, no glacier or snowmelt occurs when air temperatures are negative.

In this study, discharge observations available at the San Andrés station for the hydrological years from April 2008 to March 2014 were used for model calibration and validation, using the even years for calibration and odd years for validation, to reduce climatic dependence in the calibration [35]. The data from the on-glacier automatic weather stations, available from November 2012 to November 2014, were used to determine the precipitation and temperature gradients on the watershed. Daily precipitation and temperatures from the permanent Río Tinguiririca weather station were used to force the model.

3.1.2. Model Parameters

In watersheds such as the one studied here, the runoff modelled by SRM+G depends on air temperature, the degree-day factors for snow and glacier melt, and the snow and glacier areas in the watershed [29]. The degree-factor for snowmelt is calculated following the relationship established by Rango and Martinec (1996) [36], as shown in Equation (5):

$$ddf = 1.1\frac{\rho_s}{\rho_w} \quad (5)$$

where ρ_s is the snow density and ρ_w is water density, both in kg m^{-3}. The degree-day factors for snow varies during the ablation months (November to March), with a maximum value of 0.55 cm $°C^{-1}$ day^{-1} derived from Equation (5) in March, with a snow density of 500 kg m^{-3}, and a minimum value of 0.10 cm $°C^{-1}$ day^{-1} in November. The minimum value and the intermediate values (for December to February) are calibrated through density in order to better adjust the representation of the model during the thaw period. The seasonality in the snow ddf accounts for the unaccounted seasonal variation in incoming solar radiation and snow aging, the latter of which leads to increasing snow density, decreasing albedo and increased ablation [37–40]. The value of ddf is set to zero on the coldest winter months (April to October) to account for reduced incoming solar radiation and limited or no melting in winter. Unlike snow, glacial ice does not have significant albedo variations during the year [40]. Furthermore, this particular glacier has few debris on its surface, which are only located in the lower ablation zone); therefore, a constant degree-day factor for glacier melt (a_G) was used based on the study by Bravo et al. (2017) [31] at Universidad Glacier, who used a value of 0.8 cm °C^{-1} day^{-1} based on values for glacier ice from Hock (2005) [37]. Constant values have also been used in the Himalayas [29,41] and in other basins of central and southern Andes [42–44].

Constant runoff coefficients were used for snow, glacier, and rain areas. The values used in this work for the snow and ice runoff coefficients were taken from Ismail and Bogacki (2018) [29], for glacial watersheds in India, where $c_S = 0.8$ and $c_G = 0.7$. The runoff coefficient for rain area was adjusted as part of the model calibration. The recession coefficient k was calculated according to Rango and Martinec (1996) [36] using Equation (6):

$$k = x \cdot Q^{-y} \tag{6}$$

where k is the recession coefficient; x and y are constants that must be determined for the analyzed watershed and Q is the observed discharge. The recession coefficient varies on a daily time step, while the values of x and y are established as constants. Here, one pair of values (x, y) was used for the ablation period (November to March) and another for the precipitation period (April to October). The x and y values were set as model calibration parameters and determined through an optimization method using genetic algorithms. The automatic calibration method using genetic algorithms [45] seeks values for the calibration parameter that optimize the statistical indicators of model fit, which are presented in Section 3.2.

Air temperature was distributed to the catchment using a mean altitudinal temperature gradient of −6.8 °C km^{-1} calculated from the two weather stations located on the glacier (Figure 2). Precipitation were distributed over the catchment on each elevation band with a precipitation gradient, expressed by Equation (7):

$$PP_i = 1.02 \ln(Z_i) - 7.32 \tag{7}$$

where PP_i represents a percentage of precipitation measured at the Río Tinguiririca meteorological station in elevation band i; Z_i is the mean height in meters above sea level of the band i. This precipitation gradient was obtained through the relationship between the rainfall measured in AWS1 in summer months (Figure 2) (at 2790 m.a.s.l.) and the Río Tinguiririca meteorological station (at 1134 m.a.s.l.) for a one-year period of analysis. The obtained parameter for this study are consistent with those calculated by Ragettli et al. (2014) in the central Andes [46]. Four elevation bands were used to distribute precipitation and to calculate snow and glacier melt over the catchment, based on the precipitation and temperature gradients. A threshold temperature of 0 °C was used to separate liquid and solid precipitation. The parameters used in the model are listed in Table 1.

Table 1. SRM+G model parameters for Universidad Glacier watershed.

Parameters	Symbol	Value	Units	Remarks
Runoff Coefficient-Rain	C_R	0.6	-	Constant
Runoff Coefficient-Snow	C_S	0.8	-	Constant
Runoff Coefficient-Glacier	C_G	0.7	-	Constant
Degree-day Factor-Snow	ddf	0.10–0.55	cm °C^{-1} d^{-1}	November–March
Degree-day Factor-Glacier	aG	0.80	cm °C^{-1} d^{-1}	Constant
Recession Coefficient	x	1.0248	-	November–March
		0.9251	-	April–October
	y	0.0892	-	November–March
		0.0180	-	April–October
Temperature Lapse Rate	α	−6.8	°C km^{-1}	Constant

Glacierized areas were obtained from the public glacier inventory available from the Dirección General de Aguas (hereafter denoted DGA) in 2014 [47]. According to DGA (2014), Universidad Glacier retreated by 0.03 km^2 year^{-1} during de 1945–2011 period, reaching an area of 27.44 km^2 in April 2013. The snow cover area was determined on a daily scale using the Moderate Resolution Imaging Spectroradiometer (MODIS) Snow Cover Daily L3 Global 500-m grid satellite product

(MOD10A1) [48,49]. This product provides the normalized difference snow index (NDSI), calculated using Equation (8):

$$NDSI = \frac{MODIS_{B4} - MODIS_{B6}}{MODIS_{B4} + MODIS_{B6}} \tag{8}$$

where $MODIS_{B4,B6}$ are bands 4 and 6 of the sensor. Grid cells with a NDSI index larger than 0.4 were classified as snow [50]. A filter was applied to discard the whole daily grid when cloud cover was greater than 20%, and the missing grid was interpolated using the previous and next day with available information through nearest neighbor interpolation. Finally, the area of liquid precipitation is determined with the temperature lapse rate; if precipitation occurs within an elevation band with temperature greater than zero, then the area of that band not covered by snow or ice is contributes liquid precipitation.

3.2. Model Effiency Indicators

For both the calibration and validation periods, four statistical fit indicators were used to measure the model efficiency, based on the comparison between the observed (O) and simulated discharges (P).

The coefficient of determination (r^2) is defined as the squared Pearson's correlation coefficient [51] and is calculated according to Equation (9):

$$r^2 = \left(\frac{\sum_{i=1}^{n}(O_i - \overline{O})(P_i - \overline{P})}{\sqrt{\sum_{i=1}^{n}(O_i - \overline{O})^2}\sqrt{\sum_{i=1}^{n}(P_i - \overline{P})^2}} \right)^2 \tag{9}$$

where $\overline{O}$ is the average of the observed values and $\overline{P}$ is the average of the simulated values.

The Nash–Sutcliffe Efficiency (NSE), proposed by Nash and Sutcliffe (1970) is a normalized statistic that estimates the relative magnitude of the residual variance compared to the measured data variance [52]. A score of $NSE = 1$ indicates a perfect simulation, $NSE = 0$ indicates that the simulation is not better that using mean observations as sole predictor, and $NSE < 0$ indicates that the simulation is worse than using mean observations (Equation (10)):

$$NSE = 1 - \frac{\sum_{i=1}^{n}(O_i - P_i)^2}{\sum_{i=1}^{n}(O_i - \overline{O})^2} \tag{10}$$

The root mean square error (RMSE) is estimated through the square root of the difference between the observed and simulated values over the total number of data [52] and is calculated according to Equation (11):

$$RMSE = \sqrt{\frac{\sum_{i=1}^{n}(O_i - P_i)^2}{N}} \tag{11}$$

where N is the total amount of data.

The modified Kling–Gupta Efficiency (KGE') is an improvement on the Nash–Sutcliffe Efficiency proposed by Kling and Gupta (2009) [53], in which the correlation, deviation and variability components are weighted equally, solving systematic problems of maximum value underestimation [54]. The modified version [55] is calculated according to Equation (12):

$$KGE' = 1 - \sqrt{(r-1)^2 + (\alpha - 1)^2 + (\delta - 1)^2} \tag{12}$$

where r is the Pearson's correlation coefficient; α is the ratio of the coefficient of variation of the observed and simulated values; δ is the ratio between mean simulated and observed discharges.

The r^2, NSE and KGE' indicators have an optimum of 1, while $RMSE$ has an optimum of 0.

3.3. Climate Change Scenarios

In accordance with the Fifth Assessment Report (AR5) [10], this study focused on two of the four climate change scenarios, better known as Representative Concentration Pathways, RCP 4.5 and RCP 8.5. The first represents a stable scenario where the emission of greenhouse gas increases until the mid-21st century [56,57], while the second is a more unfavorable and high-emission scenario where emissions keep increasing exponentially to the end of the 21st century [58].

Based on the selected emission pathways, ensemble simulations from 18 global circulation models (GCMs) were selected from the Coupled Model Inter-Comparison Project Phase 5 (CMIP5) [59], providing a total of 70 future climate simulations. The ensembles are a set of simulations that characterize climate projections, where the differences in the initial conditions and the formulation of the models give rise to different evolutions of the climate system that can contribute information on the uncertainty associated with the error of each model [10]. The GCMs used in this study for climate simulations are presented in Table 2.

Table 2. Base GCMs used in this study.

GCM	Climate Modeling Center and Location	Ensemble	GCM	Climate Modeling Center and Location	Ensemble
ACCESS1	Centre for Australian Weather and Climate Research, Australia	r1i1p1	GFDL	NOAA Geophysical Fluid Dynamics Laboratory, USA	r1i1p1, r2i1p1, r3i1p1
BNU-ESM	College of Global Change and Earth System Science, Beijing Normal University, China	r1i1p1	GISS-E2	NSA Goddard Institute for Space Studies, USA	r1i1p1, r2i1p1, r3i1p1, r4i1p1, r5i1p1
CanESM2	Canadian Centre for Climate Modelling and Analysis, Canada	r1i1p1, r2i1p1, r3i1p1, r4i1p1, r5i1p1	HadGEM2-AO	National Institute of Meteorological Research, Korea Meteorological Administration, Korea	r1i1p1, r2i1p1, r3i1p1
CCSM4	National Centre for Atmospheric Research, USA	r1i1p1, r2i1p1, r3i1p1, r4i1p1, r5i1p1, r6i1p1, r7i1p1, r8i1p1	HadGEM2-CC	Met Office Hadley Centre, UK	r1i1p1, r2i1p1, r3i1p1, r4i1p1, r5i1p1
CESM1	Community Earth System Model Contributors	r1i1p1, r2i1p1, r3i1p1	IPSL-CM5A	Institut Pierre Simon Laplace, France	r1i1p1, r2i1p1, r3i1p1, r4i1p1
CMCC-CM	Centro Euro-Mediterrano per I Cambianmenti Climatici, Italy	r1i1p1, r2i1p1, r3i1p1, r4i1p1, r5i1p1	MIROC-ESM	Japan Agency for Marine Earth Science and Technology, Atmosphere and Ocean Research Institute (The University of Tokyo), and National Institute for Environmental Studies, Japan	r1i1p1
CSIRO-MK3	Commonwealth Scientific and Industrial Research Organization in collaboration with Queensland Climate Change Centre of Excellence, Australia	r1i1p1, r2i1p1, r3i1p1, r4i1p1, r5i1p1, r6i1p1, r7i1p1, r8i1p1, r9i1p1, r10i1p1	MPI-ESM	Max Planck Institute for Meteorology, Germany	r1i1p1, r2i1p1, r3i1p1
EC-EARTH	EC-EARTH consortium, Europe	r2i1p1, r8i1p1, r9i1p1, r12i1p1	MRI-CGCM3	Meteorological Research Institute, Japan	r1i1p1, r2i1p1, r3i1p1, r4i1p1, r5i1p1
FGOALS	LASG, Institute of Atmospheric Physics, Chinese Academy of Sciences and CESS, Tsinghua University, China	r1i1p1	NorESM1	Norwegian Climate Centre, Norway	r1i1p1, r2i1p1, r3i1p1

Downscaling

The GCMs have a spatial resolution that varies between approximately 1.5° and 2.5°, which is inadequate for the spatial scale of this study because the representativeness of the involved climate phenomena is lost, preventing characterizations at the watershed level [60]. As a result of this issue, there has been increasing use of regional climate models (RCMs), which transform the spatial scale of GCMs to a finer scale that is more applicable to watershed-level studies, with study areas under 40,000 km^2 [61]. The approaches used to downscale GCMs can be dynamic (use of regional climate models) or statistical. In both cases it is sought to eliminate the systematic bias in global models and transform the coarse GCM climate outputs towards at a reduced spatial resolution [62]. The dynamic methods have demanding computational requirements, such that the use of statistical methods is more common. In this study, the Quantile Mapping Bias Correction method was used to downscale the GCMs outputs to the watershed. This method uses as distinct bias correction for the percentiles of the cumulative distribution function (CDF) of the GCMs, with respect to the percentiles of the observed data. That is, for each GCM climate datum a CDF is determined, through which an occurrence probability is obtained. Subsequently, the GCM CDF is adjusted to the observed CDF, removing the bias and allowing it to be quantified in order to eliminate it in the projection phase [63]. The foregoing is defined according to Equation (13):

$$P_o = {F_o}^{-1}(F_m(P_m)) \tag{13}$$

where P_o is an observed climate variable; P_m is a variable simulated by the GMCs; F_m is the CDF of P_m and ${F_o}^{-1}$ is the inverse CDF corresponding to P_o.

In the case of precipitation, a two-parameter Gamma probability distribution was used, as expressed in Equations (14)–(16) [60], while for temperature a normal distribution was used, according to Equation (17) [64].

$$f_\gamma(X|\alpha,\beta) = X^{\alpha-1}\cdot\frac{1}{\beta^\alpha\cdot\Gamma(\alpha)}\cdot e^{\frac{-x}{\beta}} \quad ;X \geq 0; \quad \alpha,\beta > 0 \tag{14}$$

$$\Gamma(\alpha) = \int_0^\infty u^{\alpha-1}e^{-u}du \tag{15}$$

$$\alpha = \frac{\sigma^2}{\mu}, \quad \beta = \left(\frac{\mu}{\sigma}\right)^2 \tag{16}$$

$$f_N(X|\mu,\sigma) = \frac{1}{\sigma\cdot\sqrt{2\pi}}\cdot e^{\frac{-(x-\mu)^2}{2\sigma^2}} \quad ;X \in R \tag{17}$$

where X is the variable to evaluate, μ is the average of the total time series of data, and σ is the standard deviation of the series.

Three periods were defined for the downscaling process. The first period is associated with observed climate data (temperature and precipitation) from 1980 to 2005 and is referred to as the observed historical period. Historical climate data were obtained from the meteorological station Río Tinguirica. The second period covers the same period (1980–2005), but includes the climate data simulated by the GCMs, and is referred to as the simulated historical period. Lastly, the third period extends from 2006 to 2100 and corresponds to climate data projected by the GCMs and is referred to as the projected period.

3.4. Projections

Future discharge was projected by forcing the hydrological model with the downscaled GCM precipitation and temperature, extrapolating from the weather station to the rest of the basin, using the model parameters in Table 1. To estimate future glacier area, a volume-area scaling approach was used, following the empirical relation in Equation (18) [65]:

$$V = c(S)^\gamma \tag{18}$$

where V is the glacier volume in km^3, S is the glacier area in km^2, and γ and c are constants. In this case, γ was set to 1.357 following Bahr et al. (1997) [66], and k is derived from Universidad glacier's initial volume $V_{(t=2009)}$ = 1.67 km w.e.^{3} and initial area $S_{(t=2009)}$ = 29.03 km^2, which results in k = 0.017 km$^{3-2\gamma}$. These initial volume comes from an interpolation of radio-echo soundings carried out on the glacier for that year [67]. Knowing these constants, it is possible to project the area of the glacier from the simulated volume changes, through Equation (19) [68]:

$$\Delta V(t) = b_n(t)S(t) \tag{19}$$

where ΔV is the volume change, b_n is the simulated net glacier mass balance and S is the glacier area. Computed negative (positive) changes in area are removed (added) starting from the lowest elevation band in the model. To estimate the future snow cover area, multiple regression models that relate the snow cover area to temperature and precipitation were used [69], following Equation (20):

$$A_s = T{\cdot}T_{factor} + PP{\cdot}PP_{factor} \pm CI \tag{20}$$

where A_S is the snow cover area in km^2; T is the daily temperature in °C and PP is daily precipitation in mm. The values of T_{factor}, PP_{factor} and CI were determined for this study from multiple regression, using the same methodology used by Khadha et al. (2014) [69] for a two-year period and used on a daily scale. The seasonal snow area relationships were determined with Equation (20), as shown in Table 3.

Table 3. Multiple regression parameter for the snow area projections.

Season	T_Factor	PP_Factor	CI
Summer	38.375	5.586	−47.032
Autumn	−22.68	−4.667	48.575
Winter	−5.404	1.263	43.616
Spring	34.734	5.037	−3.468

This relationship, for all the seasons results in a coefficient of determination of r^2 = 0.89, where the observed snow areas are compared with those simulated using Equation (20) for the years with data availability. Finally, the distribution of the calculated snow area among the four elevation bands was carried out by filling from the highest band to the lowest.

4. Results

4.1. Hydrological Model

The hydrological model (SRM+G) results for the analyzed period (2008–2014) are presented in the hydrograph in Figure 3, while the correlations between observed and simulated discharges for the calibration and validation periods are shown in Figure 4a,b. Table 4 shows the model efficiency measures for the two subperiods. In the hydrograph in Figure 3, it can be observed that the SRM+G model better represents the observed discharge during the melting month (November–March), on runoff peaks. During the winter period, the model has more difficulties representing the observed low discharge.

The scatterplots in Figure 4 show more dispersion for the validation periods than for the calibration period; however, both cases have a correlation trend consistent with the hydrograph in Figure 3, with more dispersion for discharges greater than 20 m^3/s. The r^2, NSE and KGE are above 0.70 and in some cases close to 1 (Table 4). Usually, a value of 0.7 or more for these indicators indicates efficient modeling [70].

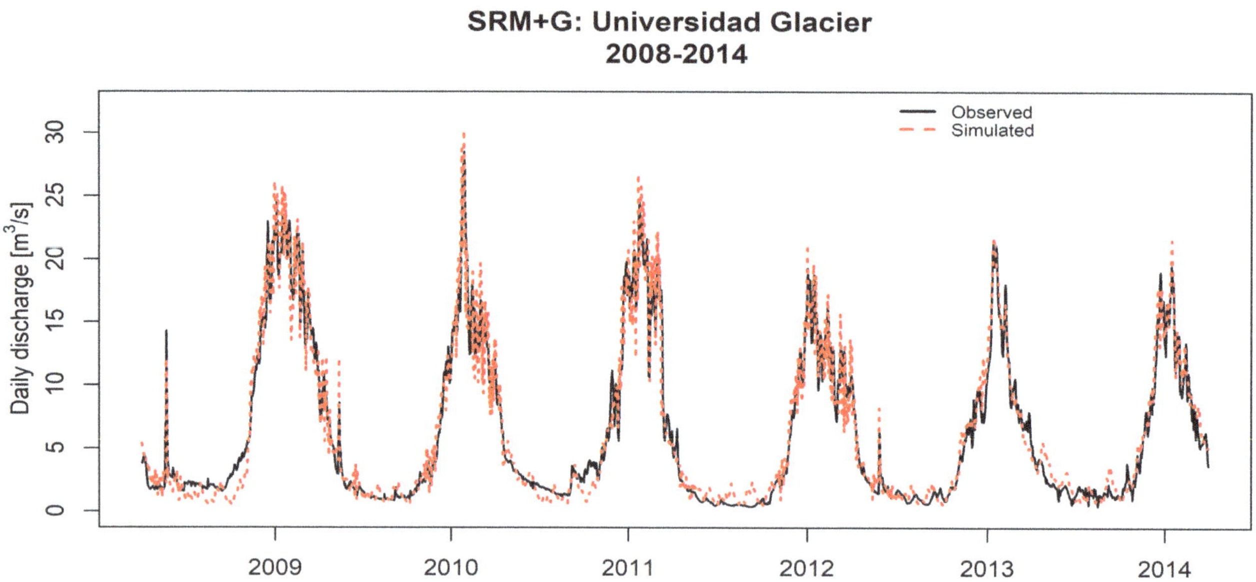

Figure 3. Hydrogram of the SRM+G hydrological model at the San Andrés discharge station. The calibration period corresponds to the even years and the validation period to the odd years. The high discharge periods extend from November to March, during which the melting period can be recognized.

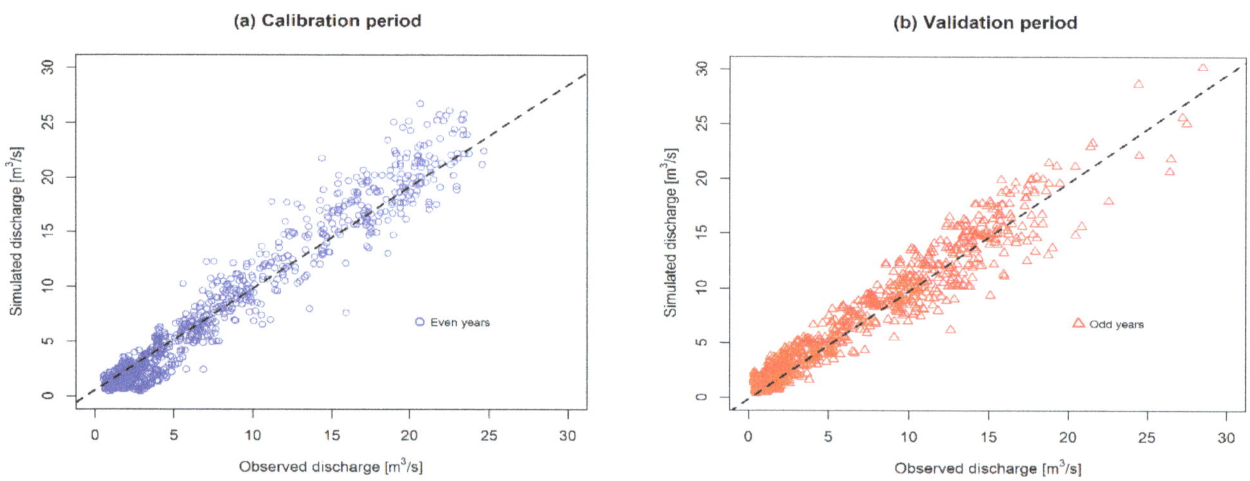

Figure 4. Comparison between observed and simulated values: (**a**) For the calibration period for even years; (**b**) For the validation period for odd years. The dashed line indicates the correlation trend.

Table 4. Statistical efficiency indicator.

Indicator	Values	
	Calibration	**Validation**
r^2	0.93	0.90
NSE	0.92	0.88
RMSE	1.32	1.15
KGE'	0.90	0.89

4.2. Climate Change

In Figure 5, the result of the statistical downscaling of the monthly averages of both precipitation (Figure 5a) and temperature (Figure 5c) are shown. The methodology was applied for both emission scenarios (RCP 4.5 and RCP 8.5) and the bias of the 70 GCMs used in this study was corrected (Figure 5b,d).

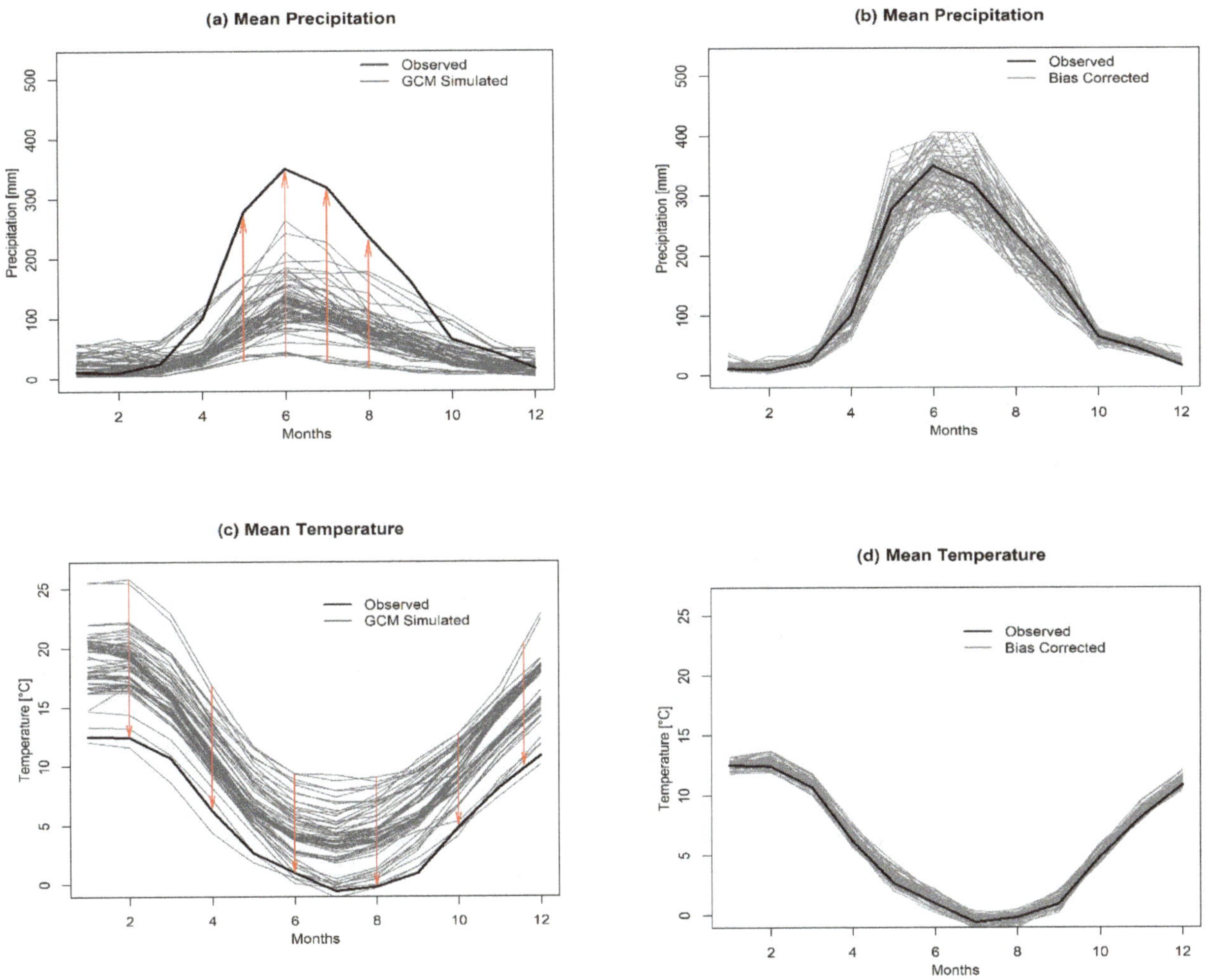

Figure 5. Statistical scaling results for precipitation and temperature, applied to the 70 GCM simulations used in this study; (**a**,**c**): Indicate the observed value curves in contrast to the values simulated by the GCMs. (**b**,**d**): Show the observed value curves along with the bias-corrected curves.

In Figure 5a, it can be observed that precipitation is overestimated in summer by some models, while it is underestimated by all models in the winter months. The red arrows indicate the expected effect of the bias correction procedure, i.e., bringing the models with greater dispersion closer to the expected mean. Figure 5b shows how the bias correction produces a better fit with the mean value of historical precipitation, although it generates a remaining dispersion around the observed values which is greatest in winter, reaching over 100 mm. In the case of temperature (Figure 5c) the same behavior is observed, with most of the models overestimating temperature, but in this case throughout the year. After the downscaling, little scatter remains and time series of temperatures correctly fit the observed historical period obtained (Figure 5d).

4.3. Projections

The simulated glacier mass balance for all GCMs analyzed shows a significant loss of glacier mass under both scenarios of climate change (Figure 6). Under the RCP 4.5 scenario, the ensemble of projected glacier volumes shows decreases ranging from 12 to 37% by 2050, while by the end of the century the ensemble volume loss trajectory reaches from 25 to 68%. The volume loss is greater under the RCP 8.5 scenario, decreasing between 20 and 48% by 2050 and reaching 51 to 84% by the end of the century.

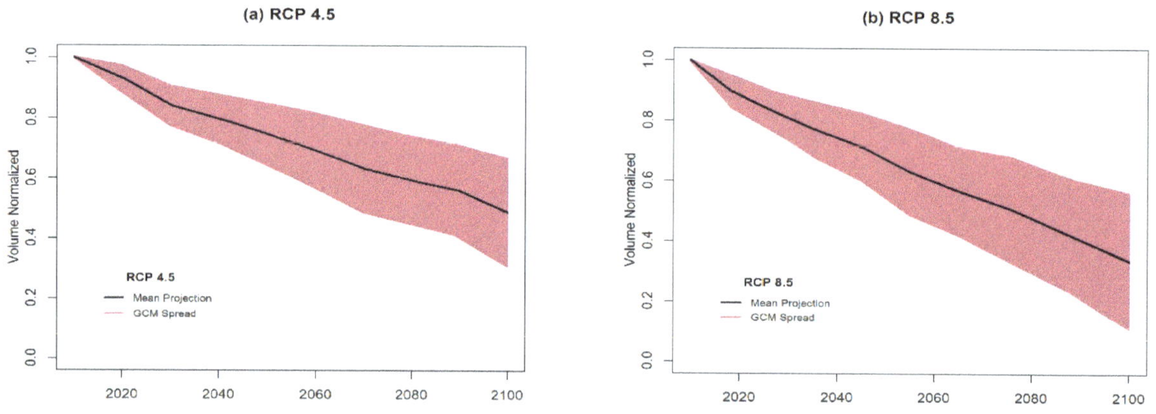

Figure 6. Volume changes on the Universidad Glacier; (**a**): Under RCP 4.5. (**b**): Under RCP 8.5. The volume is normalized by the initial glacier volume.

Table 5 shows the spread in downscaled GCM projections of climate variables for both emission scenarios in 2050 and 2100. There is a progressive decrease in precipitation and an increase in temperature for both scenarios, by 2050, with increasing differences between scenarios by the end of the century. The projected negative precipitation trend is particularly worrying in light of the mega drought that currently affects the area and which caused serious impacts on water resource availability. Based on the projections in Table 5, the simulated discharges using the downscaled precipitation and temperature under RCP 4.5 and RCP 8.5 scenarios are shown in Figure 7.

Table 5. Climate data projections for the two emission scenarios used in this study.

	Precipitation [mm]					
Emission Scenario	**2050**			**2100**		
	5%	**Mean**	**95%**	**5%**	**Mean**	**95%**
RCP 4.5	−7%	−16%	−25%	−12%	−24%	−35%
RCP 8.5	−10%	−20%	−30%	−21%	−40%	−58%
	Temperature [°C]					
Emission Scenario	**2050**			**2100**		
	5%	**Mean**	**95%**	**5%**	**Mean**	**95%**
RCP 4.5	1.5	1.8	2.1	2.3	2.9	3.4
RCP 8.5	2.3	2.9	3.4	3.2	3.9	4.6

In both cases, there are 70 discharge series estimates, associated with each GCM and their ensemble members. An increase in runoff is observed for both scenarios, which reaches a peak around 2040, after which the runoff begins to decline. These results hence suggest that the peak water in this basin has not yet been reached and that increased glacier melting in response to warming will increase discharge until the peak water is reached around 2040. The decreasing discharge trend afterward is mainly due to decreasing glacier mass and the associated reduced contribution to runoff. A comparison of the two scenarios (RCP 4.5 and RCP 8.5) shows that in the first case, there is a sustained increase in mean annual discharge (MAD) over time until 2040, after which the MAD begins to decline. The peak water is better pronounced under scenario RCP4.5 and more subdued under RCP8.5. Increased mass loss under RCP8.5 depletes glacier volume faster (Figure 6), which restricts the contribution of glacial runoff by 2040 compared to RCP4.5. The changing runoff contributions are examined in Figure 8, where the contribution of the monthly average runoff for the year 2100 is shown under both emission scenarios RCP 4.5 and RCP 8.5.

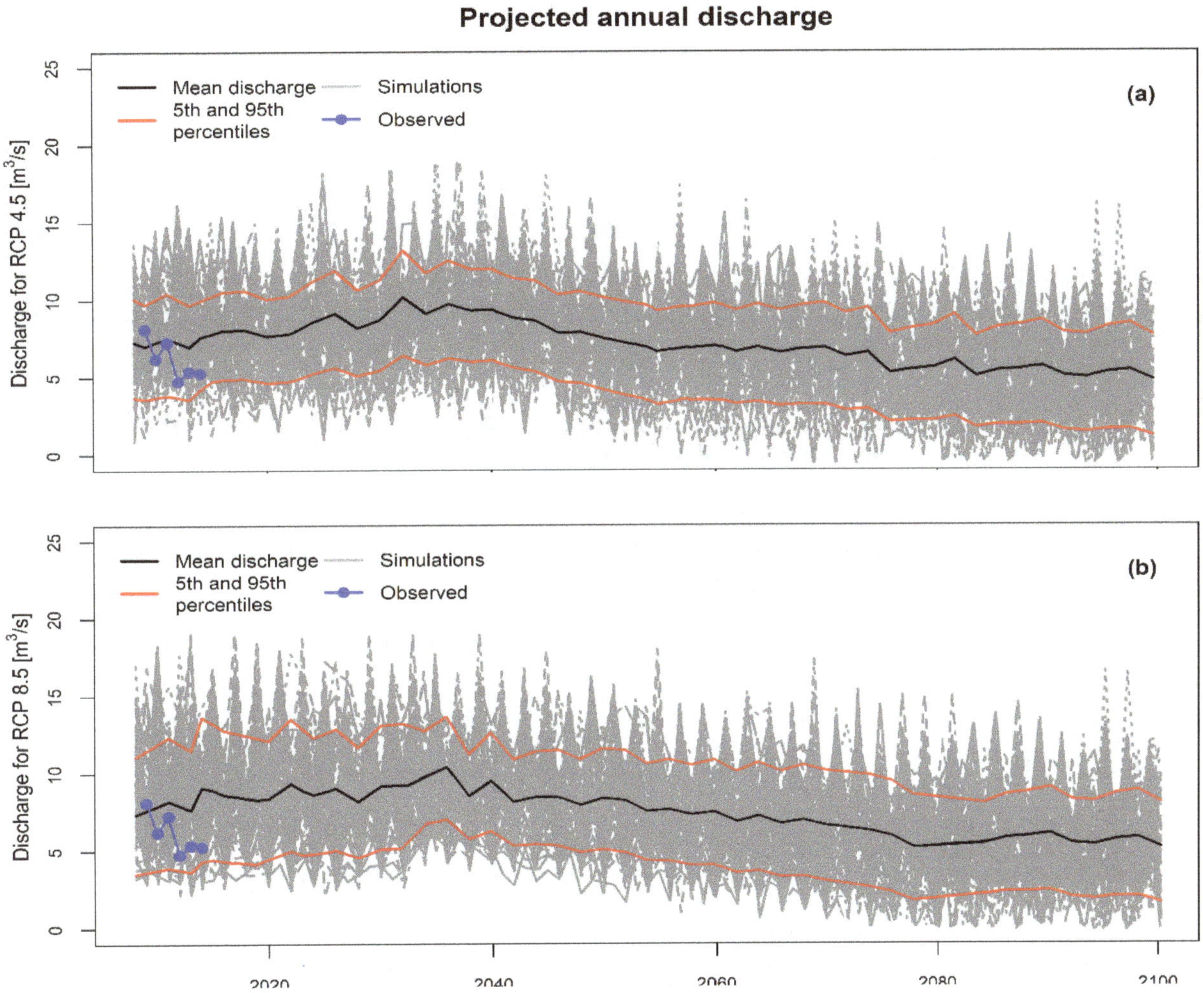

Figure 7. Results of the 70 simulations for two emission scenarios. (**a**) RCP 4.5 scenario; (**b**) RCP 8.5 scenario. The confidence intervals associated with the 5th and 95th percentiles are indicated in red, along with the mean of the discharges simulated by the 70 GCM simulations under the two scenarios (in black). Blue dots indicate the observed mean annual discharges (2008–2014).

The contribution of rainfall during the winter months (April–August) decreases by 2100 under both scenarios, due to the projected decreases in precipitation, as shown in Table 5. The reduced precipitation also explains part of the decreased snowmelt contributions in the future, which is further exacerbated by the conversion of snowfall to rain under warming temperatures. The decreasing snowmelt and rain contribution to streamflow are particularly worrying for the long-term sustainability of water resources downstream. Figure 8 shows that the bulk of discharge in this headwater catchments originates from glacier melt, and that the relative contribution of glacier melt to discharge increases in the future as runoff from snowmelt and rain decrease. As such, the flow remains dominated by glacial melt in the warmest months during the ablation period (October to March) under both scenarios but the peak contribution is much reduced an occurs earlier, by one month under RCP4.5 (in January) and by two months under RCP8.5 (in December). Hence, glacier melt constitutes a key buffer for the drying and warming trend projected in this region. However, and as seen in Figure 7, the peak water is reached around 2040 and the progressive depletion of glacier volume leads to decreasing glacier runoff onward, threatening the long-term sustainability of meltwater resources in this region.

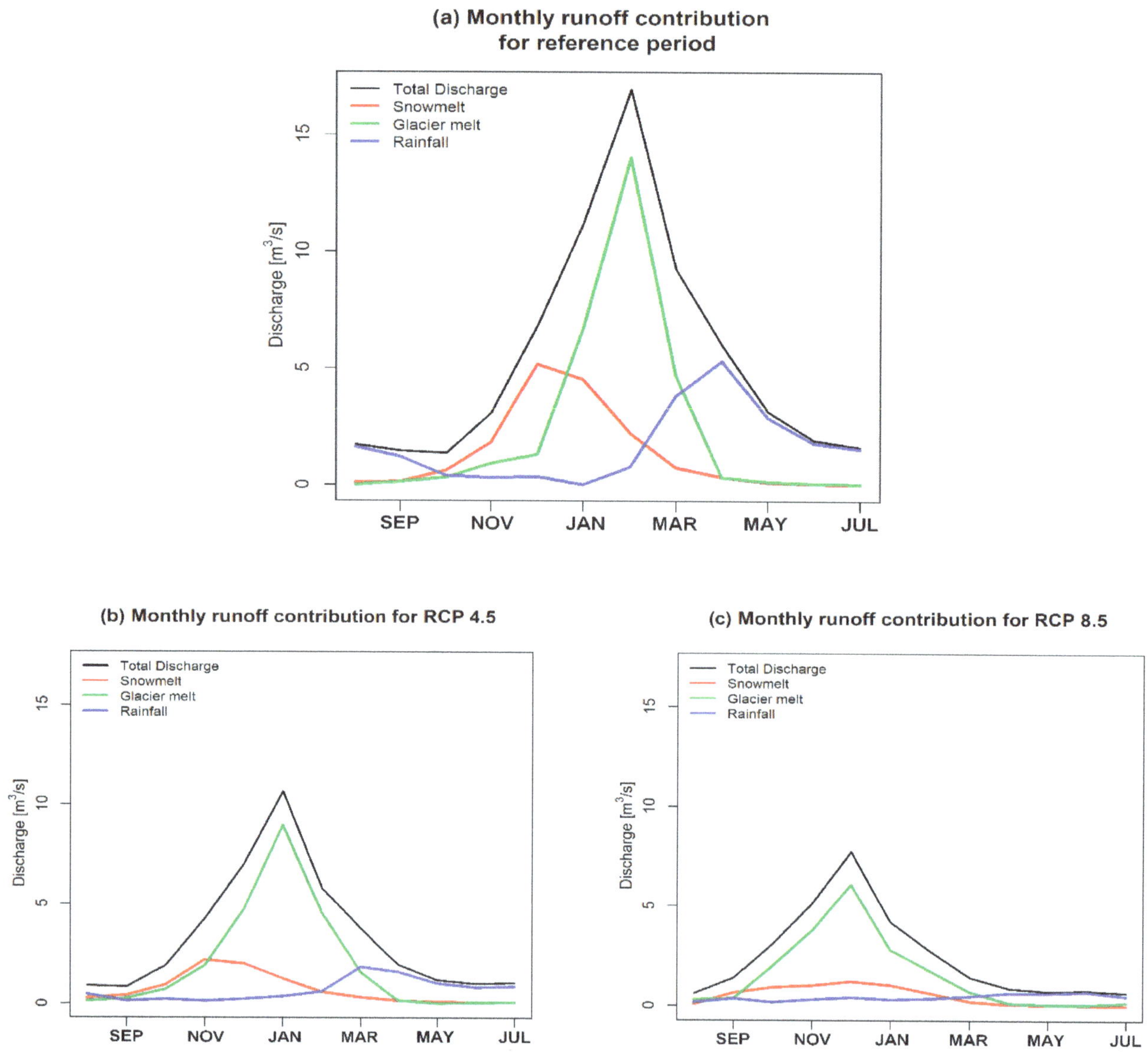

Figure 8. Monthly runoff contributions: (**a**) For the reference period (2008–2014); (**b**) For 2100 under emission scenario RCP 4.5 and (**c**) For 2100 under emission scenario RCP 8.5.

5. Discussion and Conclusions

The Snowmelt Runoff Model + Glacier hydrological model was able to efficiently simulate the hydrological response of the studied watershed. Values above 0.8 were obtained for the four statistical indicators used to measure efficiency, indicating reliable simulations [70]. In comparison, the performance of the hydrological modeling of the Tinguiririca River at Bajo Briones for 1987–1988 by Escobar [34], which had a r^2 of 0.88, was surpassed, even taking into account that the relative glacier cover is smaller within the contributing area of the Bajo Briones station. This increase in performance is attributed to the addition of the glacier module, which expectedly improves the hydrological response of the watershed. While the model has a greater percentage error during low-water periods, it correctly simulates melting periods when the water supply is greater. At an annual scale, the observed and simulated mean discharge values differ by only 7%. SRM+G is a hydrological model that can adequately represent runoff in basins with an area smaller than 500 km^2. The model is thus particularly flexible for data-scarce regions where there is insufficient information to apply a fully-distributed model. The increased error observed in winter simulated discharge is due to the sensitivity of the model in distinguishing between liquid and solid precipitation. Another aspect that could improve the representation and efficiency of the model in the future would be to incorporate the effect of the debris on the lowermost glacier of the Universidad Glacier has, which could influence the spatial–temporal variation of the degree-day factor for ice and thus better represent the associated glacier ice melt processes [71–73]. It must be stressed that, for studies related to climate change, a greater number of years to calibrate and validate the selected hydrological model is recommended [74]. However, this is often difficult or impossible to achieve in high altitude, remote glacierized catchments such as the one studied here. Still, the good performance obtained in this limited data context suggests that the parsimonious modelling approach used is useful to project future discharge in response to climate change scenarios.

The bias correction procedure based on the quantile mapping method resulted in rescaled GCM precipitation that closely approach the observed values of the historical period, but with greatest remaining dispersion during winter, over 100 mm. This remaining scatter causes large positive discharge deviations that exceed the 95% confidence range (Figure 7). According to the downscaled projections, there will be a decrease in mean annual discharges with respect to current conditions. Mean annual discharge are projected to decrease by 18.1 and 43.3% by 2050 for RCP 4.5 and RCP 8.5, respectively, while by 2100, the projected decreases reach 31.4 and 54.2%. Similar responses have been projected in Chile, but with an even greater decrease, in the upper Juncal River basin [75], where mean annual discharge was projected to decrease by 62% for the RCP 8.5 scenario for 2100. Omani et al. (2016) show similar results [76] for a glacierized basin located in the central Andes, where they report a decrease in MAD of 17% for RCP 4.5 and 30% for RCP 8.5, by mid-century.

While according to the projections there will be an increase in water supply from the catchment until around 2040, it must be stressed that this increase will occur at the cost of a progressive and irreversible depletion in glacier volume, which will drive the reducing discharge trend from 2040 onward. The results of this work on glacier volume loss are consistent with other similar studies in Southern Andes, which reported losses around 25 and 68% for RCP 4.5 and 51 to 84% for RCP 8.5, both at the end of the century [3,77,78].

It has been shown that climate change has a negative impact in Andean zones because it increases the retreat of glacier masses and thus causes the irreversible loss of natural reservoirs, putting the sustainability of ecosystems into jeopardy and intensifying conflicts among stakeholders [23,79]. The projected ephemeral increase in meltwater outflux until 2040 simulated in this study thus represents a nonrenewable water source, as glacier depletion leads to reduced glacial runoff afterward. It has been estimated that the mega drought currently affecting central Chile will intensify by 2100 [80]. As such, the heavy dependence on water from the cryosphere may become a conflict scenario in various parts of the country near the end of this century. The temporary increase in water availability resulting from increased glacier melting could also have other consequences such as an increase in erosion and a reduction in infiltration time, which would affect aquifer recharge and the hydrological regime of downstream rivers [22]. Adaptation plans are highly recommended amid changes in the hydrological regime in these types of watersheds [81], especially those with intensive agricultural and industrial uses.

Author Contributions: Conceptualization, R.E.-M. and H.A.; Methodology, R.E.-M., H.A., C.K.; software, R.E.-M.; Formal Analysis, R.E.-M. and H.A.; Investigation, R.E.-M., H.A., M.S.-A., C.K.; Data Curation, R.E.-M. and R.U.; writing—original draft preparation, R.E.-M. and H.A.; writing—review and editing, H.A., R.U. and C.K.; visualization, R.E.-M., C.K.; supervision, H.A. All authors have read and agreed to the published version of the manuscript.

Acknowledgments: This research was supported by project ANID/FONDAP/15130015, Water Research Center for Agriculture and Mining (CRHIAM).

References

1. Migliavacca, F.; Confortola, G.; Soncini, A.; Senese, A.; Diolaiuti, G.A.; Smiraglia, C.; Barcaza, G.; Bocchiola, D. Hydrology and potential climate changes in the Rio Maipo (Chile). *Geogr. Fis. E Din. Quat.* **2015**, *32*, 155–168.
2. Huh, K.; Baraër, M.; Mark, B.; Ahn, Y. Evaluating Glacier Volume Changes since the Little Ice Age Maximum and Consequences for Stream Flow by Integrating Models of Glacier Flow and Hydrology in the Cordillera Blanca, Peruvian Andes. *Water* **2018**, *10*, 1732. [CrossRef]
3. Huss, M.; Hock, R. Global-scale hydrological response to future glacier mass loss. *Nat. Clim. Chang.* **2018**, *8*, 135–140. [CrossRef]
4. Ayala, A.; Pellicciotti, F.; MacDonell, S.; McPhee, J.; Vivero, S.; Campos, C.; Egli, P. Modelling the hydrological response of debris-free and debris-covered glaciers to present climatic conditions in the semiarid Andes of central Chile. *Hydrol. Process.* **2016**, *30*, 4036–4058. [CrossRef]
5. Bown, F.; Rivera, A.; Acuña, C. Recent glacier variations at the Aconcagua basin, central Chilean Andes. *Ann. Glaciol.* **2008**, *48*, 43–48. [CrossRef]
6. Gascoin, S.; Kinnard, C.; Ponce, R.; Lhermitte, S.; MacDonell, S.; Rabatel, A. Glacier contribution to streamflow in two headwaters of the Huasco River, Dry Andes of Chile. *Cryosphere* **2011**, *5*, 1099–1113. [CrossRef]
7. Aitken, D.; Rivera, D.; Godoy-Faúndez, A.; Holzapfel, E. Water scarcity and the impact of the mining and agricultural sectors in Chile. *Sustainability* **2016**, *8*, 128. [CrossRef]
8. Valdés-Pineda, R.; Pizarro, R.; García-Chevesich, P.; Valdés, J.B.; Olivares, C.; Vera, M.; Balocchi, F.; Pérez, F.; Vallejos, C.; Fuentes, R.; et al. Water governance in Chile: Availability, management and climate change. *J. Hydrol.* **2014**, *519*, 2538–2567. [CrossRef]
9. Rössler, O.; Fischer, A.M.; Huebener, H.; Maraun, D.; Benestad, R.E.; Christodoulides, P.; Soares, P.M.M.; Cardoso, R.M.; Pagé, C.; Kanamaru, H.; et al. Challenges to link climate change data provision and user needs: Perspective from the COST-action VALUE. *Int. J. Climatol.* **2019**, *39*, 3704–3716. [CrossRef]
10. IPCC. *Fifth Assessment Report Synthesis Report: Climate Change*; Intergovernmental Panel on Climate Change: Geneva, Switzerland, 2014; ISBN 978-92-9169-343-6.
11. Araya-Osses, D.; Casanueva, A.; Román-Figueroa, C.; Uribe, J.M.; Paneque, M. Climate change projections of temperature and precipitation in Chile based on statistical downscaling. *Clim. Dyn.* **2020**, *54*, 4309–4330. [CrossRef]
12. Pellicciotti, F.; Ragettli, S.; Carenzo, M.; McPhee, J. Changes of glaciers in the Andes of Chile and priorities for future work. *Sci. Total Environ.* **2014**, *493*, 1197–1210. [CrossRef] [PubMed]
13. Hayat, H.; Akbar, T.A.; Tahir, A.A.; Hassan, Q.K.; Dewan, A.; Irshad, M. Simulating Current and Future River-Flows in the Karakoram and Himalayan Regions of Pakistan Using Snowmelt-Runoff Model and RCP Scenarios. *Water* **2019**, *11*, 761. [CrossRef]
14. Garreaud, R.D.; Boisier, J.P.; Rondanelli, R.; Montecinos, A.; Sepúlveda, H.H.; Veloso-Aguila, D. The Central Chile Mega Drought (2010–2018): A climate dynamics perspective. *Int. J. Climatol.* **2020**, *40*, 421–439. [CrossRef]
15. Falvey, M.; Garreaud, R.D. Regional cooling in a warming world: Recent temperature trends in the southeast Pacific and along the west coast of subtropical South America (1979–2006). *J. Geophys. Res.* **2009**, *114*, D04102. [CrossRef]
16. Carrasco, J.F.; Casassa, G.; Quintana, J. Changes of the 0 °C isotherm and the equilibrium line altitude in central Chile during the last quarter of the 20th century/Changements de l'isotherme 0 °C et de la ligne d'équilibre des neiges dans le Chili central durant le dernier quart du 20ème siècle. *Hydrol. Sci. J.* **2005**, *50*. [CrossRef]
17. Kinnard, C.; Ginot, P.; Surazakov, A.; MacDonell, S.; Nicholson, L.; Patris, N.; Rabatel, A.; Rivera, A.; Squeo, F.A. Mass Balance and Climate History of a High-Altitude Glacier, Desert Andes of Chile. *Front. Earth Sci.* **2020**, *8*. [CrossRef]
18. Le Quesne, C.; Acuña, C.; Boninsegna, J.A.; Rivera, A.; Barichivich, J. Long-term glacier variations in the Central Andes of Argentina and Chile, inferred from historical records and tree-ring reconstructed precipitation. *Palaeogeogr. Palaeoclimatol. Palaeoecol.* **2009**, *281*, 334–344. [CrossRef]
19. Masiokas, M.H.; Christie, D.A.; Le Quesne, C.; Pitte, P.; Ruiz, L.; Villalba, R.; Luckman, B.H.; Berthier, E.; Nussbaumer, S.U.; González-Reyes, Á.; et al. Reconstructing the annual mass balance of the Echaurren Norte

glacier (Central Andes, 33.5° S) using local and regional hydroclimatic data. *Cryosphere* **2016**, *10*, 927–940. [CrossRef]

20. Dussaillant, I.; Berthier, E.; Brun, F.; Masiokas, M.; Hugonnet, R.; Favier, V.; Rabatel, A.; Pitte, P.; Ruiz, L. Two decades of glacier mass loss along the Andes. *Nat. Geosci.* **2019**, *12*, 802–808. [CrossRef]
21. Ragettli, S.; Immerzeel, W.W.; Pellicciotti, F. Contrasting climate change impact on river flows from high-altitude catchments in the Himalayan and Andes Mountains. *Proc. Natl. Acad. Sci. USA* **2016**, *113*, 9222–9227. [CrossRef]
22. Gleick, P.H.; Palaniappan, M. Peak water limits to freshwater withdrawal and use. *Proc. Natl. Acad. Sci. USA* **2010**, *107*, 11155–11162. [CrossRef] [PubMed]
23. Casassa, G.; López, P.; Pouyaud, B.; Escobar, F. Detection of changes in glacial run-off in alpine basins: Examples from North America, the Alps, central Asia and the Andes. *Hydrol. Process.* **2009**, *23*, 31–41. [CrossRef]
24. Hrachowitz, M.; Clark, M.P. HESS Opinions: The complementary merits of competing modelling philosophies in hydrology. *Hydrol. Earth Syst. Sci.* **2017**, *21*, 3953–3973. [CrossRef]
25. Arsenault, R.; Brissette, F.P. Continuous streamflow prediction in ungauged basins: The effects of equifinality and parameter set selection on uncertainty in regionalization approaches. *Water Resour. Res.* **2014**, *50*, 6135–6153. [CrossRef]
26. Ficklin, D.L.; Barnhart, B.L. SWAT hydrologic model parameter uncertainty and its implications for hydroclimatic projections in snowmelt-dependent watersheds. *J. Hydrol.* **2014**, *519*, 2081–2090. [CrossRef]
27. Nemri, S.; Kinnard, C. Comparing calibration strategies of a conceptual snow hydrology model and their impact on model performance and parameter identifiability. *J. Hydrol.* **2020**, *582*, 124474. [CrossRef]
28. Martinec, J.; Rango, A.; Roberts, R. Snowmelt runoff model (SRM) user's manual. *Agric. Exp. Stn. Spec. Rep.* **2008**, *100*, 180. [CrossRef]
29. Ismail, M.F.; Bogacki, W. Scenario approach for the seasonal forecast of Kharif flows from the Upper Indus Basin. *Hydrol. Earth Syst. Sci.* **2018**, *22*, 1391–1409. [CrossRef]
30. Kinnard, C.; MacDonell, S.; Petlicki, M.; Mendoza Martinez, C.; Abermann, J.; Urrutia, R. Mass Balance and Meteorological Conditions at Universidad Glacier, Central Chile. In *Andean Hydrology*; CRC Press: Boca Raton, FL, USA, 2018; pp. 102–123.
31. Bravo, C.; Loriaux, T.; Rivera, A.; Brock, B.W. Assessing glacier melt contribution to streamflow at Universidad Glacier, central Andes of Chile. *Hydrol. Earth Syst. Sci.* **2017**, *21*, 3249–3266. [CrossRef]
32. Podgórski, J.; Kinnard, C.; Pętlicki, M.; Urrutia, R. Performance Assessment of TanDEM-X DEM for Mountain Glacier Elevation Change Detection. *Remote Sens.* **2019**, *11*, 187. [CrossRef]
33. EULA. *Establecimiento de Red de Estaciones Nivo-Glaciales de Cordillera de la Región de O´Higgins y Desarrollo de un Modelo para la Gestión Integrada de los Recursos Hídricos de la Cuenca del Río Rapel*; Universidad de Concepción: VI Región, Chile, 2017.
34. Escobar, C. *Aplicacion del Modelo "SRM 3-11" en Cuencas de los Andes Centrales. Segudas Jornadas de Hidráulica Francisco Javier Domínguez*; Department of Hydrology, Ministry of Public Works: Santiago, Chile, 1992; pp. 283–295.
35. Arsenault, R.; Brissette, F.; Martel, J.-L. The hazards of split-sample validation in hydrological model calibration. *J. Hydrol.* **2018**, *566*, 346–362. [CrossRef]
36. Rango, A.; Martinec, J. Revisiting the Degree-Day Method for Snowmelt Computations. *JAWRA J. Am. Water Resour. Assoc.* **1995**, *31*, 657–669. [CrossRef]
37. Hock, R. Glacier melt: A review of processes and their modelling. *Prog. Phys. Geogr. Earth Environ.* **2005**, *29*, 362–391. [CrossRef]
38. Xie, S.; Du, J.; Zhou, X.; Zhang, X.; Feng, X.; Zheng, W.; Li, Z.; Xu, C.-Y. A progressive segmented optimization algorithm for calibrating time-variant parameters of the snowmelt runoff model (SRM). *J. Hydrol.* **2018**, *566*, 470–483. [CrossRef]
39. Adnan, M.; Nabi, G.; Kang, S.; Zhang, G.; Adnan, R.M.; Anjum, M.N.; Iqbal, M.; Ali, A.F. Snowmelt Runoff Modelling under Projected Climate Change Patterns in the Gilgit River Basin of Northern Pakistan. *Pol. J. Environ. Stud.* **2017**, *26*, 525–542. [CrossRef]
40. Hock, R. Temperature index melt modelling in mountain areas. *J. Hydrol.* **2003**, *282*, 104–115. [CrossRef]

41. Ismail, M.F.; Bogacki, W.; Muhammad, N.; Bagacki, W. Degree Day Factor Models for Forecasting the Snowmelt Runoff for Naran Watershed. *Sci. Int.* **2015**, *27*, 1961–1969.
42. Schneider, C.; Kilian, R.; Glaser, M. Energy balance in the ablation zone during the summer season at the Gran Campo Nevado Ice Cap in the Southern Andes. *Glob. Planet. Chang.* **2007**, *59*, 175–188. [CrossRef]
43. Brock, B.W.; Burger, F.; Rivera, A.; Montecinos, A. A fifty year record of winter glacier melt events in southern Chile, 38°–42° S. *Environ. Res. Lett.* **2012**, *7*, 045403. [CrossRef]
44. Fernández, A.; Mark, B.G. Modeling modern glacier response to climate changes along the Andes Cordillera: A multiscale review. *J. Adv. Model. Earth Syst.* **2016**, *8*, 467–495. [CrossRef]
45. Pino Vargas, E.; Cornejo Navarretty, L.; Román Arce, C. Uso de Algoritmos Genéticos para la Calibración de un Modelo Hidrológico Precipitación Escorrentia en la Cuenca del Caplina. *Cienc. Desarro.* **2019**, *18*, 45–50. [CrossRef]
46. Ragettli, S.; Cortés, G.; McPhee, J.; Pellicciotti, F. An evaluation of approaches for modelling hydrological processes in high-elevation, glacierized Andean watersheds. *Hydrol. Process.* **2014**, *28*, 5674–5695. [CrossRef]
47. Dirección General de Aguas. *Inventario Nacional de Glaciares*; Dirección General de Aguas: Santiago, Chile, 2014.
48. Arsenault, K.R.; Houser, P.R.; De Lannoy, G.J.M. Evaluation of the MODIS snow cover fraction product. *Hydrol. Process.* **2014**, *28*, 980–998. [CrossRef]
49. Hall, D.K.; Riggs, G.A. *MODIS/Terra Snow Cover Daily L3 Global 500 m SIN Grid, Version 6. [NDSI]*; NASA National Snow and Ice Data Center Distributed Active Archive Center: Boulder, CO, USA, 2016.
50. Salomonson, V.V.; Appel, I. Development of the Aqua MODIS NDSI fractional snow cover algorithm and validation results. *IEEE Trans. Geosci. Remote Sens.* **2006**, *44*, 1747–1756. [CrossRef]
51. Krause, P.; Boyle, D.P.; Bäse, F. Comparison of different efficiency criteria for hydrological model assessment. *Adv. Geosci.* **2005**, *5*, 89–97. [CrossRef]
52. Ritter, A.; Muñoz-Carpena, R. Performance evaluation of hydrological models: Statistical significance for reducing subjectivity in goodness-of-fit assessments. *J. Hydrol.* **2013**, *480*, 33–45. [CrossRef]
53. Gupta, H.V.; Kling, H.; Yilmaz, K.K.; Martinez, G.F. Decomposition of the mean squared error and NSE performance criteria: Implications for improving hydrological modelling. *J. Hydrol.* **2009**, *377*, 80–91. [CrossRef]
54. Thiemig, V.; Rojas, R.; Zambrano-Bigiarini, M.; De Roo, A. Hydrological evaluation of satellite-based rainfall estimates over the Volta and Baro-Akobo Basin. *J. Hydrol.* **2013**, *499*, 324–338. [CrossRef]
55. Kling, H.; Fuchs, M.; Paulin, M. Runoff conditions in the upper Danube basin under an ensemble of climate change scenarios. *J. Hydrol.* **2012**, *424–425*, 264–277. [CrossRef]
56. Wise, M.; Calvin, K.; Thomson, A.; Clarke, L.; Bond-Lamberty, B.; Sands, R.; Smith, S.J.; Janetos, A.; Edmonds, J. Implications of Limiting CO2 Concentrations for Land Use and Energy. *Science* **2009**, *324*, 1183–1186. [CrossRef]
57. Smith, S.J.; Wigley, T.M.L. Multi-Gas Forcing Stabilization with Minicam. *Energy J.* **2006**, SI2006. [CrossRef]
58. Riahi, K.; Rao, S.; Krey, V.; Cho, C.; Chirkov, V.; Fischer, G.; Kindermann, G.; Nakicenovic, N.; Rafaj, P. RCP 8.5—A scenario of comparatively high greenhouse gas emissions. *Clim. Chang.* **2011**, *109*, 33–57. [CrossRef]
59. Taylor, K.E.; Stouffer, R.J.; Meehl, G.A. An Overview of CMIP5 and the Experiment Design. *Bull. Am. Meteorol. Soc.* **2012**, *93*, 485–498. [CrossRef]
60. Teutschbein, C.; Seibert, J. Bias correction of regional climate model simulations for hydrological climate-change impact studies: Review and evaluation of different methods. *J. Hydrol.* **2012**, *456–457*, 12–29. [CrossRef]
61. Teutschbein, C.; Seibert, J. Regional Climate Models for Hydrological Impact Studies at the Catchment Scale: A Review of Recent Modeling Strategies. *Geogr. Compass* **2010**, *4*, 834–860. [CrossRef]
62. Li, H.; Sheffield, J.; Wood, E.F. Bias correction of monthly precipitation and temperature fields from Intergovernmental Panel on Climate Change AR4 models using equidistant quantile matching. *J. Geophys. Res.* **2010**, *115*, D10101. [CrossRef]
63. Gudmundsson, L.; Bremnes, J.B.; Haugen, J.E.; Engen Skaugen, T. Technical Note: Downscaling RCM precipitation to the station scale using quantile mapping—A comparison of methods. *Hydrol. Earth Syst. Sci. Discuss.* **2012**, *9*, 6185–6201. [CrossRef]

64. Fang, G.H.; Yang, J.; Chen, Y.N.; Zammit, C. Comparing bias correction methods in downscaling meteorological variables for a hydrologic impact study in an arid area in China. *Hydrol. Earth Syst. Sci.* **2015**, *19*, 2547–2559. [CrossRef]
65. Chen, J.; Ohmura, A. Estimation of Alpine glacier water resources and their change since the 1870s. *Hydrol. Mt. Reg.* **1990**, *193*, 127–135.
66. Bahr, D.B.; Meier, M.F.; Peckham, S.D. The physical basis of glacier volume-area scaling. *J. Geophys. Res. Solid Earth* **1997**, *102*, 20355–20362. [CrossRef]
67. Científicos, C.; De Aguas, D.G. Dirección General de Aguas. In *Estimación de Volúmenes de Hielo Mediante Radio Eco Sondaje en Chile Central*; General Water Authority: Santiago, Chile, 2012.
68. Radić, V.; Hock, R. Modeling future glacier mass balance and volume changes using ERA-40 reanalysis and climate models: A sensitivity study at Storglaciären, Sweden. *J. Geophys. Res. Earth Surf.* **2006**, *111*. [CrossRef]
69. Khadka, D.; Babel, M.S.; Shrestha, S.; Tripathi, N.K. Climate change impact on glacier and snow melt and runoff in Tamakoshi basin in the Hindu Kush Himalayan (HKH) region. *J. Hydrol.* **2014**, *511*, 49–60. [CrossRef]
70. Moriasi, D.N.; Arnold, J.G.; van Liew, M.W.; Bingner, R.L.; Harmel, R.D.; Veith, T.L. Model Evaluation Guidelines for Systematic Quantification of Accuracy in Watershed Simulations. *Trans. ASABE* **2007**, *50*, 885–900. [CrossRef]
71. Parajuli, A.; Chand, M.B.; Kayastha, R.B.; Shea, J.M.; Mool, P.K. Modified temperature index model for estimating the melt water discharge from debris-covered Lirung Glacier, Nepal. *Proc. Int. Assoc. Hydrol. Sci.* **2015**, *368*, 409–414.
72. Chand, M.B.; Kayastha, R.B.; Parajuli, A.; Mool, P.K. Seasonal variation of ice melting on varying layers of debris of Lirung Glacier, Langtang Valley, Nepal. *Proc. Int. Assoc. Hydrol. Sci.* **2015**, *368*, 21–26. [CrossRef]
73. Braithwaite, R.J. Encyclopedia of snow, ice and glaciers. In *Encyclopedia of Snow, Ice and Glaciers*; Springer: Dordrecht, The Netherlands, 2011; pp. 186–190.
74. Klein Tank, A.; Zwiers, F. *Guidelines on Analysis of Extremes in a Changing Climate in Support of Informed Decisions for Adaptation*; Climate Data and Monitoring, WCDMP-No. 72, WMO-TD No. 1500; World Meteorological Organization: Geneva, Switzerland, 2009.
75. Ragettli, S.; Pellicciotti, F.; Immerzeel, W. Contrasting response of glacierized catchments in the Central Himalaya and the Central Andes to climate change. In Proceedings of the EGU General Assembly, Vienna, Austria, 12–17 April 2015.
76. Omani, N.; Srinivasan, R.; Karthikeyan, R.; Venkata Reddy, K.; Smith, P.K. Impacts of climate change on the glacier melt runoff from five river basins. *Trans. ASABE* **2016**, *59*, 829–848.
77. Radić, V.; Hock, R. Regionally differentiated contribution of mountain glaciers and ice caps to future sea-level rise. *Nat. Geosci.* **2011**, *4*, 91–94. [CrossRef]
78. Radić, V.; Bliss, A.; Beedlow, A.C.; Hock, R.; Miles, E.; Cogley, J.G. Regional and global projections of twenty-first century glacier mass changes in response to climate scenarios from global climate models. *Clim. Dyn.* **2014**, *42*, 37–58. [CrossRef]
79. Aldunce, P.; Araya, D.; Sapiain, R.; Ramos, I.; Lillo, G.; Urquiza, A.; Garreaud, R. Local Perception of Drought Impacts in a Changing Climate: The Mega-Drought in Central Chile. *Sustainability* **2017**, *9*, 2053. [CrossRef]
80. Penalba, O.C.; Rivera, J.A. Regional aspects of future precipitation and meteorological drought characteristics over Southern South America projected by a CMIP5 multi-model ensemble. *Int. J. Climatol.* **2016**, *36*, 974–986. [CrossRef]
81. Carey, M.; Huggel, C.; Bury, J.; Portocarrero, C.; Haeberli, W. An integrated socio-environmental framework for glacier hazard management and climate change adaptation: Lessons from Lake 513, Cordillera Blanca, Peru. *Clim. Chang.* **2012**, *112*, 733–767. [CrossRef]

4

Potential Impacts of Climate Change on Water Resources in the Kunhar River Basin, Pakistan

Rashid Mahmood [1,*], Shaofeng Jia [1] and Mukand S. Babel [2]

Academic Editors: Yingkui Li and Michael A. Urban

1 Institute of Geographic Science and Natural Resources Research/Key Laboratory of Water Cycle and Related Land Surface Processes, Chinese Academy of Sciences, Beijing 100101, China; jiasf@igsnrr.ac.cn
2 Water Engineering and Management, Asian Institute of Technology, Pathumthani 12120, Thailand; msbabel@gmail.com
* Correspondence: rashi1254@gmail.com.

Abstract: Pakistan is one of the most highly water-stressed countries in the world and its water resources are greatly vulnerable to changing climatic conditions. The present study investigates the possible impacts of climate change on the water resources of the Kunhar River basin, Pakistan, under A2 and B2 scenarios of HadCM3, a global climate model. After successful development of the hydrological modeling system (HEC-HMS) for the basin, streamflow was simulated for three future periods (2011–2040, 2041–2070, and 2071–2099) and compared with the baseline period (1961–1990) to explore the changes in different flow indicators such as mean flow, low flow, median flow, high flow, flow duration curves, temporal shift in peaks, and temporal shifts in center-of-volume dates. From the results obtained, an overall increase in mean annual flow was projected in the basin under both A2 and B2 scenarios. However, while summer and autumn showed a noticeable increase in streamflow, spring and winter showed decreased streamflow. High and median flows were predicted to increase, but low flow was projected to decrease in the future under both scenarios. Flow duration curves showed that the probability of occurrence of flow is likely to be more in the future. It was also noted that peaks were predicted to shift from June to July in the future, and the center-of-volume date—the date at which half of the annual flow passes—will be delayed by about 9–17 days in the basin, under both A2 and B2 scenarios. On the whole, the Kunhar basin will face more floods and droughts in the future due to the projected increase in high flow and decrease in low flow and greater temporal and magnitudinal variations in peak flows. These results highlight how important it is to take cognizance of the impact of climate change on water resources in the basin and to formulate suitable policies for the proper utilization and management of these resources.

Keywords: Climate change; downscaling; temperature; precipitation; hydrologic modelling; water resources; Kunhar basin; Pakistan

1. Introduction

The concentration of greenhouse gases (GHGs) has dramatically increased during the last few decades because of anthropogenic forces such as burning of fossil fuels and biomass, land use changes, rapid industrialization, and deforestation. This increased GHGs concentration has resulted in global warming and a global energy imbalance [1,2]. According to the Fifth Assessment Report of the Intergovernmental Panel on Climate Change (IPCC-AR5), the global average temperature has increased by 0.85 °C (0.65 °C–1.06 °C) over the period of 1800–2012, relative to 1961–1990 [3], and 0.74 °C ± 18 °C has been detected during the last hundred years (1906–2005) [4]. There is a very high likelihood of this trend in global warming to only exacerbate. The global average temperature is projected to increase by 1.7 °C–4.8 °C in 2081–2100 (relative to 1986–2005) under different Representative Concentration Pathways (RCPs).

The projected increase in global warming is likely to intensify the hydrological cycle of the world, and hence, disturb the existing hydrological system. As a consequence of global warming, hydrological systems are likely to experience changes in the average availability of water, as well as changes in extreme events such as increase in precipitation intensity, frequency, and/or amount of heavy precipitation [5–7]. For example, the average annual precipitation is likely to increase in the high latitudes and the equatorial Pacific Ocean by the end of the 21st century under RCP8.5. However, in several mid-latitude areas and subtropical dry regions, mean precipitation is likely to decrease. The areas of the globe with increasing precipitation are much more than those with decreasing precipitation [3]. This disturbance in a hydrological system can pose problems for public health, industrial and municipal water demand, water energy exploitation, and the ecosystem. However, as stated above, the impact of climate change on hydrological systems may vary from region to region [2,3,5,8].

Hydrological systems are of great importance as they greatly affect the environmental and economic development of a region and are highly complex because they comprise the atmosphere, cryosphere, hydrosphere, biosphere, and geosphere. The hydrological cycle of a basin (catchment) is mainly influenced by the physical characteristics of the basin, climatic conditions in the basin, and human activities. Most studies on climate change have focused on temperature, precipitation, and evaporation [9] since these are considered to be the key indicative factors of climate change and variability in a river basin [10]. There is an increasing consensus that changing trends in climatic variables, especially temperature and precipitation, can change the hydrological and ecological conditions of a river basin [11,12]. Given all these conditions, it is of great importance to assess the possible impacts of climate change on the water resources of a region, which can help in the proper utilization and management of these water resources.

The economy of Pakistan relies greatly on agriculture, which is heavily dependent on the Indus River Irrigation System. Water issues in Pakistan are crucial challenges for the policymakers and managers of water resources in the country [13]. Today, Pakistan is one of the most water-stressed countries in the world as water availability in the country has reduced from about 5000 m^3 per capita per year in 1952 to 1100 m^3 per capita per year in 2006, which is an alarming situation [14]. According to the United Nations report, a country with water availability of less than 1000 m^3 per capita per year is considered a water scarce country [15]. The reduction in water availability has created tensions among different provinces due to the increasingly unequal distribution of water. Climate change and variability are likely to affect water availability for, and magnitude of, irrigation and hydropower production in the country. These changing water conditions are likely to increase the tensions among the provinces, especially at the downstream (Sindh province) [13]. Therefore, a clear estimation of future water resources under changing climatic conditions is of great importance for the planning, operation, and management of hydrological installations in any watershed in Pakistan.

In the last few years, outputs from General Circulation Models (GCMs)—the most advanced and numerical based coupled models—are fed into hydrological models to explore the impacts of climate change on water resources in various parts of the world in the future. However, these models are coarse in spatial resolution (about 200–500 km) [16] and might not be suitable at the basin level, especially small basins which require very fine spatial resolution [17,18]. To bridge GCM resolution and basin scales, downscaling—dynamical and statistical—techniques have been developed. In dynamical downscaling, a high-resolution numerical based Regional Climate Model (RCM), with a resolution of about 5-50 km, uses the course outputs of a GCM and provides detailed information or high resolution outputs at the basin level [2]. On the other hand, Statistical downscaling (SD) methods (*i.e.*, stochastic weather generator, regression, and weather typing) create empirical/statistical relationships between the GCM scale and basin scale variables (e.g., temperature and precipitation). Compared to dynamical downscaling, SD methods are faster and computationally inexpensive, and thus offer approaches that have been rapidly adopted by a wider community of scientists [17]. For the present study, downscaled data (maximum temperature, minimum temperature, and precipitation) was obtained from [1]. They used a Statistical Downscaling Model (SDSM), a combination of multiple linear regression and a stochastic weather generator, to downscale temperature and precipitation over the period of 2011–2099 under A2 and B2 scenarios of HadCM3, a global climate model.

Different studies such as Akhtar *et al.* [13], Ahmad *et al.* [19], Shrestha *et al.* [20], and Bocchiola *et al.* [21] have assessed the impacts of climate change on the water resources of Pakistan [13,19–21]. These studies were mostly conducted in the Upper Indus basin using hydrological models such as Snowmelt Runoff model (SRM), Hydrologiska Byråns Vattenbalansavdelning (HBV), Soil and Water Assessment Tool (SWAT), and the WEB-DHM-S model. However, to the best of our knowledge, no studies have been conducted to explore the potential impacts of climate change on the water resources of the Kunhar River basin, which is one of the main tributaries of the transboundary Jhelum River and is located entirely in Pakistan. The Kunhar River originates from the Greater Himalayas and contributes to the Mangla Reservoir after joining the Jhelum River. The Mangla Reservoir's water is used for irrigation and hydropower production. It is the second largest reservoir of Pakistan. Although HEC-HMS, a well-known hydrological model, has been successfully used for small to large and plains to mountainous areas of the world [22–27], almost no studies have been reported in Pakistan that use HEC-HMS for the assessment of climate change impacts on water resources.

Thus, the main objectives of the present study were: (1) the development of HEC-HMS, a hydrological model, in the mountainous Kunhar River basin which is greatly influenced by winter snowfall; and (2) to assess the possible impacts of climate change on the water resources of the Kunhar River basin, located in Pakistan, under A2 and B2 scenarios of HadCM3. Description of the data and the study area are given in Sections 2 and 3 of this paper. A brief introduction of the hydrological model used in the study, and the methods used for the analysis of streamflow, is given in Section 4. Sections 5 and 6 include the results/discussion and conclusions respectively. This study will be very useful for policies that govern the utilization and management of the water resources of the Kunhar basin, which contributes significantly to the Mangla reservoir, under the effects of climate change.

2. Data Description

2.1. Hydro-Meteorological Data

Observed daily historical data of three climate stations (maximum temperature, minimum temperature, and precipitation) were collected from the Pakistan Meteorological Department (PMD) and the streamflow data of two hydrometric stations, spanning 1961 to 2000, from the Water and Power Development Authority (WAPDA). The geographical and basic information of the hydro-meteorological stations is presented in Table 1 and Figure 1.

Table 1. Basic information about hydro-climatic stations in the Kunhar River basin.

SR	Station	Lat	Long	Altitude	Period	PP	TX	TN
	Meteorological	(° N)	(° E)	(m ASL)	(year)	Annual (mm)	Mean (°C)	Mean (°C)
1	Balakot	34.55	73.35	995	1961–2000	1514	24.9	12.4
2	Muzaffarabad	34.37	73.48	702	1961–2000	1477	28.4	13.5
3	Naran	34.9	73.65	2421	1961–2000	1522	12.3	3.2
						1504	**21.8**	**9.7**
	Hydrometric					**Mean Discharge** Annual (mm)	m^3/s	**Catchment Area** km^2
1	Naran	34.90	73.65	2362	1961–2000	1441	47	1011
2	Gari Habibullah	34.40	73.38	810	1961–2000	1349	101	2335

PP, precipitation; *TX*, maximum temperature; *TN*, minimum temperature; *ASL*, above sea level.

2.2. HadCM3 Downscaled Data

Downscaled data of maximum temperature (TX), minimum temperature (TN), and precipitation (PP) for the period of 2011-2099 was taken from [1]. They downscaled TX, TN, and PP with a well-known downscaling model, Statistical Downscaling Model (SDSM), under A2 and B2 scenarios of HadCM3 in the Jhelum River basin. They selected HadCM3 for two reasons: (a) it showed better results during the evaluation of various GCMs (CGCM2, CSISRO, CCSR/NIES, and HadCM3); and (b) it has been used by a majority of studies for statistical downscaling of climate variables around the Jhelum basin [1]. They also applied the bias correction technique on the downscaled data to get closer to realistic results.

SDSM is a hybrid of multiple linear regression and a stochastic generator [28] and is widely used for downscaling climate variables [1]. HadCM3 is a Global Climate Model (GCM) developed by the Hadley Center of the UK Meteorological Office. It has a spatial resolution of 2.5° × 3.75° (latitude × longitude), with a surface spatial resolution of about 278 × 417 km, decreasing to 278 × 295 km at 45 degrees North and South [29]. A2 and B2 are the emission scenarios developed by the IPCC. The A2 scenario describes a very heterogeneous world: very slow fertility patterns across regions, continuous increase in global population, regionally oriented economic development, more fragmented and slower per capita economic growth, and more rapid technological changes as compared to other scenarios. The B2 scenario presents the world with an emphasis on social and environmental sustainability and local solutions to economic issues. Under this scenario, the population of the world increases at a rate lower than under A2 and economic development is intermediate. B2 also posits less rapid and more diverse technological changes relative to the B1 and A1 scenarios [30].

Mahmood and Babel [1] have shown that TX, TN, and PP are projected to increase by 0.91 °C–3.15 °C, 0.93 °C–2.63 °C, and 6%–12% respectively under A2, and 0.69 °C–1.92 °C, 0.56 °C–1.63 °C, and 8%–14% respectively under B2. According to them, in autumn, winter, summer, and spring, precipitation is projected to increase by 24%–32%, 11%–17%, 9%–12% and 4%–5% under A2 and 12%–25%, 12%–15%, 10%–14%, 0.2%–4% respectively under B2. The projected increases in temperature (TX and TN) are 2.7 °C–3.0 °C, 2.5 °C–3.2 °C, and 2.0 °C–2.6 °C, 3.2 °C–3.9 °C under A2 and 1.6 °C–1.73 °C, 2.0 °C–2.5 °C, 1.4 °C–1.7 °C, 1.4 °C–1.8 °C under B2 by the end of this century.

2.3. DEM Data

Elevation data is one of the primary sources used in hydrological studies. The most common form of elevation data is the digital elevation model (DEM) that is widely used to extract the topographical characteristics of a terrain [31,32]. In this study, a DEM of 30 m, created by the Advanced Spaceborne Thermal Emission and Reflection Radiometer (ASTER), was obtained from the U.S. Geological Survey (USGS) (http://gdex.cr.usgs.gov/gdex/). This DEM is shown in Figure 1. ASTER-DEM was developed by the National Aeronautics and Space Administration (NASA) of USA, the Japan Space Systems (JSS), and the Ministry of Economy, Trade and Industry (METI) of Japan. The coverage area of ASTER-DEM ranges between 83° N and 83° S, covering 99% of the Earth's landmass.

In the present study, DEM was processed with HEC-GeoHMS, an extension of GIS, to extract basin parameters like slopes of rivers and sub-basins, areas of sub-basins, river lengths, longest flow paths, elevations in the basin, flow direction, streamlines, contour lines, and aspects during the process of watershed delineation. Among them, contours, aspects, slope, and the delineated sub-basin are shown in Figure 2.

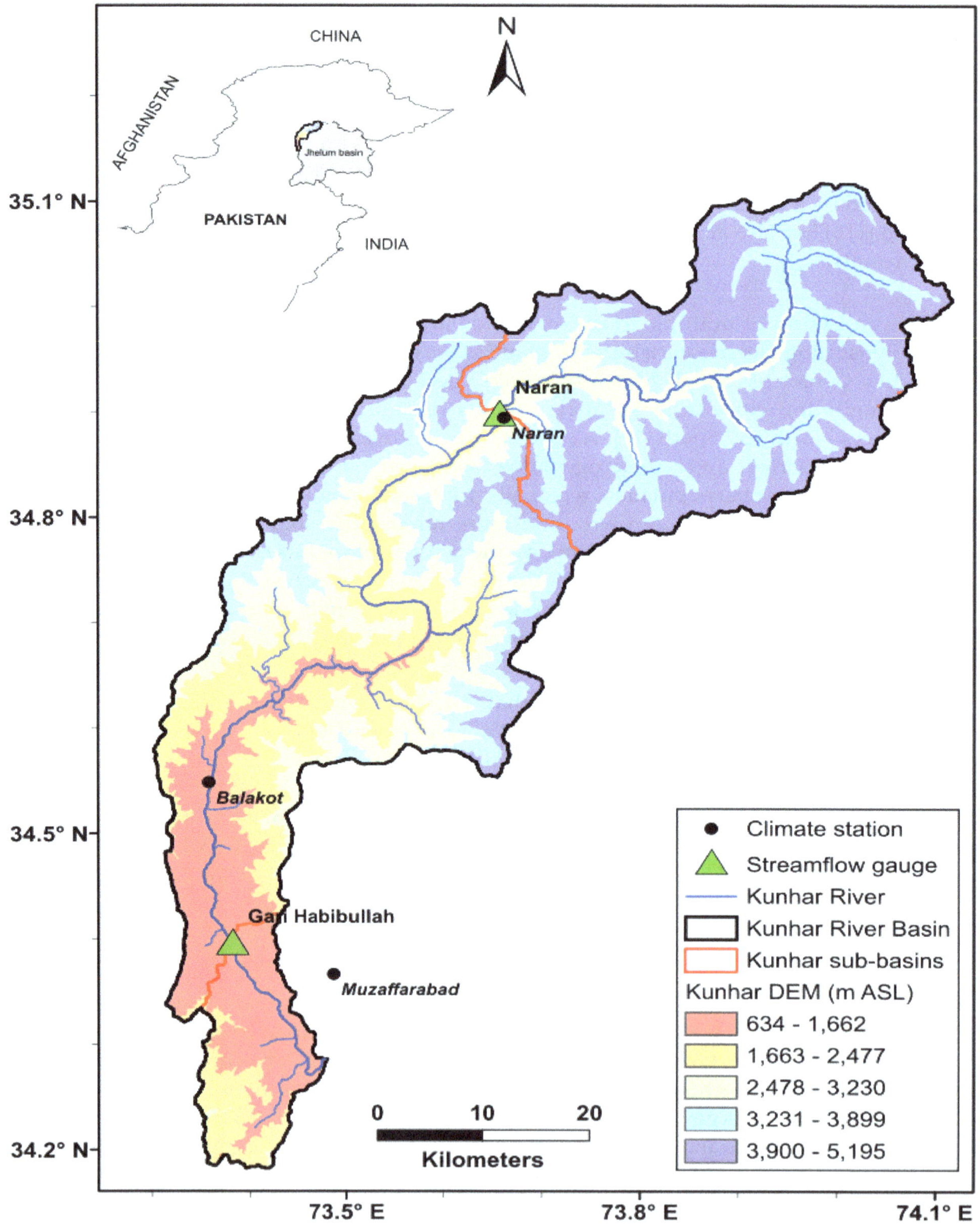

Figure 1. Location map of the Kunhar River basin showing hydro-climatic stations, streamlines, and altitude.

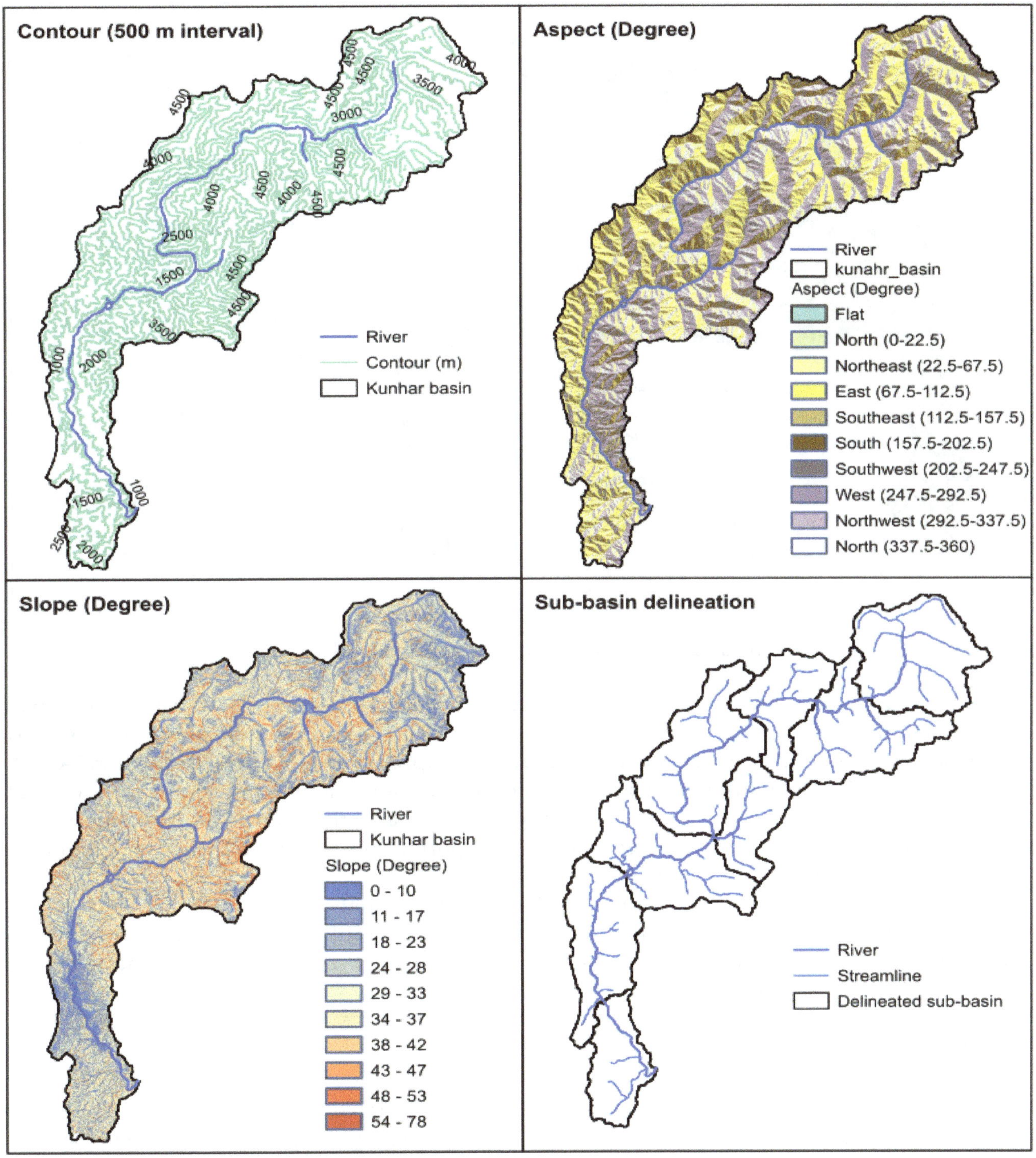

Figure 2. Topographical features and streamline delineation of the Kunhar River basin.

2.4. *Land Cover and Soil Data*

Land cover can strongly affect the hydrological processes in a region. These processes are mainly influenced by the density of plant cover and the morphology of plant species [33]. The land cover data for the Kunhar River basin was derived from the global land cover data (1 km resolution) developed by the Joint Research Center (JRC) of the European Commission (http://eusoils.jrc.ec.europa.eu/data.html). This data is shown in Figure 3.

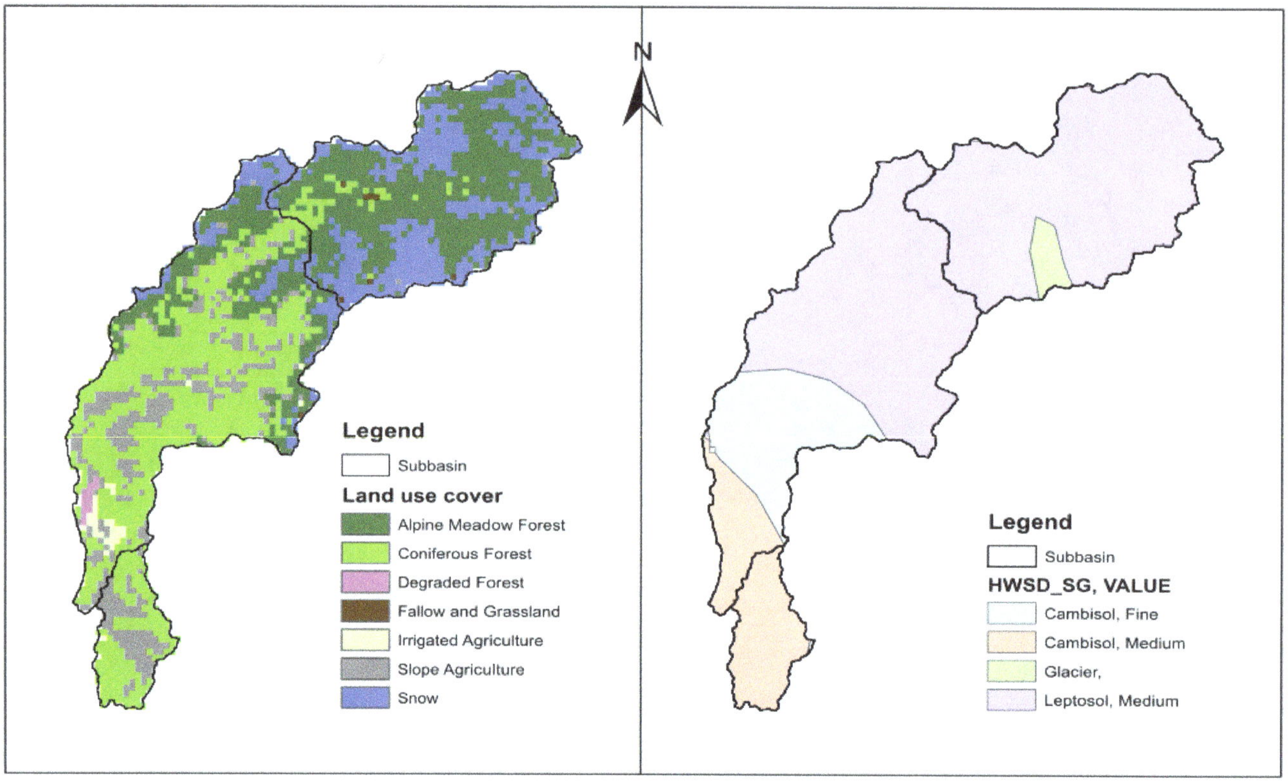

Figure 3. Land use and soil map of the Kunhar River basin.

The soil data for the Kunhar basin was obtained from the Harmonized World Soil Database, with a resolution of 1 km, as is shown in Figure 3. This database was developed by FAO in collaboration with the International Institute Of Applied Systems Analysis (IIASA), the International Soil Reference and Information Centre (ISRIC) of World Soil Information, the Institute of Soil Science of Chinese Academy of Sciences (ISSCAS), and the Joint Research Centre of the European Commission (JRC) (http://www.fao.org/soils-portal/soil-survey/soil-maps-and-databases/harmonized-world-soil-database-v12/en/).

The soil and land cover data were used to extract the initial estimations of the hydrological properties of the basin such as maximum moisture deficit, hydraulic conductivity, crop coefficient, percentage of imperviousness, and basin lag. However, the exact estimates of these parameters for each sub-basin were obtained only during the calibration process.

3. Study Area

The Kunhar River basin is located in the northern side of Pakistan and stretches between 34.2°–35.1° N and 73.3°–74.1° E, as is shown in Figure 1. The Kunhar drains the southern slope of the Greater Himalayas, located in the Khyber Pakhtunkhwa (KPK) province of Pakistan. It originates from the Lulusar Lake in the Kaghan valley of KPK. It passes through Jalkhand, Bata Kundi, Naran, Kaghan, Kawai, Balakot, and Gari Habibullah and finally joins the Jhelum River at Rara. The Kunhar's water is rich in algal flora, resulting in a great diversity of the aquatic life it harbors [34]. It has a drainage area of 2535 km^2, with elevation ranging from 600 to 5000 m (Figure 1).

The Kunhar is one of the biggest tributaries of the transboundary Jhelum River basin. This is the only main tributary which is entirely situated in Pakistan's territory and therefore, has great importance from the perspective of hydrological monitoring by the WAPDA of Pakistan. The Kunhar contributes about 11% of the total flow to the Mangla Reservoir constructed in the Jhelum River basin. The Mangla Reservoir is the second largest water storage site in the country. The water stored here is primarily used for irrigating 6 million hectares of land in the country and for generating hydropower. The installed capacity of the Mangla power plant is 1000 MW, and electricity is generated as a byproduct [1,35]. Snowmelt from the Kunhar basin contributes about 65% to the total discharge of the Kunhar River and 20%-40% to the Jhelum River at Manga [35]. The population in the Kunhar basin is almost entirely rural and their economy is generally agro-pastoral based. The principal occupation of the population is agriculture, although rearing livestock is also practiced in the adjacent mountainous areas. A small portion of the population is involved in trade, local labor, and employment in the bigger cities of the country [36]. In Pakistan, 93% of the total annual flow is used for agricultural purposes, 5% for industrial use, and 2% for domestic use [37]. The water of the Kunhar River is mostly used for irrigation, municipal use, power generation, and recreation [35]. Since no industries are located in the basin, most of the water of the Kunhar Basin (about 98%) is used for agriculture and the rest for domestic use.

The data of other major topographical characteristics such as slope, contours lines, aspects, and delineated sub-basins, which were extracted from DEM, are presented in Figure 2. This figure shows that the basin has undulating relief ranging from 0° to 78°. The plains along the course of the Kunhar River are located on a gentle slope (0°-10°). However, most parts of the basin have moderate (>10° and <30°) to steep slopes (>30°).

The basin has a great diversity of vegetation such as temperate coniferous forests, subtropical coniferous forests, alpine meadows, agricultural cover, and snow, as described in Table 2 and shown in Figure 3. These varied land covers are reclassified into seven main classes to explore the major land use covers in the basin. Forest, agriculture, and snow cover the maximum area of the basin with about 65%, 14%, and 20% respectively (Table 2). This information was derived from the land cover data of 1 km resolution in the basin.

Table 2. Basic characteristics of soil and land use data in the Kunhar River basin.

	HWSD-Soil Group	Area		Texture	Top Soil Fraction			TSGC
		%	km²		Sand (%)	Silt (%)	Clay (%)	Gravel (%)
1	Cambisol	14	344	Medium	42	36	22	9
2	Cambisol	13	342	Fine	22	30	48	8
3	Leptosol	71	1801	Medium	43	34	23	26
4	Glacier	2	48	-	0	0	0	0
	Land use cover	**Area**		**Land Use Cover**				
	Re-classes	**%**	**km²**	**Original**				
1	Coniferous Forest	33.1	838	Temperate Conifer/Subtropical Conifer/Tropical Moist Deciduous				
2	Degraded Forest	0.4	11	Degraded Forest				
3	Fallow and Grassland	0.3	7	Slope Grasslands/Sparse vegetation (cold)/Gobi/Desert (cold)				
4	Alpine Meadow Forest	31.8	805	Alpine Meadow				
5	Irrigated Agriculture	1.0	25	Irrigated Agriculture				
6	Slope Agriculture	13.0	329	Slope Agriculture				
7	Snow	20.5	521	Snow				

HWSD, Harmonized World Soil Database; *TSGC*, Top Soil Gravel Content.

There are three main groups of soils in the Kunhar River basin, as shown in Figure 3: 1) cambisol fine, 2) cambisol medium, and 3) leptosol. Cambisol characterizes weak to moderately developed soils and it (both fine and medium varieties) covers 27% of the basin, while leptosol is very shallow soil over hard rock and is unconsolidated and very gravelly material. Leptosol covers around 71% area of the basin. Some glacier patches also exist in the upper part of the basin. The basic information about soil types found in the basin is given in Table 2. The properties of these soils were derived from the soil data of 1 km resolution in the basin.

The Kunhar basin is located in a humid, subtropical zone. In the present study, hydro-climatic data was processed for the period of 1961–2000 to extract some basic information about the hydro-climatic conditions in the basin. This information is presented in Figure 4. The average annual temperature in the basin is about 13 °C (2 °C–23 °C). February is the coldest and July is the warmest month here.

This was calculated from the data of the Naran and Balakot climate stations available for the period of 1961-2000. At Balakot, located in the lower part of basin (Figure 1), TN ranges from 2 °C (January) to 23 °C (July) and TX from 14 °C (January) to 35 °C (June). At Naran, located in the upper part of the basin, TN ranges from −9 °C (February) to 13 °C (July) and TX from −1 °C (February) to 23 °C (July) (Figure 4a). The Kunhar basin has an annual precipitation of about 1500 mm (Table 1) with two peaks (Figure 4b). The first peak occurs in the upper part of the basin in the month of March because of the Western Disturbances (WDs) system in winter. Most parts of Pakistan, and its northwestern parts primarily, obtain precipitation due to WDs. WDs are caused by depressions over Mediterranean regions, resulting in precipitation over central and southwest Asia in the months of December to March [38,39]. The second peak happens in the month of July and in the lower parts of the basin due to the summer monsoons, which are the result of the saturated south western winds from the Bay of Bengal and Arabian Sea. It can be concluded that the monsoons do not reach the upper part of the basin, although WDs affect the whole basin. On the other hand, there is only one big streamflow peak, both in the upper (Naran) and lower parts (Gari Habibullah) of the basin, that occurs in the month of July (Figure 4b). This means the precipitation from December to March (winter) accumulates as snow cover, especially in the upper parts of the basin, and then starts melting after March and lasts till July when it overlaps with monsoon precipitation and results in one big peak. An average flow of about 1350 mm (103 m^3/s) has been measured at Gari Habibullah near the mouth of the basin for the period of 1961-2000. The same was reported by de-Scally as well [39].

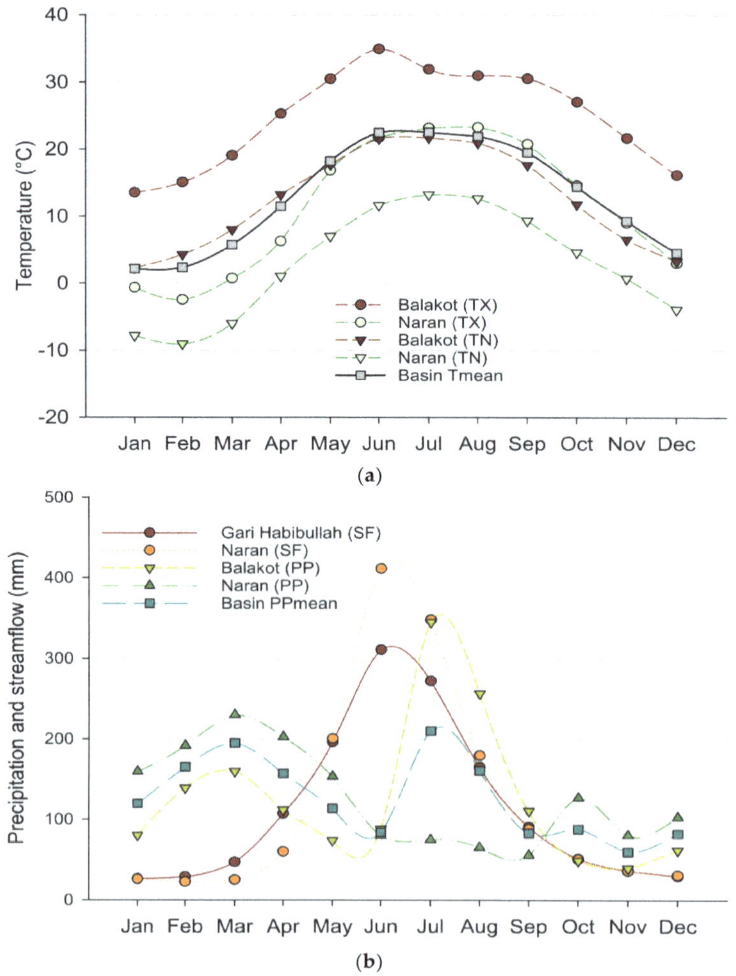

Figure 4. Monthly (**a**) max temperature (TX), min temperature (TN), (**b**) precipitation (PP), and streamflow (SF) in the Kunhar River basin for the period of 1961–2000.

4. Methodology

4.1. Description of HEC-HMS

The Hydrological Modeling system (HEC-HMS) is a rainfall-runoff simulation software used for a wide range of watersheds from large river basins to small urban areas. The model was formulated by the U.S. Army Corps of Engineers at the Hydrologic Engineering Center (HEC). HEC-HMS comprises different loss techniques such as SCS curve number, initial and constant, Green Ampt, one-layer deficit-constant, Smith Parlange, and five-layer soil moisture accounting. These techniques are used to estimate excess precipitation in fairly simple to very complex infiltration and evapotranspiration environments. This model can be used for both event and continuous modeling.

In order to calculate direct runoff from excess precipitation, seven methods including SCS, Clark, Snyder, and ModClark are available in the system. This model also consists of five base flow methods including recession method, constant monthly method, and linear reservoir method, and six channel routing methods including Muskingum, and modified pulse methods. It also has six kinds of meteorological models like Thiessen and inverse distance methods to analyze meteorological data such as precipitation, evapotranspiration, and snowmelt. The meteorological model extracts the precipitation for each sub-basin in the watershed. However, currently, only two methods (Temperature Index and Gridded Temperature Index) are available to compute runoff from snowfall in this modeling system [40,41].

A complete basin model setup for rainfall-runoff processes comprises a basin model, a meteorological model, control specification, and input time series. The basin model describes the physical characteristics of the study region such as the areas of sub-basins and river lengths of a watershed. Each basin model in HEC-HMS consists of a loss method, a transforming method, a base flow method, and a channel routing method [23]. Control specification is one of the main components of this model's setup and it controls the simulation period. For example, it controls when the model is to start and stop, and what the time interval for simulation should be. The input time series encompasses precipitation, temperature, evapotranspiration, and observed streamflow *etc.*, which have a direct link with the basin's model and the meteorological model. A detailed description of the model's formulation and its various processes is provided in the User's Manual and Technical Reference Manual of HEC-HMS [40] (p. 318 and p.157).

In the present study, the basin model was developed using the deficit and constant loss (DCL) method for calculating excess precipitation, the SCS unit hydrograph for transforming direct runoff, Muskingum for channel routing, and the recession method for base flow. The meteorological model was established using the Thiessen polygon gauge weight method for precipitation calculation, the temperature index method for snowmelt modeling, and the monthly evapotranspiration method. The Thiessen polygons were created and their weights were calculated by HEC-GeoHMs in accordance with the precipitation gauges. The same combination has been used in different studies [15,17,18].

The DCL model computes excess precipitation in a watershed. It is a single layer continuous method used for calculating the changes in soil moisture content. It is similar to the initial and constant loss (ICL) method but this method recovers the initial losses after a long period of no precipitation. This method contains four main parameters: maximum deficit, initial deficit, constant rate, and impervious percentage. These parameters can be initially estimated using soil and land cover data as initial inputs for the model but are finalized only during the calibration process.

The excess precipitation calculated from DCL was transformed into direct surface runoff by the SCS unit hydrograph method. Basin lag is the only parameter of the SCS method, and it needs to be determined during calibration. It can also be estimated as an initial value for calibration by multiplying the time of concentration by 0.6. In this study, the recession method was used to calculate the base flow which contributes to the total flow from the watershed. Three parameters—initial discharge, recession constant, and threshold—were determined during calibration.

To transfer the total flow from one point to other, the Muskingum method was used. This method is a simple mass conservation scheme for routing flow through channels. There are two main parameters for this method: travel time (K), and the Muskingum coefficient (X). The Muskingum coefficient ranges between 0 and 0.5.

The Thiessen polygon method was used to assign weights to each gauge in the watershed during the development of the meteorological model. For snowmelt modeling, different elevation bands were used for each sub-basin in the temperature index method (TIM). TIM is an extension of the degree-day technique to calculate flow from snowpack. In the degree-day approach, a fixed amount of snowmelt is assigned for each degree above freezing point. This method is a conceptual representation of the cold energy stored in the snowpack. This also takes care of past conditions and some other climatic factors during the calculation of snowmelt. Different parameters such as base temperature, wet melt rate, rain rate limit, melt rate pattern, lapse rate, and antecedent temperature index are required for this method [32]. A lapse rate of $-7.0\ ^{\circ}/\text{km}$ was calculated for the study area and kept constant for the entire Kunhar River basin. The FAO Penman-Monteith method—recommended as the standard method for computing potential evapotranspiration [24]—was carried out to calculate the potential evapotranspiration in the basin. A schemetic diagram for the setup of HEC-HMS in the Kunhar River basin is shown in Figure 5.

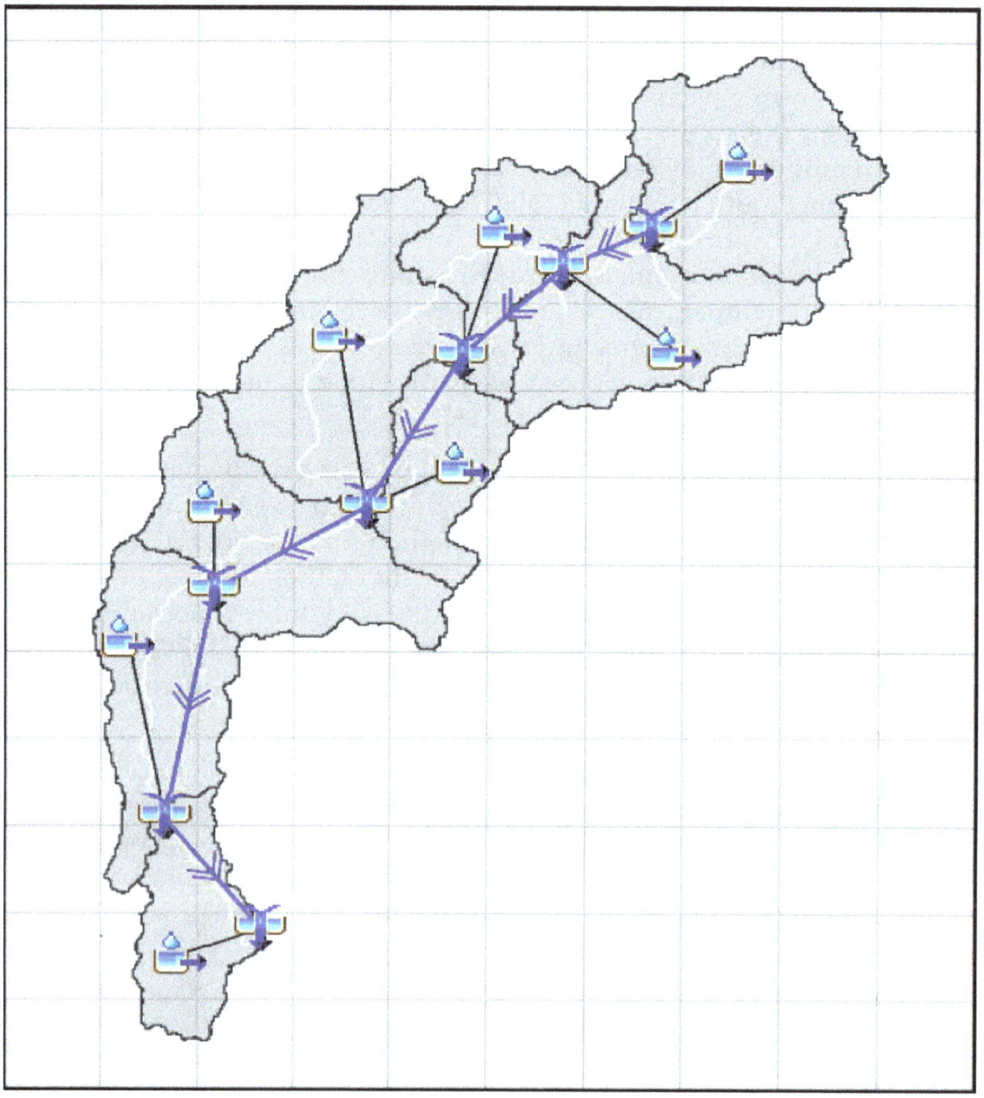

Figure 5. Schemetic diagram for the setup of HEC-HMS hydrological modeling system.

4.2. The Model's Calibration and Validation

The calibration of a model is a process in which the model's parameters are adjusted in such a way that the simulated flow captures the variations of the observed flow [25]. In this study, a split sample method was used for calibration and validation. In this method, the calibration period does not overlap with the validation period. A data period of eight years, from 1982 to 1989, was chosen as the calibration period and the period from 1978 to 1981 for validation because these periods had minimum missing values of both precipitation and streamflow. The physical properties—land use cover and soil properties—of the watershed were considered as constant during the simulation period.

Nelder Mead and Univariate Gradient are the two main algorithms to optimize the objective function. There are seven different kinds of objective functions in HEC-HMS. In this study, the sum of the squared residual was chosen and was minimized using the Nelder-Mead algorithm to explore the optimized model parameters in order to get the best results of the simulation. The simulated flows were compared with the observed flow using the coefficient of determination (R^2), percent deviation (D), and Nash-Sutcliffe efficiency (E). The R^2 values indicate how well the variations in the observed data are captured by the simulated data, D describes the mean percent deviation between observed and simulated flow, and E shows how well the observed plot fits with the simulated plot [22]. For more illustrative purposes, the simulated data was also compared with observed data graphically to explore how well the low and high observed flows were captured by simulated flow. In the present study, the model was calibrated and validated at both Naran and Gari Habibullah gauging stations.

The model's performance parameters—R^2, D, and E—were calculated using the following equations:

$$R^2 = \frac{\sum \left(Q_{obs} - \overline{Q_{obs}}\right) \times \left(Q_{sim} - \overline{Q_{sim}}\right)}{\sqrt{\sum \left(Q_{obs} - \overline{Q_{obs}}\right)^2 \times \left(Q_{sim} - \overline{Q_{sim}}\right)^2}} \quad (1)$$

Q_{obs} and Q_{sim} are observed and simulated values respectively. If the value of R^2 is close to 1, it indicates a good correlation between simulated and observed flows. The correlation is considered optimum if R^2 is exactly equal to 1 [22].

$$D\,(\%) = 100 \times \frac{\sum \left(Q_{sim} - Q_{obs}\right)}{\sum Q_{obs}} \quad (2)$$

The value of D should ideally be close to 0%. Positive and negative values are respectively indicators of over- and under-estimation by the model [22].

$$E = 1 - \frac{\sum \left(Q_{sim} - Q_{obs}\right)^2}{\sum \left(Q_{obs} - \overline{Q_{obs}}\right)^2} \quad (3)$$

The value of E lies between 0 and 1. A positive value close to 1 implies good calibration while a negative value close to 0 is not acceptable. If the value of E is greater than 0.75, then the results are considered to be good, and if it is between 0.36 and 0.75, the results are satisfactory [41].

4.3. Projected Changes in Streamflow

After successful calibration and validation, the downscaled daily time series (A2 and B2) of precipitation and temperature for the period of 2011 to 2099 were used as input for HEC-HMS to simulate daily flow data at both gauges (Naran and Gari Habibullah). The physical characteristics of the Jhelum basin were kept constant throughout the simulation period. However, it cannot be ignored that these characteristics do vary with time. The simulated data were divided into three periods: the 2020s (2011–2040), the 2050s (2041–2070), and the 2080s (2071–2099) and all three were compared with the baseline period (1961–1990) to assess changes in flow in the future. Different indicators such as mean flow, low flow, median flow, high flow, flow duration curves, temporal shift in peaks, and temporal shifts in center-of-volume dates were calculated for the three periods and the results were compared to the baseline period's data so as to explore the impact of climate change on the streamflow in the basin.

When analyzing streamflow to construct an installation such as a reservoir and headwork on a river, two questions are frequently asked: (1) how often will the streamflow occur in the future? and (2) what will be the magnitude of the streamflow? Flow duration curves are the main tools to deal with these two questions. These curves present the percentage of times that the flow in a stream is likely to exceed or be equal to a specified value of the flow. These curves can be applied in different kinds of studies such as hydropower management, reservoir sedimentation, water quality management, and low and high flow studies [42]. The following equation is used to construct the flow duration curves:

$$P\,(\%) = \frac{M}{(n+1)} \times 100 \tag{4}$$

P or the probability of flow is equal to or exceeds a specified value (% of time), M is the rank of events, and N is the number of events in a specified period of time.

In the present study, the daily time series were used to construct the flow duration curves for the base period (1961–1990) and for the three future periods: the 2020s, 2050s, and 2080s.

5. Results and Discussion

5.1. Calibration and Validation

Table 3 shows the model's performance parameters (E, D, and R^2) that were calculated using the daily and monthly observed and simulated streamflows for calibration (1982–1989) and validation (1978–1981) at Naran and Gari Habibullah. In the case of daily data, the values of E and R^2 ranged from 0.55 to 0.74 during calibration and validation at both stations, and the D values lay between −9% and 12%. The D values show that the model underestimated values during the calibration and validation periods at Naran and also underestimated them at Gari Habibullah during calibration. However, the model overestimated the values during validation at the Gari Habibullah station. The results were greatly improved when these parameters were calculated from the monthly time series. The values of E and R^2 were improved to 0.63–0.88. These results are quite satisfactory and complement well some previous studies such as Meenu *et al.* [22] in India, Verma *et al.* [23] in India, Yimer *et al.* [24] in Ethiopia, and Garcia *et al.* [25] in Spain. All these studies also used HEC-HMS to simulate streamflow for climate change studies, with E ranging from 0.48–0.83 and R^2 from 0.63–0.84.

Table 3. Nash-Sutcliffe efficiency (E), coefficient of determination (R^2), and percent deviation (D) for calibration (1982–1989) and validation (1978–1981) for different stream gauges, calculated from daily and monthly data, in the Kunhar River basin.

Hydrometric Station	Calibration			Validation		
	E	R^2	D (%)	E	R^2	D (%)
From daily streamflow						
Naran	**0.71**	0.74	−7.78	0.71	0.73	−3.13
Gari Habibullah	0.66	0.67	−8.69	0.55	0.65	12.17
From monthly streamflow						
Naran	0.83	0.84	−11.02	0.83	0.84	−3.14
Gari Habibullah	0.88	0.85	−9.92	0.63	0.70	12.18

The graphical comparisons of observed against simulated flow for the calibration and validation periods are shown in Figures 6 and 7 respectively. At both Naran and Gari Habibullah gauging stations, patterns of observed flow were well matched by patterns of simulated flow during the calibration period. However, the peak and low flows were not captured well by the model. At Naran, in some years, peak and low flows were underestimated by the model but it was the reverse in the case of

Gari Habibullah. This underestimation of flow might be due to the lack of enough precipitation gauges available in the basin or it might be due to the use of a simple temperature index model to calculate flow from snowmelt because about 65% of the total flow comes from snowmelt. Over- and under-estimation of flow must be considered during the discussion of this study's results. For example, at Gari Habibullah, the model overestimated the flow by 12%. This means that if the projected increase at Gari Habibullah is 50%, then the final projected increase will be 38%.

In the case of validation, the peaks were comparatively better captured by the simulated flow at both stations, especially at Naran. On the whole, the calibration and validation results were satisfactory and comparable with some previous studies like Meenu *et al.* [22], Verma *et al.* [23], Yimer *et al.* [24], and Garcia *et al.* [25].

The present study's results could be much more exact if the number of precipitation gauges were more than are currently available. In the present study, the vegetation cover (where 20% is snow cover) was considered constant for the entire calibration and validation periods. Vegetation cover, especially snow cover, could not have been the same for all the years of the calibration and validation periods. It can be concluded that taking into consideration changes in vegetation cover will perhaps improve the results.

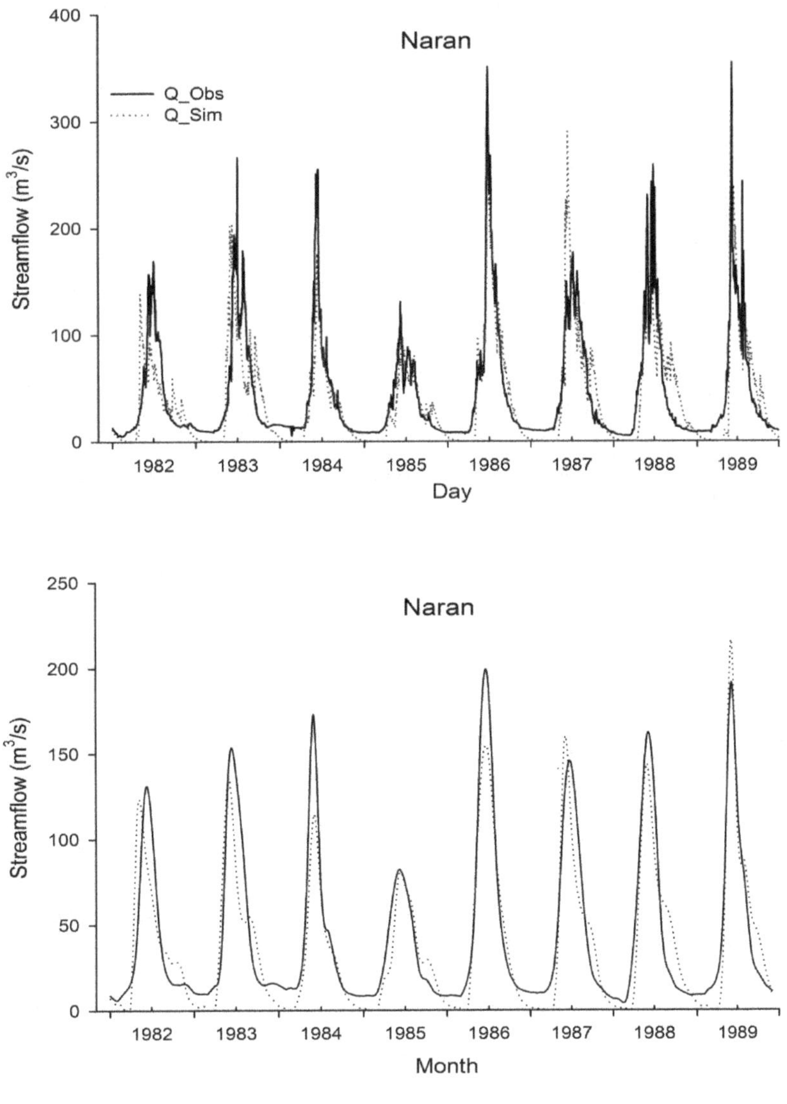

Figure 6. *Cont.*

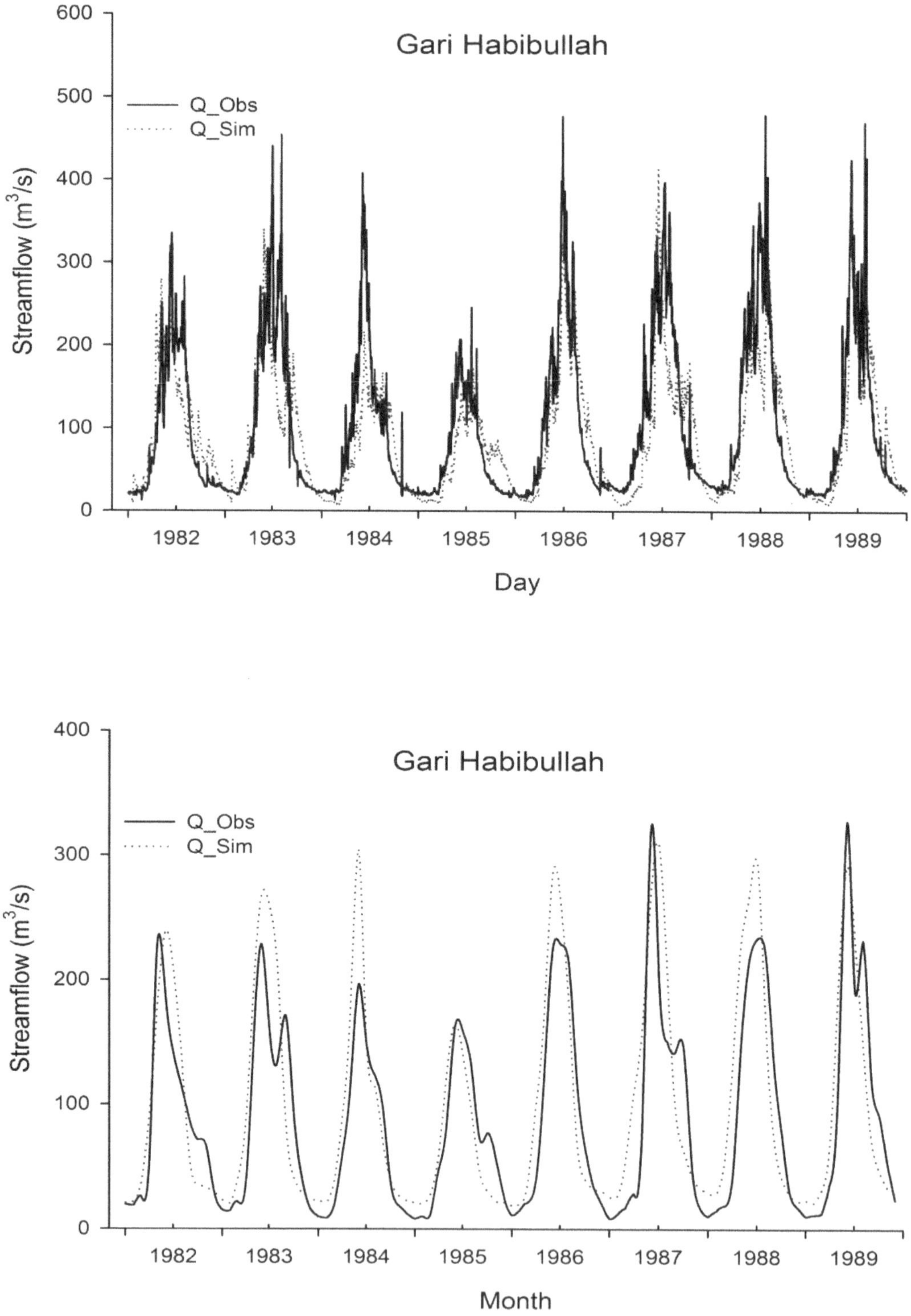

(**b**)

Figure 6. Comparison of daily and monthly observed and simulated streamflow at (**a**) Naran and (**b**) Gari Habibullah for calibration (1982–1989) in the Kunhar River basin.

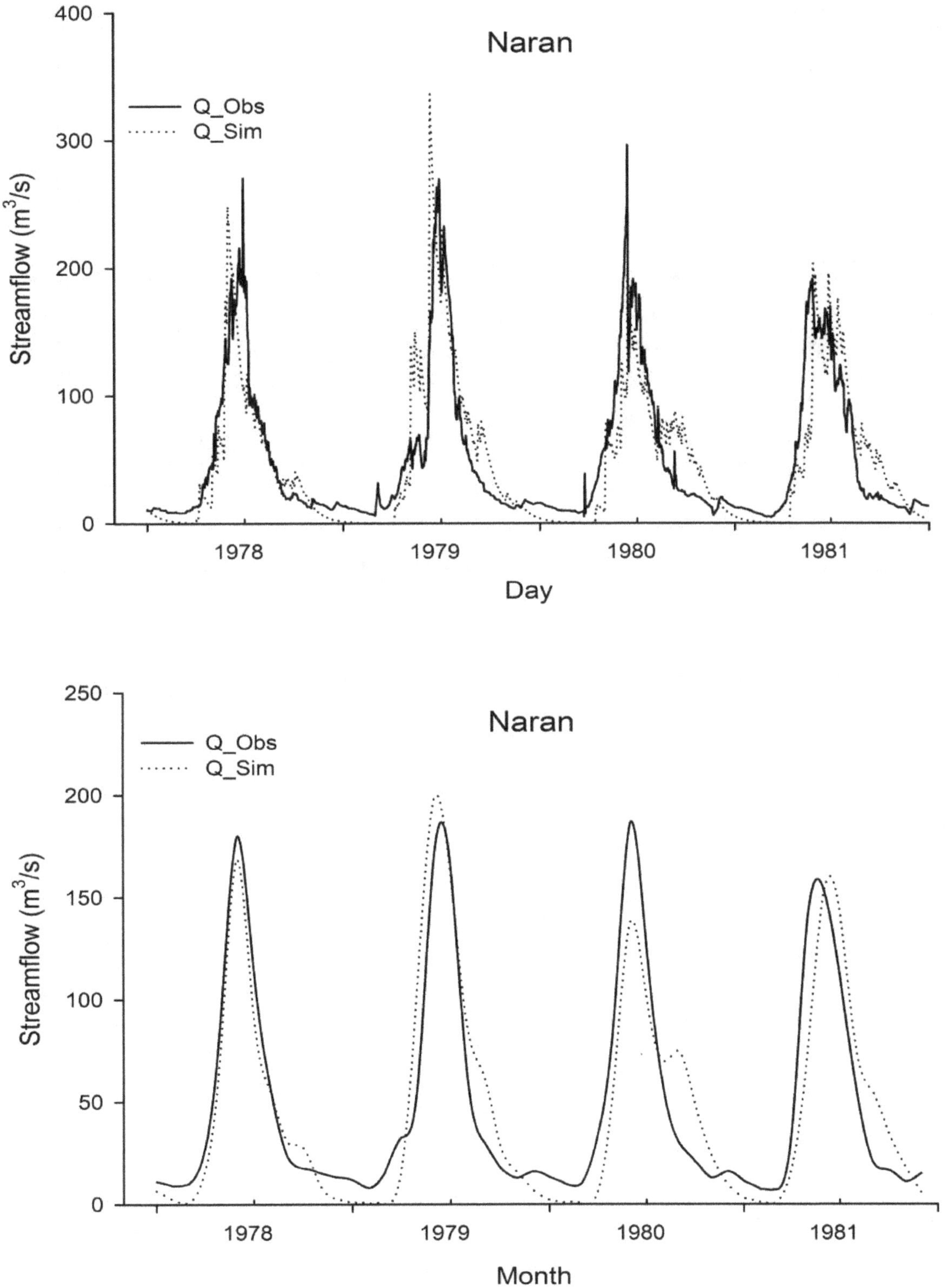

(a)

Figure 7. *Cont.*

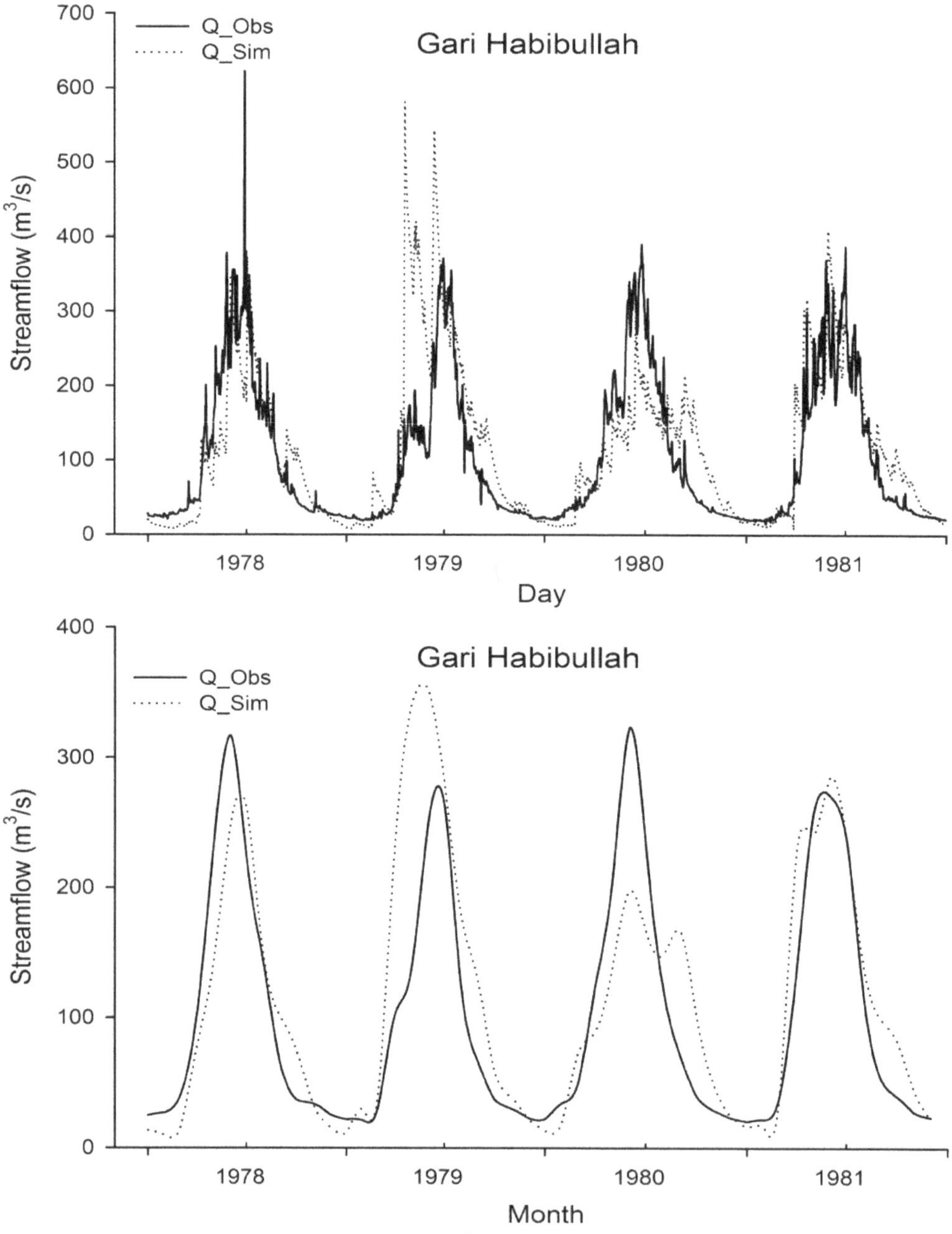

(**b**)

Figure 7. Comparison of daily and monthly observed and simulated streamflow at (**a**) Naran and (**b**) Gari Habibullah for validation (1978-1981) in the Kunhar River basin.

5.2. Projected Changes in Mean Streamflow

Table 4 outlines, in percentage, the projected changes in seasonal and annual flows in the 2020s, 2050s, and 2080s with respect to the baseline period (1961–1990) under A2 and B2 scenarios. In all three periods and under both scenarios, the flow in winter (DJF) and spring (MAM) seasons was projected to decrease at both gauges except in winter at Gari Habibullah in the 2080s. In both seasons, the negative changes at Naran (ranging from 40%-73%) were greatly higher than Gari Habibullah, (ranging from 6% to 28%) under both A2 and B2. Mahmood and Babel [43] conducted a study about extreme temperature events in the Jhelum River basin under A2 and B2, and they found that the intensity of cold days and cold nights will increase in the future. The reduction in flow (as indicated by the results of this study) is likely to be the case due to more accumulation of snowfall that would result from the increased intensity of cold days and cold nights in winter (based on the projections of Mahmood and Babel that indicate that precipitation will increase in winter and decrease in spring) [1].

Conversely, the summer (JJA) and autumn (SON) seasons were projected to have increased flow in all three future periods and under both scenarios but with different magnitudes. Autumn showed highest projected increase at both stations and under both scenarios, ranging from 91% to 131%. This high increase in autumn flow is most likely due to the most projected increase of 25%-32% in precipitation indicated by Mahmood and Babel [1]. The projected changes in this study appear much more than the projected changes in precipitation explored by Mahmood and Babel [1]. This could be because the vegetation cover was considered to be constant throughout the simulation period in this study. In the present state, snow cover (20%) dominates the basin and we used the same percentage of snow cover and constant proportions of other vegetation types for the simulation of projected flow. It is quite possible that snow cover will reduce in the future because of increasing temperature in the basin. So, along with validation results wherein the model overestimated by about 12%, if we also reduce snow cover, then the changes would be less than the projected changes.

On the whole, the mean annual flow was projected to increase by 42%–43% at the end of this century in the basin (at both sites) under both scenarios. The high increase in flow in summer and autumn and the decrease in winter and spring shows that it is hard to rely on only one GCM. So, a group of GCMs is required to give a clearer picture about these results.

The projected decrease in winter and spring can cause water shortage during these seasons not only for the agriculture sector but also for the domestic sector. On the other hand, the projected increase in summer (monsoon season) and autumn can result in flooding in the basin, and that can cause economic losses in the basin. These losses can be reduced by building reservoirs in the basin which can be used to store water during the flooding season and the stored water can be used during the period of shortage.

Table 4. Future changes (%) in streamflow at different gauges relative to the baseline period (1961-1990) under A2 and B2 scenarios in the Kunhar River basin.

A2 Scenario	**1961–1990 (m^3/s)**		**2020s**		**2050s**		**2080s**	
	Naran	**G. Habib**	**Naran**	**G. Habib**	**Naran**	**G. Habib**	**Naran**	**G. Habib**
Winter	10	25	−70	0	−71	−6	−70	8
Spring	37	110	−56	−14	−52	−26	−40	−12
Summer	121	226	53	38	51	46	61	49
Autumn	24	58	104	109	98	107	111	126
Annual	48	105	8	33	6	30	16	43
B2 Scenario								
Winter	10	25	−73	−2	−73	−11	−70	2
Spring	37	110	−62	−16	−58	−28	−49	−10
Summer	121	226	58	38	56	43	54	46
Autumn	24	58	88	97	92	91	97	131
Annual	48	105	3	29	5	24	8	42

5.3. Changes in Flow Duration Curve as Well as Low, Medium, and High Flows

Figure 8 displays the comparison of flow duration curves during the baseline period (1961–1990) and the three future periods (2020s, 2050s and 2080s) under A2 and B2 scenarios at the Gari Habibullah hydrometric station. This comparison shows that the probability of occurrence of flow and magnitudes of flow could be higher in the future in the Kunhar basin under both scenarios. Table 5 shows the projected changes in high flow (Q_5), median flow (Q_{50}), and low flow (Q_{95}) in the 2020s, 2050s, and 2080s with respect to the baseline under both scenarios. Under both scenarios, Q_5 and Q_{50} were projected to increase in all three future periods in the Kunhar basin, with 17%–52% (Q_5) and 43%–84% (Q_{50}). This increase in mean flow is most likely due to the increase in mean annual precipitation and increase in high flow is most likely due to the increase in summer precipitation (monsoon season), as also projected by Mahmood and Babel [1]. However, Q_{95} was predicted to decrease by 18%–99% in the future in the basin. This might be due to the increase in the intensity of cold days and cold nights (mentioned by Mahmood and Babel [43]) which may cause precipitation as snow accumulation in winter (low flow season).

These changes in high flow and flow duration curve show that the frequency of floods and their magnitudes will increase in the future which will create a lot of management problems in the basin. Flooding can not only cause economic losses but also loss of life. Nonetheless, with proper utilization and management of the increased flow, Pakistan can actually increase food and hydropower production in the basin.

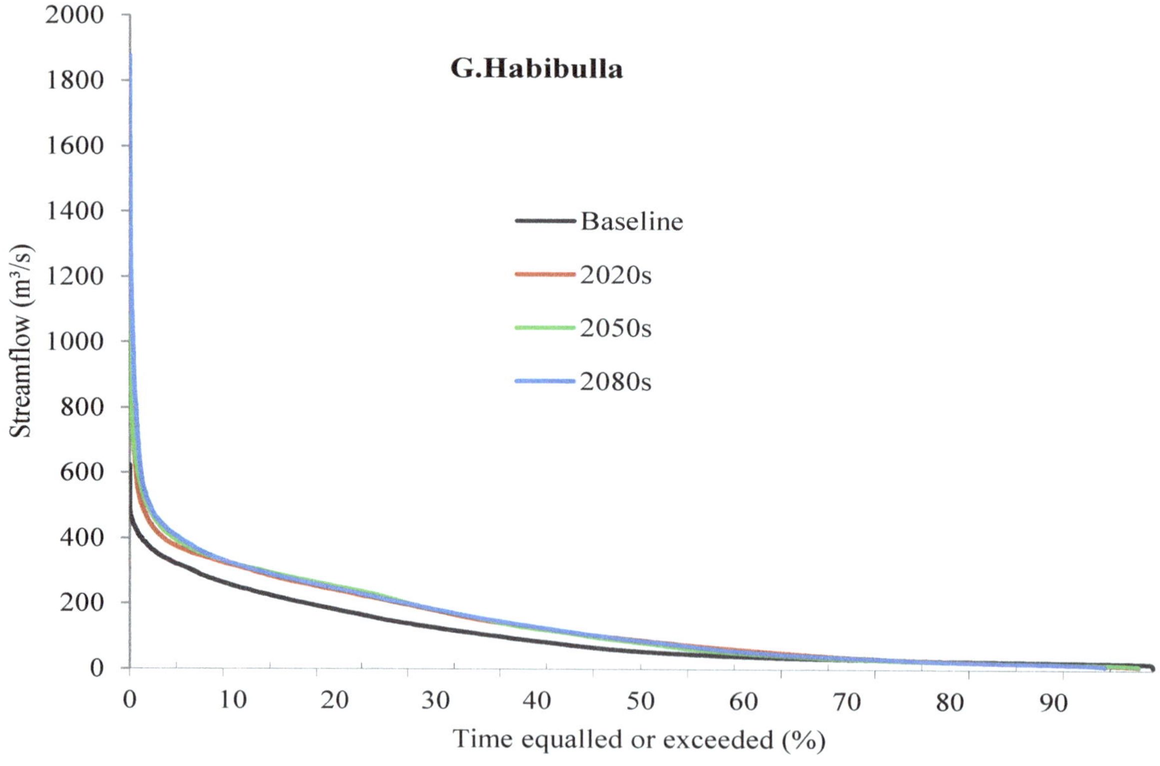

(a)

Figure 8. *Cont.*

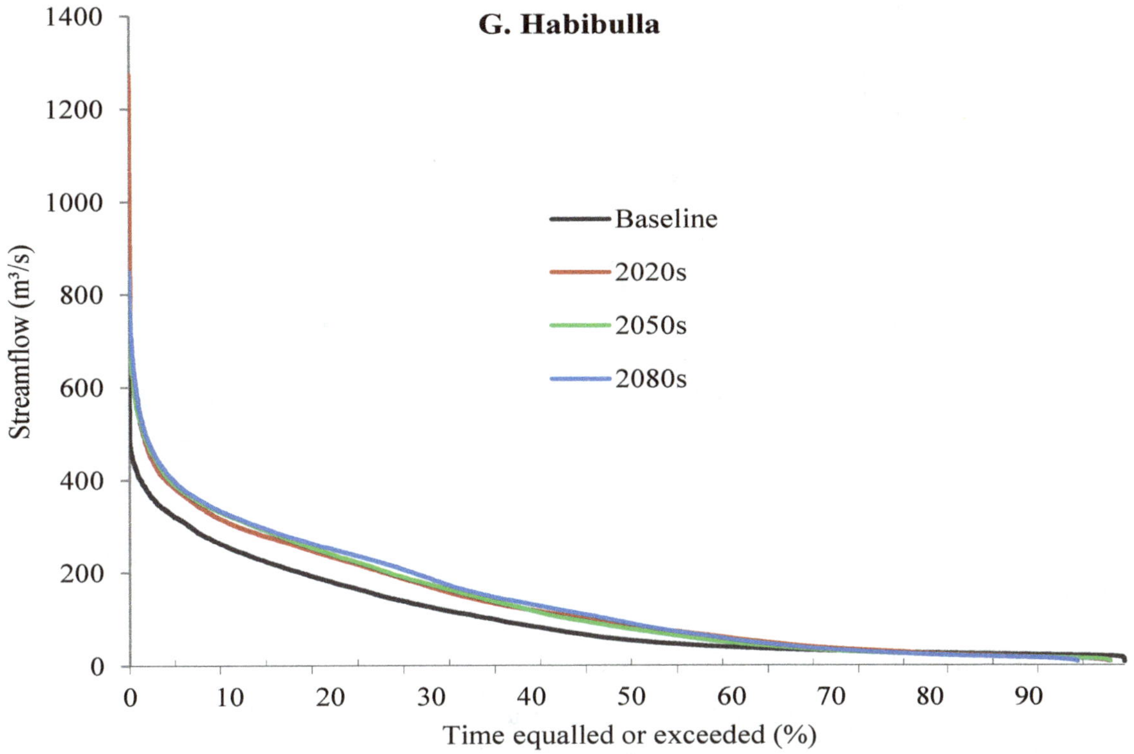

(b)

Figure 8. Flow duration curves under (**a**) A2 and (**b**) B2 scenarios at Gari Habibullah in the Kunhar River basin.

Table 5. Percent future changes in low, median, and high flows with respect to the baseline (1961–1990) under A2 and B2 scenarios in the Kunhar River basin.

A2 Scenario	1961–1990 (m^3/s)		2020s		2050s		2080s	
	Naran	G. Habib	Naran	G. Habib	Naran	G. Habib	Naran	G. Habib
Q_5	170	316	41	17	39	22	52	28
Q_{50}	19	54	43	67	44	55	61	76
Q_{95}	8	21	−99	−25	−99	−22	−99	−18
B2 Scenario								
Q_5	170	316	42	19	46	21	43	24
Q_{50}	19	54	31	65	43	51	45	86
Q_{95}	8	21	−99	−23	−99	−25	−99	−21

Q_5, high flow; Q_{50}, median flow; Q_{95}, low flow.

5.4. Temporal Shifts in Peak Flows

In Figure 9, the mean monthly discharge of the baseline period (1961–1990) is plotted against the mean monthly discharge in the future periods (2020s, 2050s, and 2080s) to explore temporal shifts and magnitudes of peak flows at Gari Habibullah. At this site, a definite delay and increase in peak flows were projected in all three future periods under both scenarios. The peak flows were projected to shift from June to July, with about a 10%–25% increase under both scenarios. This shows that the basin will not only face an increase in frequency and magnitude of floods (mentioned in previous sections) but will also face the shift of these floods from June to July.

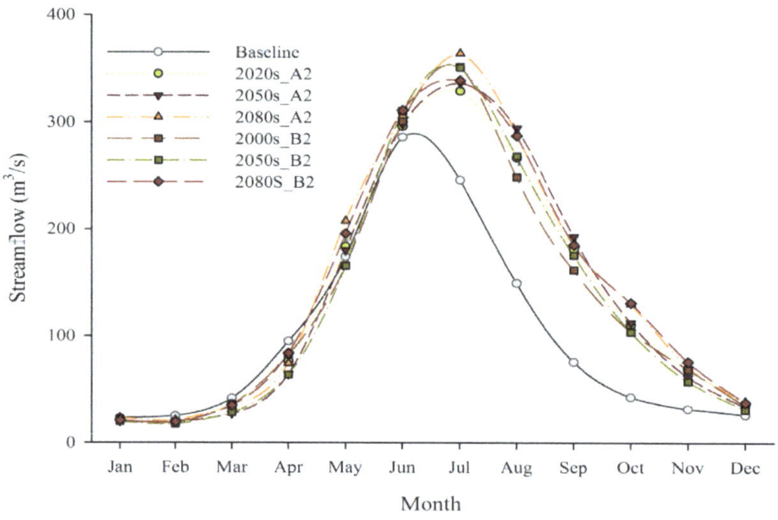

Figure 9. Future shift of peak flows in timing and magnitude under A2 and B2 in the Kunhar River basin.

5.5. Temporal Shifts in Center-of-Volume Date (CVD)

To determine the impact of climate change on the timing of streamflows, an indicator such as center-of-volume date (CVD)—a date at which half of the total volume of streamflow passes through at a gauging station for a specific time period—was used in the present study. CVD was calculated according to the equation described in [44]. Table 6 shows the changes in CVD under A2 and B2, with respect to the baseline period, in the three future periods at Naran and Gari Habibullah. The positive values show delay in flow, and the negative values show earlier in flows. The delay flows were projected at both stations in the Kunhar River basin under A2 and B2 in all three future periods, with 9–17 days delay. Table 6 shows that about half of the flow, on an average, of each year, in the Kunhar basin, passed through by 2–6 July for the baseline period (1961–1990) at both sites. However, this is predicted to shift to mid-July in the future.

Table 6. Future changes in center-of-volume dates (CVD) with respect to the baseline period (1961–1990) at different streamflow stations under both scenarios, A2 and B2, in the Kunhar River basin.

River	Station	CVD of Baseline		2020s	2050s	2080s
		Day of Year		A2 (B2) Day		
Kunhar	G. Habib	183	2 July	15(14)	17(16)	9(16)
Kunhar	Naran	187	6 July	13(14)	13(12)	11(10)

6. Conclusions

Pakistan is one of the most water-stressed countries in the world and its water resources are greatly vulnerable to changing climate. In the present study, the possible impacts of climate change on the water resources of the Kunhar River basin, Pakistan, were assessed under A2 and B2 scenarios of HadCM3. The Kunhar River originates in the Greater Himalayas and is one of the main tributaries of the transboundary Jhelum River. It is located entirely in Pakistan and contributes to the Mangla Reservoir after joining the Jhelum River.

The HEC-HMS hydrological model was used to simulate streamflow in the basin for the future. The model was calibrated and validated for the periods of 1982–1989 and 1978–1981 respectively, at two hydrometric stations (Naran and Gari Habibullah). Three indicators (*i.e.*, Nash Efficiency, coefficient of determination, and percentage deviation), and graphical representations of differences between observed and simulated data were used to check the performance of the model. Downscaled temperature and precipitation data for the period of 2011–2099 under A2 and B2 scenarios of HadCM3 were obtained from Mahmood and Babel [1] and fed into HEC-HMS to simulate the streamflow for the future. Mahmood and Babel projected an overall increase of 1.92 °C–3.15 °C in temperature and 5%–11% in precipitation in the basin. In this study, the simulated streamflow data was divided into three future periods (2011–2040, 2041–2070, and 2071–2099) and was compared with the baseline period (1961–1990). Different indicators like changes in mean flow, low flow, median flow, high flow, flow duration curves, temporal shift in peaks, and temporal shifts in center-of-volume dates were used to investigate the changes in streamflow under A2 and B2. The main conclusions of the study are the following:

1. Mean annual flow was projected to increase in the basin under both A2 and B2 scenarios. Noticeable increase in streamflow was predicted for summer and autumn. However, spring and winter showed decrease in flow.
2. High and median flows were predicted to increase but low flows were projected to decrease in the future under both scenarios. Flow duration curves showed that the probability of occurrence of flow will be more in the future, relative to the baseline.
3. Peaks were predicted to shift from June to July in the future. Similarly, center-of-volume date, a date at which half of the annual water passes, might get delayed by about 9–17 days in the basin under both A2 and B2.

The overall conclusion of the study is that the Kunhar basin is likely to face more floods and droughts in the future due to the projected increase in high flows and decrease in low flows. Many temporal and magnitudinal variations in peak flows would be faced in the basin. This can create many problems for the policy makers and managers of water resources if they do not consider the impacts of changing climate in the basin. For further studies of the basin, we recommend the use of different GCMs so as to cover the range of uncertainties related to GCMs and to explore a wider range of possible impacts of climate change on water resources in the basin.

Limitations of the Study

In the present study, the impact of climate change on the water resources of the Kunhar River basin were assessed by using only single GCM (HadCM3), although it is suggested that more than one GCM should be used to cover the uncertainties related to GCM outputs. Only two meteorological stations are available in the basin, an indication of the data scarcity of the basin. The scarcity of data can cause a lower level of performance by a hydrological model during the calibration and validation processes. Land cover and soil properties were considered constant throughout the simulation period; such an assumption can affect the projections of streamflow in the basin.

Acknowledgments: This study is part of the first author's doctoral research work, conducted at the Asian Institute of Technology (AIT), Thailand. The authors wish to acknowledge and offer gratitude to the Pakistan Meteorological Department (PMD) and the Water and Power Development Authority (WAPDA) of Pakistan for providing important and valuable data for the research. Heartfelt gratitude is also extended to the Higher Education Commission (HEC) of Pakistan for providing financial support to the first author for his doctoral studies at AIT. The authors are also grateful to the Natural Sciences Fund of China (41471463) and CAS (Chines Academy of Sciences) for providing financial support to complete this research.

Author Contributions: The first author conducted this research during his doctoral program under the supervision of the third author. The second author supervised the writing of this article and also reviewed it.

References

1. Mahmood, R.; Babel, M.S. Evaluation of sdsm developed by annual and monthly sub-models for downscaling temperature and precipitation in the jhelum basin, pakistan and india. *Theor. Appl. Climatol.* **2013**, *113*, 27–44. [CrossRef]

2. Chu, J.; Xia, J.; Xu, C.Y.; Singh, V. Statistical downscaling of daily mean temperature, pan evaporation and precipitation for climate change scenarios in haihe river, China. *Theor. Appl. Climatol.* **2010**, *99*, 149–161. [CrossRef]
3. Intergovernmental Panel on Climate Change (IPCC). *Climate Change 2013: The Physical Science Basis*; Contribution of Working Group I to the Fifth Assessment Report of the Intergovernmental Panel on Climate Change; Cambridge University Press: Cambridge, UK; New York, NY, USA, 2013; p. 1535.
4. Solomon, S., Qin, D., Manning, M., Chen, Z., Marquis, M., Averyt, K.B., Tignor, M., Miller, H.L., Eds.; *Climate Change 2007: The Physical Science Basis*; Contribution Of Working Group I to the Fourth Assessment Report of the Intergovernmental Panel on Climate Change; Cambridge University Press: Cambridge, UK, 2007.
5. Khattak, M.S.; Babel, M.S.; Sharif, M. Hydro-meteorological trends in the upper indus river basin in Pakistan. *Clim. Res.* **2011**, *46*, 103–119. [CrossRef]
6. Ma, Z.Z.; Wang, Z.J.; Xia, T.; Gippel, C.J.; Speed, R. Hydrograph-based hydrologic alteration assessment and its application to the Yellow River. *J. Environ. Inform.* **2014**, *23*, 1–13. [CrossRef]
7. Yang, W.; Yang, Z.F. Evaluation of sustainable environmental flows based on the valuation of ecosystem services: A case study for the Baiyangdian Wetland, China. *J. Environ. Inform.* **2014**, *24*, 90–100. [CrossRef]
8. Zhang, X.C.; Liu, W.Z.; Li, Z.; Chen, J. Trend and uncertainty analysis of simulated climate change impacts with multiple GCMs and emission scenarios. *Agric. Forest Meteor.* **2011**, *151*, 1297–1304. [CrossRef]
9. Wang, H.; Zhang, M.; Zhu, H.; Dang, X.; Yang, Z.; Yin, L. Hydro-climatic trends in the last 50 years in the lower reach of the Shiyang River Basin, NW China. *Catena* **2012**, *95*, 33–41. [CrossRef]
10. Dhital, Y.; Tang, Q.; Shi, J. Hydroclimatological changes in the Bagmati River Basin, Nepal. *J. Geogr. Sci.* **2013**, *23*, 612–626. [CrossRef]
11. Gleick, P. Regional hydrologic consequences of increases in atmospheric CO_2 and other trace gases. *Climatic Change* **1987**, *10*, 137–160. [CrossRef]
12. Jordan, Y.C.; Ghulam, A.; Chu, M.L. Assessing the impacts of future urban development patterns and climate changes on total suspended sediment loading in surface waters using geoinformatics. *J. Environ. Inform.* **2014**, *24*, 65–79. [CrossRef]
13. Akhtar, M.; Ahmad, N.; Booij, M.J. The impact of climate change on the water resources of Hindukush–Karakorum–Himalaya region under different glacier coverage scenarios. *J. Hydrol.* **2008**, *355*, 148–163. [CrossRef]
14. Jetly, R. *Pakistan in Regional and Global Politics*; Taylor & Francis: New Dehli, India, 2012.
15. Connor, R. *The United Nations World Water Development Report 2015: Water for a Sustainable World*; UNESCO Publishing: Paris, France, 2015.
16. Gebremeskel, S.; Liu, Y.B.; de Smedt, F.; Hoffmann, L.; Pfister, L. Analysing the effect of climate changes on streamflow using statistically downscaled gcm scenarios. *Int. J. River Basin Manag.* **2005**, *2*, 271–280. [CrossRef]
17. Wilby, R.L.; Hay, L.E.; Gutowski, W.J.; Arritt, R.W.; Takle, E.S.; Pan, Z.; Leavesley, G.H.; Martyn, P.C. Hydrological responses to dynamically and statistically downscaled climate model output. *Geophys. Res. Lett.* **2000**, *27*, 1199–1202. [CrossRef]
18. Hay, L.E.; Wilby, R.L.; Leavesley, G.H. A comparison of delta change and downscaled GCM scenarios for three mountainous basins in The United States. *J. Am. Water Resour. Assoc.* **2000**, *36*, 387–397. [CrossRef]
19. Ahmad, Z.; Hafeez, M.; Ahmad, I. Hydrology of mountainous areas in the upper Indus Basin, Northern Pakistan with the perspective of climate change. *Environ. Monit. Assess.* **2012**, *184*, 5255–5274. [CrossRef] [PubMed]
20. Shrestha, M.; Koike, T.; Hirabayashi, Y.; Xue, Y.; Wang, L.; Rasul, G.; Ahmad, B. Integrated simulation of snow and glacier melt in water and energy balance-based, distributed hydrological modeling framework at Hunza River Basin of Pakistan Karakoram region. *J. Geophys. Res. Atmos.* **2015**. [CrossRef]
21. Bocchiola, D.; Diolaiuti, G.; Soncini, A.; Mihalcea, C.; D'Agata, C.; Mayer, C.; Lambrecht, A.; Rosso, R.; Smiraglia, C. Prediction of future hydrological regimes in poorly gauged high altitude basins: The case study of the upper Indus, Pakistan. *Hydrol. Earth Syst. Sci.* **2011**, *15*, 2059–2075. [CrossRef]
22. Meenu, R.; Rehana, S.; Mujumdar, P.P. Assessment of hydrologic impacts of climate change in Tunga-Bhadra river basin, India with HEC-HMS and SDSM. *Hydrol. Process.* **2012**, *27*, 1572–1589. [CrossRef]

23. Verma, A.; Jha, M.; Mahana, R. Evaluation of HEC-HMS and WEPP for simulating watershed runoff using remote sensing and geographical information system. *Paddy Water Environ.* **2010**, *8*, 131–144. [CrossRef]
24. Yimer, G.; Jonoski, A.; Griensven, A.V. Hydrological response of a catchment to climate change in the Upper Beles River Basin, Upper Blue Nile, Ethiopia. *Nile Basin Water Eng. Sci. Mag.* **2009**, *2*, 49–59.
25. García, A.; Sainz, A.; Revilla, J.A.; Álvarez, C.; Juanes, J.A.; Puente, A. Surface water resources assessment in scarcely gauged basins in the north of Spain. *J. Hydrol.* **2008**, *356*, 312–326. [CrossRef]
26. Munyaneza, O.; Mukubwa, A.; Maskey, S.; Uhlenbrook, S.; Wenninger, J. Assessment of surface water resources availability using catchment modelling and the results of tracer studies in the mesoscale Migina Catchment, Rwanda. *Hydrol. Earth Syst. Sci.* **2014**, *18*, 5289–5301. [CrossRef]
27. Babel, M.; Bhusal, S.; Wahid, S.; Agarwal, A. Climate change and water resources in the Bagmati River Basin, Nepal. *Theor. Appl. Climatol.* **2014**, *115*, 639–654. [CrossRef]
28. Wilby, R.L.; Dawson, C.W.; Barrow, E.M. SDSM—A decision support tool for the assessment of regional climate change impacts. *Environ. Model. Softw.* **2002**, *17*, 145–157. [CrossRef]
29. Gordon, C.; Cooper, C.; Senior, C.A.; Banks, H.; Gregory, J.M.; Johns, T.C.; Mitchell, J.F.B.; Wood, R.A. The simulation of sst, sea ice extents and ocean heat transports in a version of the hadley centre coupled model without flux adjustments. *Clim. Dyn.* **2000**, *16*, 147–168. [CrossRef]
30. IPCC. *Emissions Scenarios: A Special Report of Working Group III*; Cambridge University Press: Cambridge, UK, 2000; p. 570.
31. Agarwal, C.S.; Garg, P.K. *Textbook on Remote Sensing: In Natural Resources Monitoring and Management*; A H Wheeler Publishing Co Ltd: New Delhi, India, 2002.
32. Singh, D.; Gupta, R.D.; Jain, S. Assessment of impact of climate change on water resources in a hilly river basin. *Arab. J. Geosci.* **2015**, *8*, 10625–10646. [CrossRef]
33. Ghoraba, S.M. Hydrological modeling of the simly dam watershed (Pakistan) using GIS and SWAT model. *Alex. Eng. J.* **2015**, *54*, 583–594. [CrossRef]
34. Leghari, M.K.; Waheed, S.B.; Leghari, M.Y. Ecological study of algal flora of Kunhar River, Pakistan. *Pak. J. Bot.* **2001**, *33*, 629–636.
35. De Scally, F.A. Relative importance of snow accumulation and monsoon rainfall data for estimating annual runoff, Jhelum Basin, Pakistan. *Hydrol. Sci. J.* **1994**, *39*, 199–216. [CrossRef]
36. Hussain, M.; Shah, G.M.; Khan, M.A. Traditional medicinal and economic uses of Gymnosperms of Kaghan valley, Pakistan. *Ethnobot. Leafl.* **2006**, *10*, 72–81.
37. Pakistan Ministry of Water and Power (PMWP). *Executive Summary, Pakistan Water Sector Strategy*; PMWP: Islamabad, Pakistan, 2002; p. 24.
38. Ahmad, I.; Ambreen, R.; Sun, Z.; Deng, W. Winter-spring precipitation variability in Pakistan. *Am. J. Clim. Chang.* **2015**, *4*, 115–139. [CrossRef]
39. Mahmood, R.; Babel, M.S.; Jia, S. Assessment of temporal and spatial changes of future climate in the Jhelum river basin, Pakistan and India. *Weather Clim. Extrem.* **2015**, *10*, 40–55. [CrossRef]
40. *Hydrologic Modeling System HEC-HMS*; Institute for Water Resources: Davis, CA, USA, 2010.
41. Van Liew, M.W.; Garbrecht, J. Hydrologic simulation of the Little Washita river experimental watershed using SWAT1. *J. Am. Water Resour. Assoc.* **2003**, *39*, 413–426. [CrossRef]
42. *HEC-ResSim Reservoir System Simulation User's Manual*; Hydrologic Engineering Center: Davis, CA, USA, 2007; p. 512.
43. Mahmood, R.; Babel, M.S. Future changes in extreme temperature events using the statistical downscaling model (SDSM) in the trans-boundary region of the Jhelum river basin. *Weather Clim. Extrem.* **2014**, *5–6*, 56–66. [CrossRef]
44. Stewart, I.T.; Cayan, D.R.; Dettinger, M.D. Changes toward earlier streamflow timing across western North America. *J. Clim.* **2005**, *18*, 1136–1155. [CrossRef]

5

Stressors and Strategies for Managing Urban Water Scarcity: Perspectives from the Field

Vivek Shandas [1,*], Rosa Lehman [1], Kelli L. Larson [2], Jeremy Bunn [3] and Heejun Chang [4]

Academic Editor: Yingkui Li

1 Toulan School of Urban Studies and Planning, Portland State University, Portland, OR 97201-0751, USA; rlehman@pdx.edu
2 Schools of Geographical Sciences and Urban Planning and Sustainability, Arizona State University, AZ 85287-5302, USA; kelli.larson@asu.edu
3 Herrera Inc., Seattle, WA 98121, USA; jbunn@herrerainc.com
4 Department of Geography, Portland State University, Portland, OR 97201-0751, USA; changh@pdx.edu
* Correspondence: vshandas@pdx.edu.

Abstract: Largely because water resource planning in the U.S. has been separated from land-use planning, opportunities for explicitly linking planning policies to water availability remain unexamined. The pressing need for better coordination between land-use planning and water management is amplified by changes in the global climate, which will place even greater importance on managing water supplies and demands than in the past. By surveying land and water managers in two urbanizing regions of the western United States—Portland, Oregon and Phoenix Arizona—we assessed the extent to which their perspectives regarding municipal water resource management align or differ. We specifically focus on characterizing how they perceive water scarcity problems (*i.e.*, stressors) and solutions (*i.e.*, strategies). Overall, the results show a general agreement across both regions and professions that long-term drought, population growth, and outdoor water use are the most important stressors to urban water systems. The results of the survey indicated more agreement across cities than across professions with regard to effective strategies, reinforcing the idea that land-use planners and water managers remain divided in their conception of the solutions to urban water management. To conclude, we recommend potential pathways for coordinating the fields of land and water management for urban sustainability.

Keywords: water managers; land-use planners; climate change; stressors; strategies; Portland; phoenix

---

1. Introduction

Despite the breadth and depth of the literatures on water supply and demand management, the operation of these systems have been viewed largely in isolation from land-use planning [1]. Residential development patterns, which in some cases are the direct and intentional outcome of land-use planning initiatives, are a potentially important exogenous influence on water management [2–4]. In fact, largely because water resource planning in the U.S. has traditionally been separated from land-use planning, opportunities to explicitly link planning policies with water availability have gone unexamined [5]. Further, the planning community has neglected the role of water supply, assuming it is readily available for urban (re)development. In short, integrated planning across the land and water sectors is rarely if ever practiced. To better coordinate planning for land and water resources, information about professional perspectives and practices is required. Such information can reveal areas of converging and diverging viewpoints, thereby identifying the potential for collaborations as well as for conflicts as efforts to integrate across sectors continue.

This paper seeks to characterize and compare perceptions among land-use and water-resource planners in two cities of the American West. Using humid Portland, Oregon and arid Phoenix, Arizona as case study samples, we surveyed professionals about various stressors and strategies for their local community water systems. In addition to identifying converging and diverging views, we also examined similarities and differences across the two metropolitan study areas, which may be important in signaling the unique effects of each regional context.

1.1. Pressing Challenges

The pressing need for better coordination between land-use planning and water management is amplified by changes in population and the global climate, which present even greater challenges and uncertainties than in the past. Urban water managers—defined here as those public sector employees responsible for the allocation, administration, and budgetary aspects of water management—in Portland and Phoenix face the complexity of burgeoning growth, with the populations of both urban regions expected to double in coming decades [6,7]. This comes on top of the increasing growth rates of the last decade. Compounding the macro-scale pressures faced in both cities, recent evidence suggests that global climate change will have profound impacts on water resources in the West [8]. The Pacific Northwest (PNW) is experiencing more frequent winter floods and summer warming [9,10], while the American Southwest is currently experiencing a multi-year drought [11].

Long-term climate models project increasing annual temperatures and high precipitation variability (with generally less precipitation in summer) for the PNW [9]. Higher winter temperatures combined with more precipitation will lead to higher snow-line elevations, which in turn will affect the timing of snowmelt and summer flow, ultimately impacting the availability of water resources for the Portland region [12].

With respect to Phoenix, global climate models increasingly conclude the Southwest will likely be warmer and drier in this century than in the last. Regional climate models have recently indicated reductions in surface water runoff in regional watersheds serving metropolitan Phoenix [6]. A report by the National Research Council [13] shows that a warmer and drier future will reduce snowpack, Colorado River flows, and urban water supplies. According to the 5th IPCC assessment report, there is a broad consensus among models that the region will be drier during the 21st Century, and indeed, the transition to a more arid climate is already underway [14]. Although per capita water consumption has decreased since the 1990s in most municipalities in the American west [15–18], the expected combination of reduced water supply and increasing urban populations will likely create an unprecedented state of potential water scarcity.

1.2. Potential Strategies

The process of coordinating water management still involves several policy actions enacted independently and jointly by multiple agents with different goals and objectives, as well as varying levels of influence on outcomes. Water managers have multiple options for adapting to the effects of water supply shortfalls, including taking actions to increase supplies or to decrease demands, both of which can have either short-term or long-term reach. The long-term options, while generally effective in engineering terms, are not always feasible for political and/or financial reasons. Building dams and reservoirs can be cost-prohibitive and requires localities and states to shoulder the burden alone, especially since the U.S. government appears disinterested in funding such projects. The communities designated as hosts often resist large-scale projects. As a result, augmenting supply increasingly does not provide a viable option for the long-term [19]. Rather, U.S. water managers are increasingly forced to use short-term measures that either augment supply or diminish demand [20,21]. These measures include tapping emergency water sources, trucking in water from outside locales, and imposing water-use restrictions on non-essential uses (e.g., lawn irrigation, car washing). We note that such short term measures can prove challenging due to the overall lack of precipitation, and the uncertainties about future rain events.

In the face of new and pressing challenges and a diminished ability to continue adding to the water supply, new solutions must be sought. One area of untapped potential is in collaborating with land-use planners. Earlier studies suggest that the nature of the built environment, landscape treatments, and short-term consumption behavior can also impact residential water use in urban areas [16,22–26]. This is particularly relevant to arid cities, such as Phoenix and Los Angeles, where an estimated 74% of residential water use is for outdoor purposes [27]. Wentz and Gober [22] demonstrated that residential outdoor water use increased with large lots, turf grass (as opposed to drought-tolerant landscaping), and the presence of pools. Similarly, Chang *et al.* [3] found that residential water use was higher in denser older neighborhoods close to downtown than in newer peripheral neighborhoods in the City of Portland, Oregon. House-Peters *et al.* [28] further reported that highly affluent newer neighborhoods had more seasonal water use than denser older low-income neighborhoods in Hillsboro, a suburban city of Portland. Sauri [29] found that demand management policies such as conservation and water pricing may not be sufficient for controlling water consumption levels when low-density urban development continues or income gains occur. In a study of annual residential water use in the Portland metropolitan area, Shandas *et al.*, [23] stressed the importance of investigating behavioral and land-use density barriers that limit the responsiveness of water use to varying climate conditions. Automated timers, for example, generally do not appear to be adjusted in response to weather conditions. The built environment of cities, which is largely controlled by land-use planning, combines with individual and group behaviors to mediate the extent of water consumption [30,31].

1.3. Integration Across Sectors

To address calls for greater integration between water and land managers, many national organizations, professional journals, trade publications, and international conferences are bringing attention to highly inter-connected nature of water and land systems. The National Science Foundation, for example, recently released a call for proposals examining the water–land nexus, and the European Commission is hotly debating the future of this field [32]. Amidst the challenges for integrating water and land systems, little is known about the extent to which the people who are responsible for managing these systems view the same challenges and opportunities. To what degree do land managers agree with water managers in terms of the stressors on the urban water system? What are the similarities and differences between the two groups in terms of strategies for addressing future changes in water supply? These and other questions help to frame a critical issue in the management of water supplies because, arguably, if these two constituencies do not view the problem as the same, then their solutions will also face formidable challenges to implement.

By surveying both land and water managers in two urbanizing regions of the western United States, we sought to assess the extent to which their perspectives align or differ. We specifically focused on their perceptions of the problem (e.g., stressors) and potential solutions (e.g., strategies). While previous studies have looked at single regions [31,33] we sought to examine two regions—Phoenix, Arizona, and Portland, Oregon—that have different land-use policies, climate regimes, and population growth trends. By engaging water managers and land-use planners in these regions, we were able to address two research questions: (1) To what extent do urban water use managers and land-use planners share perspectives regarding the stressors to and strategies for water management both within and between the two locations and professional fields? (2) What factors might help to explain any observed differences and similarities in perspectives about the water management system?

2. Materials and Methods

Portland and Phoenix share the dual challenges of rapid growth and climatic uncertainty. The challenges however occur under different physical geographies, political cultures, and growth-management and land-use policies. These differences allowed us to study perceptions of land-use planners and water-resource managers in two distinct areas, by comparing factors that may exacerbate or mitigate future water shortages, and strategies for ameliorating potential water challenges.

2.1. Survey Design

We used a cross-sectional survey design that was informed by existing literature on the challenges facing the regions [31,34]. An online survey of municipal water managers and land-use planning professionals in the Phoenix and Portland regions constituted the primary vehicle for the empirical assessment, resulting in four samples: Phoenix water managers, Phoenix land planners, Portland water managers, and Portland land planners. The survey was administered in the spring of 2010 to municipal professionals in both regions—32 in metropolitan Phoenix and 25 in the Portland area. By identifying all the municipal land use and water professionals in both regions, we identified nearly all of the potential recipients of the survey. For land planners, our priority was to survey planning directors/managers or the closest equivalent for each city within our study regions. If none could be found, the city/town manager was contacted instead. For water managers, our priority was to target water resource managers/directors or the closest equivalent. While the small number of municipalities in each region did not allow us to conduct statistical tests of inference, they did provide a descriptive assessment of the water scarcity strategies employed in each region, and patterns among survey respondents, comparing the land and water managers and the two study regions.

To address our research questions, we developed survey questions that consisted of two main points—stressors and strategies—with respondents rating a set of options on a scale of one to ten (from 1 = not at all significant to 10 = greatly significant). Twelve stressors and 13 strategies were developed for both potential risk factors and solutions based on a review of literature and discussions with professionals and among our research team. Following from Larson and colleagues [35,36] these items present a variety of natural and anthropogenic risk factors as well as potential voluntary and regulatory-based strategies in both the land and water sectors. For each of the possible factors representing a stressor, respondents were asked, "To what extent are each of the following factors important when considering the future of water supply." For each factor that represented a strategy, respondents were asked, "To what extent are each of the following factors important when considering strategies for addressing shortages in future water supply." Respondents had 12 stressors and 13 strategies to consider, relating to three sources of influence: environment and population, infrastructure and management, and end-user behavior, as shown in Tables 1 and 2. The three categories correspond to the generally accepted challenges facing water resource management, including a growing population and environment, combined with strategies facing the future, such as climate change, technological applications, and human or individual actions. We distinguish two aspects of climate change, namely natural climate variability and human-induced climate change in part because American public opinion continues to be divided about the causes for changes in climate conditions.

Table 1. Stressors listed for consideration in survey with descriptive statistics (Minimum, Maximum, Median, Standard Deviation).

Category of Stressor	Stressor	Min.	Max.	Med.	SD
Environment and Population	Population growth	1	10	7	2.89
	Long-term drought	1	10	7	3.08
	Natural climate variability	1	10	5	2.88
	Human-induced climate changes	1	10	4	2.68
Infrastructure and Management	Inadequate access to water sources	1	10	4	3.02
	Infrastructure to store, treat, deliver water	1	10	4	2.69
	Land-use planning or development	1	10	5	2.46
	Water-resource planning or management	1	10	5	2.66
End-User Behavior	Indoor water uses	1	8	4	2.06
	Outdoor water uses	1	10	6	2.57
	Household or residential water use	1	10	5	2.36
	Industrial or other business water use	1	10	5	2.52

Table 2. Strategies listed for consideration in survey with descriptive statistics (Minimum, Maximum, Median, Standard Deviation).

Category of Strategy	Strategy	Min.	Max.	Med.	SD
Environment and Population	Planning for future climate changes	1	10	5	2.74
	Restricting new building permits	1	10	4	2.56
	Limiting new growth or development	1	10	6	2.97
Infrastructure and Management	More compact or dense communities	1	10	5	3.08
	Retiring agricultural land	1	10	4	2.90
	Acquiring new sources of water	1	10	8	2.73
	Building structures to store water	1	10	4	2.94
	Upgrading water delivery infrastructure	1	10	7	2.53
	Wholesale contracts with other providers	1	10	5	2.93
End-User Behavior	Restrictions or bans to limit water use	1	10	6	2.50
	Increasing the price of water	1	10	7	2.96
	Water conservation education	1	10	8	2.49
	New water efficiency technology	1	10	8	2.53

2.2. Methods of Analysis

Due to the small sample representing diverse municipalities in each region, we developed analytical techniques that emphasize the descriptive and qualitative patterns in the data. These included graphic depictions of the data, information from a series of workshops, and where available, using qualitative responses to flesh out responses to the survey. Specifically, for the purposes of data interpretation, we considered a rating of 7 or above to indicate the respondent considered the strategy or stressor to be important or very important. By looking only at those responses with a rating of 7 or higher we can effectively parse out the differences between the two professions and the two regions using descriptive means.

Across both regions and sectors, 57 surveys were completed and returned for a response rate of 50%—some municipalities had more than one respondent. For Portland, both surveys were returned from 7 towns, either the land or water survey was returned by 10 towns, and 7 towns did not respond at all. For Phoenix, both surveys were returned by 8 towns, one or the other was returned by 17 towns, and 7 towns were not represented in this study. While the response rate was similar across regions, land planners responded at a higher rate. For the water survey, respondents included Public Works Directors, City Engineers, Environmental Managers, and a Town Manager. The respondents from the land-use survey were mostly planners, but also included City Administrators and Community Development Directors from smaller towns. As a whole, the surveys represent a variety of towns across the study regions, ranging from small to large. In accordance with policies governing the ethical treatment of research participants, the survey results are presented anonymously.

3. Results and Discussion

Overall, the results of the surveys show significantly more agreement in identifying the most significant stressors—of which only two stood out—whereas there were many strategies identified as important and a wider range of opinion. There was also somewhat more agreement across cities than across professions, particularly for strategies.

3.1. Stressors

Three stressors emerged as the most widely perceived as important in both cities and by both professions. These were, in ranked order, long term drought, population growth, and outdoor water use (Figure 1). The remaining possible stressors were not viewed as important by the majority of respondents, and there was general consensus that indoor water use is the least important stressor. Phoenix professionals were generally more focused on long-term drought than Portland professionals; 83% of Phoenix water managers identified this as an important contributor to water stress, and 62% of Phoenix land-use planners agreed (Figure 2). Two most widely identified stressors were related to the environment and population, while the third—outdoor water use—was an end-user behavior. Notably, professionals in both locations rated human-induced climate change low—and water professionals rated it very low—as a cause of stress, despite the direct connection between climate change and summer drought in both areas that is projected by current climate models [11,12,37]. While we are not able to discern the reasons for the low rating to human-induced climate change, we note that the term 'human-induced' may reflect the discomfort with the causal connection or attribution of drought events to human activities. These results would be consistent with the earlier point regarding the fact that U.S. public opinion is still divided regarding the causes for climate change.

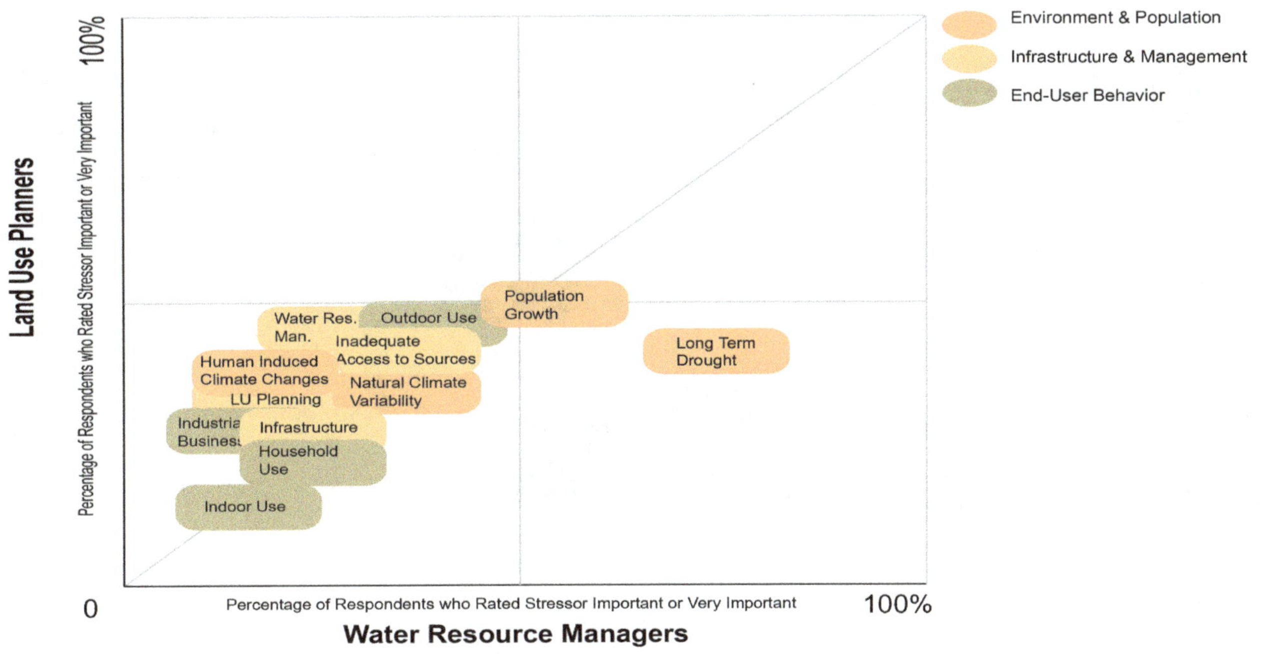

Water Resource Stressors rated at least 7/10 on a scale of importance, by profession.

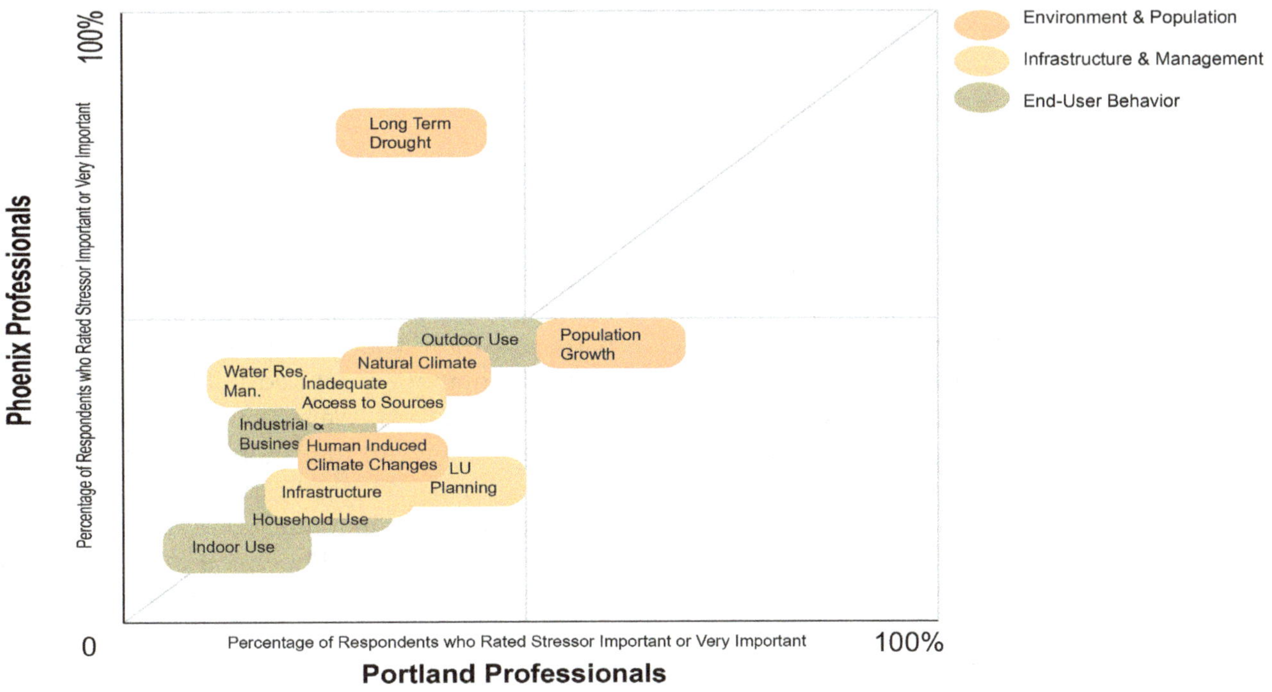

Figure 2. Water Resource Stressors rated at least 7/10 on a scale of importance, by region.

3.2. Strategies

Across both regions and professions, the results suggest greater variation in perspective on the effectiveness of potential strategies than stressors. Between both professions, four strategies were identified as somewhat effective by over 50% of respondents (as compared to only two stressors): water efficiency technology, water conservation education, acquiring new sources of water, and limiting new growth (Figure 3). Although the two biggest stressors are in the environment and population category, namely long-term drought and population growth, the corresponding strategies that directly addressed these stressors were generally not among the highest ranked.

Figure 3. Water Resource Strategies rated at least 7/10 on a scale of importance, by profession.

The strategies relating to end-user behavior generated the most support. Specifically, new water-efficient technology had the most support and least variability of all the strategies in the survey across all categories of respondents. Among water managers, there was consensus that increasing the price of water would be effective, in Portland 70% of water managers ranked it 7/10 or above, and in Phoenix, almost 60% of water managers ranked it 10/10 (Figure 4). Water conservation education was also a highly ranked strategy, although land planners lagged water managers in their enthusiasm for its effectiveness. Outright restrictions to ban or limit water-use was split in among the Portland respondents and were ranked moderately in Phoenix. Although inadequate access to water sources was not rated highly as a cause of stress, there was general agreement in Phoenix that acquiring new sources of water would be a highly effective strategy: half of Phoenix water managers ranked this a 10/10. We note that while many of these strategies were supported, recent research from California suggest that applications of water efficient technology, increasing the prices of water and education [38], may not be highly effective.

Figure 4. Water Resource Strategies rated at least 7/10 on a scale of importance, by region.

Generally, land-use planners were relatively more supportive than water managers of strategies in their field, such as limiting new growth, and more dense and compact communities. Similarly, water managers were more favorable towards upgrading water delivery systems and increasing the price of water. The main regional difference is that professionals in Phoenix were more likely to support acquiring new sources of water and limiting new growth than those in Portland. Generally, restricting new building permits was not identified but by a few land-use planners and water managers, while limiting new growth was somewhat more highly valued, particularly in Phoenix where a clear majority (75%) of water managers rated it 7/10 or above in effectiveness.

4. Discussion

Despite the sample size of the survey, we observed that across region and profession, survey respondents widely agreed on the sources of stress to urban water systems, while opinions on strategies were much less uniform and a wider array of strategies was identified as effective. The fact that only two stressors were identified by 50% or more of respondents as important while many strategies emerged could suggest that while stress to water resources comes from one or two causes, the solutions are less clear. This is logical because a suite of strategies for each specific region will be necessary, rather than a single silver bullet that fits every region. Alternatively, these results could suggest that the research team more successfully presented options for stressors with which practitioners agreed, while failing to include other important stressors.

While the top two stressors were related to environment and population, the majority of highly ranked strategies were related to end-user behavior. While understandable due to the challenge in the physical and political feasibility of other strategies, research by Lach and colleagues suggest that water managers are highly conservative when attempting to regulate end-user behavior [39]. Accordingly, while a strategy such as acquiring new sources of water might be highly effective, it is not worth consideration if there are no such new sources available or if this strategy would create negative environmental and social outcomes (e.g., on source ecosystems or communities from which water is withdrawn). Further, experiences in California suggest that water agencies are required to balance the cost of water to match what is required to address the demand, and the raising water rates must occur within limits. Local government bodies are much more likely to have the ability to enact policies related to end-use. A more positive finding is that the third most highly rated stressor was outdoor water use, which is highly likely to be impacted by land-use decisions, as described in Section 1.2. Indeed, water provides in both regions have introduced incentives for water conservation through landscape manipulation [40].

Distinctive perspectives were identified along both professional and regional lines. The biggest professional division relates to limiting new growth, which was much more highly ranked by land-use planners, and increasing the price of water, which was much more highly ranked by water managers. This is perhaps due to familiarity level—in that planners use growth measures (e.g., urban growth boundary in Portland) to reduce the physical growth of the urban area—while water manger in Phoenix use instruments such a pricing to mediate demand. The largest regional difference was that Phoenix professionals are much more worried about long-term drought, which is logical as their already arid region was in the midst of a significant drought period at the time of the survey. Professionals in Phoenix are also more optimistic about acquiring new sources of water as a solution.

The practicality of some of the highest ranked solutions is not yet clear. The two top strategies in both regions and professions are education and new technology. While these are important tools, hoped for technological solutions to environmental problems do not always materialize and often create negative, unintended consequences (e.g., dams, canalized water transfer systems, excessive ground water pumping). Conversely, some of the most aggressive strategies were not as highly ranked, perhaps because of perceived political feasibility issues. In particular, raising water prices and limiting growth may not be seen as popular choices by elected officials. Although it is important to remain critical of the practicality and feasibility of any solution, innovative ideas are already being generated, such as tiered water pricing, which was recently implemented in Santa Fe [41] This can also be seen in significant opposition to increasing the price of water in an earlier Phoenix-based survey, which demonstrated that planning professionals opposed those strategies more so than scientists as well as residents [42].

The practitioners' approach to global climate change is of particular importance for the long-term health of the water systems. Human-induced climate change was identified as an important stressor by a minority of water managers and land-use planners. Despite the fact that the two most commonly identified stressors are greatly exacerbated by climate change—long-term drought in Phoenix, and population growth in Portland—climate still is not considered unequivocally as the primary factor in addressing urban water scarcity. The related strategies showed slightly higher interest levels: planning for future climate changes was identified as an effective strategy by a third of Phoenix water and land practitioners, a third of Portland water managers, and almost half of Portland land-use planners.

The lack of recognition of climate change as an important stressor, and only moderate interest in planning for future climate changes as a strategy suggests a serious problem in resource management decision-making in Portland and Phoenix. The professionals most concerned with climate change are land-use planners in the Portland area, suggesting that they are a group to watch and may provide a model. This further suggests that there could be potential opportunities to use land planning to adapt to climate change as it relates to urban water resource management [4,23,24]. The recent experience of California's water management challenges, and the emerging field of 'climate attribution science' may offer many opportunities for further research into the linkages among climate, land use and water recourses planning.

To that end, four plausible directions emerge as a way to address the challenges facing water use in the midst of climate change. First, the gap must be closed between academic and federal concerns over climate change and the lack of concern reported by on the ground practitioners. Meeting mandates to provide water in the short run seems to be central to the administration of urban water resources. Yet, if the frequency of drought and magnitude of droughts are expected to increase in the future as projected by several studies [12,37,40] then engaging practitioners—both land-use planners and water managers—in developing climate sensitive strategies can provide effective guidance to reduce short and long term impacts. Second, a realistic understanding of the scale of the issue is required. Quantification, even as an estimate would help provide a realistic understanding of the improvements in conservation education and technology that would be necessary to offset the large-scale environmental and population challenges that are already emerging. Third, while the communities we surveyed had a high level of consensus regarding the problem (*i.e.*, stressors) and on the top strategies, assessing the effectiveness of those strategies is an area that requires further research and consideration. The consensus itself is a strong foundation for advancing greater coordination between the two fields, and provides an opportunity to build further cross-sector integration.

Finally, the continued rapid population growth that is forecasted for both study areas presents an opportunity to use land-use planning tools to lessen future stresses on urban water systems. The opportunity for linking existing land-use, housing, and transportation models that are common place in urban areas, with water demand models can help in cross-sector communication, particularly early in the land-use planning process in growing cities [43]. Integrating such models with downscaled climate system can also help to inform communities about the potential to greatly impact the stressor and strategies for managing urban water scarcity.

5. Conclusions

We set out to answer two questions. First, to what extent do urban water use managers and land-use planners share perspectives regarding the stressors to and strategies for water management systems both within and between the two locations and professional fields? Overall, survey responses were mostly consistent for the stressors (*i.e.*, problems) facing water resource management, while responses to the effectiveness of strategies were more varied by both region and profession. Professionals expressed the greatest concern about the impact of environmental and population stressors, while strategies related to end-user behavior were found to be the most effective, and there was general consensus that improved water efficiency technology and water conservation education would be effective strategies. Concomitant concerns about long-term drought seems to be pervasive among the respondents from Phoenix, perhaps since the region continues to undergo severe water shortages along with much of the Southwest. The general lack of concern about climate pressures on water resources by Portland respondents is also note worthy, and may suggest a need to further examine why climate concerns may not be central to current water management strategies.

The second research question was: what factors might help to explain any observed differences and similarities in perspectives about the water management system? The general consensus was that environmental and population factors are stressors, but that related strategies are not as effective, likely because these influences are difficult or impossible to control at the level of local government. Conversely, local government bodies may have greater ability and some certainty to influence end-user behavior, which is reflected in these strategies' high ratings. There were categorical divisions in assessing one stressor (long-term drought) and variation in the perceptions of many strategies, notably source acquisition and limiting new growth. The clear consensus that drought is a greater stressor in Phoenix than Portland is likely due to that region's significantly more arid climate with ongoing drought and to the projected impacts of climate change across the southwest. The professional divide between water managers and land planners concerning limiting new growth (favored by land-use planners) and increasing the price of water (favored by water resource managers) might be explained by their differential levels of involvement and knowledge of related programs. Fostering greater interaction between these two groups may be a timely, cost-effective, and prudent means of improving the opportunities for more effective management of this scarce resource.

Taken together, the present paper provides one perspective into the state of water resource management in two geographically diverse urban areas of the U.S. However, the local politics of water management is complex and is often based on accepted behaviors and organizational cultures [38]. The pressing challenges of climate change and precipitation uncertainty pose formidable challenges to traditional approaches. While our approach could not address the reason for specific responses, the summary of responses suggest aligned and divergence regarding the stressors and strategies by region and profession. Now more than ever we need research that articulates how current challenges to water resource management are created, in part, through the social systems we have created and enabled. Our research is one step in that direction, but additional research is needed to assess the extent to which administrative fragmentation of natural resource management is creating water scarcity. The future will require at once the need for collaboration among the diverse groups of interest managing water and land resources, and the sharing of effective strategies across jurisdictions.

Acknowledgments: Financial assistance for this paper comes from the Sector Applications Research Program (SARP), which is part of the Climate Program Office of the U.S. Department of Commerce, National Oceanic and Atmospheric Administration (NOAA) pursuant to NOAA Award No. NA09OAR4310140. In addition, the material is based upon work supported by the National Science Foundation under Grant No. SES-0951366, DMUU: Decision Center for a Desert City II: Urban Climate Adaptation; and Grant No. SES-1462086. Additional financial support was provided by the Institute for Sustainable Solution at Portland State University (James F. and Marion L. Miller Foundation sustainability grant). The statements, findings, conclusions, and recommendations expressed in this material are those of the research team and do not necessarily reflect the views of NOAA, U.S. Department of Commerce, the National Science Foundation (NSF) or the U.S. Government.

Author Contributions: Vivek Shandas, Kelli L. Larson, and Heejun Chang designed the research, including the development of survey questionnaire. Vivek Shandas and Kelli L. Larson implemented the survey. Rosa Lehman and Jeremy Bunn compiled survey questionnaire and analyzed the data. Vivek Shandas, Rosa Lehman, Kelli L. Larson, Heejun Chang and Jeremy Bunn wrote the manuscript.

References

1. Gober, P.; Larson, K.; Quay, R.; Polsky, C.; Chang, H.; Shandas, V. Why Land Planners and Water Managers Don't Talk to One Another and Why They Should. *Soc. Nat. Resour.* **2013**, *26*, 356–364. [CrossRef]
2. Shandas, V.; Parandvash, G.H. Integrating Urban Form and Demographics in Water Demand Management: An Empirical Case Study of Portland Oregon (US). *Environ. Plan. B: Plan. Des.* **2010**, *37*, 112–128. [CrossRef]
3. Chang, H.; Parandvash, G.H.; Shandas, V. Spatial Variations of Single Family Residential Water Use in Portland, Oregon. *Urban Geogr.* **2010**, *31*, 953–972. [CrossRef]
4. Chang, H.; House-Peters, L. Cities as place for climate mitigation and adaptation: A case study of Portland, Oregon, USA. *J. Korean Geogr. Soc.* **2010**, *45*, 49–74.
5. Macleod, C.J.A.; Blackstock, K.L.; Haygarth, P.M. Mechanisms to improve integrative research at the science-policy interface for sustainable catchment management. *Ecol. Soc.* **2008**, *13*, 48.
6. Gober, P.; Kirkwood, C.W. Climate Change and Water in Southwestern North America Special Feature: Vulnerability assessment of climate-induced water shortage in Phoenix. *Proc. Natl. Acad. Sci. USA* **2010**, *107*, 21289–21294. [CrossRef] [PubMed]
7. Hoyer, W.; Chang, H. Development of Future Land Cover Change Scenarios in the Metropolitan Fringe, Oregon, U.S., with Stakeholder Involvement. *Land* **2014**, *3*, 322–341. [CrossRef]
8. Georgakakos, A.; Fleming, P.; Dettinger, M.; Peters-Lidard, C.; Terese, T.C.; Richmond, K.; Reckhow, K.; White; Yates, D. Ch. 3: Water Resources. In *Climate Change Impacts in the United States: The Third National Climate Assessment*; Melillo, J.M., Richmond, T.C., Yohe, G.W., Eds.; U.S. Global Change Research Program: Washington, DC, USA, 2014. [CrossRef]
9. Mote, P.W.; Salathé, E.P., Jr. Future climate in the Pacific Northwest. *Clim. Chang.* **2010**, *102*, 29–50. [CrossRef]
10. Chang, H.; Jung, I.-W.; Steele, M.; Garnett, M. Spatial Patterns of March and September streamflow trends in Pacific Northwest Streams. *Geogr. Anal.* **2012**, *44*, 177–201. [CrossRef]
11. MacDonald, G.M. Water, climate change and sustainability in the southwest. *Proc. Natl. Acad. Sci. USA* **2010**, *107*, 21256–21262. [CrossRef] [PubMed]
12. Jung, I.-W.; Chang, H. Assessment of future runoff trends under multiple climate change scenarios in the Willamette River Basin, Oregon, USA. *Hydrol. Process.* **2011**, *25*, 258–277. [CrossRef]
13. National Research Council. *Colorado River Basin Water Management: Evaluating and Adjusting to Hydroclimatic Variability*; National Academies Press: Washington, DC, USA, 2007; p. 210.

14. Seager, R.; Ting, M.; Li, C.; Naik, N.; Cook, B.; Nakamura, J.; Liu, H. Projections of declining surface-water availability for the southwestern United States. *Nat. Clim. Chang.* **2013**, *3*, 482–486. [CrossRef]
15. Gutzler, D.S.; Nims, J.S. Interannual variability of water consumption and summer climate in Albuquerque, New Mexico. *J. Appl. Meteorol.* **2005**, *44*, 1777–1787. [CrossRef]
16. Breyer, B.; Chang, H.; Parandvash, H. Land use, temperature and single family residential water use patterns in Portland, Oregon and Phoenix, Arizona. *Appl. Geogr.* **2012**, *35*, 142–151. [CrossRef]
17. Chang, H.; Praskievicz, S.; Parandvash, H. Sensitivity of urban water consumption to weather climate variability at multiple temporal scales: The case of Portland. *Int. J. Geospatial Environ. Res.* **2014**, *1*, 7.
18. Mini, C.; Hogue, T.; Pincetl, S. The effectiveness of water conservation measures on summer residential water use in Los Angeles, California. *Landsc. Urban Plan.* **2015**, *94*, 134–146. [CrossRef]
19. Platt, R.H. The 2020 Water Supply Study for Metropolitan Boston: The Demise of Diversion. *J. Am. Plan. Assoc.* **1995**, *61*, 185–200. [CrossRef]
20. Elliott, C.; Knight, J.; White, G.F. Domestic Water supply: Right or Good? In *Ciba Foundation Symposium on Human Rights in Health*; John Wiley & Sons, Ltd.: Chichester, UK, 1974. [CrossRef]
21. Wescoat, J.L., Jr.; White, G.F. *Water for Life: Water Management and Environmental Policy*; Cambridge University Press: Cambridge, UK, 2003.
22. Wentz, E.A.; Gober, P. Factors Influencing Water Consumption for the City of Phoenix, Arizona. *Water Resour. Manag.* **2007**, *21*, 1849–1863. [CrossRef]
23. Shandas, V.; Rao, M.; McGrath, M.M. The implications of climate change on residential water use: A micro-scale analysis of Portland (OR). *J. Clim. Water* **2012**, *3*, 225–238. [CrossRef]
24. House-Peters, L.; Chang, H. Modeling the Impact of Land Use and Climate Change on Neighborhood-scale Evaporation and Nighttime Cooling: A Surface Energy Balance Approach. *Landsc. Urban Plan.* **2011**, *103*, 139–155. [CrossRef]
25. Runfola, D.M.; Polsky, C.; Nicolson, C.; Giner, N.; Pontius, R.G., Jr.; Decatur, A. A growing concern? Examining the influence of lawn size on residential water use in suburban Boston, MA, USA. *Landsc. Urban Plan.* **2013**, *119*, 113–123. [CrossRef]
26. Halper, E.B.; Dall'erba, S.; Rosalind, H.B.; Scott, C.A.; Yool, S.R. Effects of irrigated parks on outdoor residential water use in a semi-arid city. *Landsc. Urban Plan.* **2015**, *134*, 210–220. [CrossRef]
27. Mayer, P.W.; DeOreo, W.B.; Opitz, E.M.; Kiefer, J.C.; Davis, W.Y.; Dziegielewski, B.; Nelson, J.O. *Residential End Uses of Water. Final Report*; AWWA Research Foundation: Denver, CO, USA, 1999.
28. House-Peters, L.; Pratt, B.; Chang, H. Effects of urban spatial structure, socio-demographics, and climate on residential water consumption in Hillsboro, Oregon. *J. Am. Water Resour. Assoc.* **2010**, *46*, 461–472.
29. Saurí, D. Lights and shadows of urban water demand management: The case of the metropolitan region of Barcelona. *Eur. Plan. Stud.* **2003**, *11*, 229–243. [CrossRef]
30. Ouyang, Y.; Wentz, E.A.; Ruddell, B.L.; Harlan, S.L. A Multi-Scale Analysis of Single-Family Residential Water Use in the Phoenix Metropolitan Area. *J. Am. Water Resour. Assoc.* **2013**, *50*, 448–467. [CrossRef]
31. Hong, C.Y.; Chang, H. Uncovering the Influence of Household Sociodemographic and Behavioral Characteristics on Summer Water Consumption in the Portland Metropolitan Area. *Int. J. Geospatial Environ. Res.* **2014**, *1*, 2.
32. European Environment Agency. *Water Resources across Europe—Confronting Water Scarcity and Drought*; Technical Report TH-AL-09-002-EN-C; European Environment Agency, Office for Official Publications of the European Communities: Copenhagen, Denmark, 2009; pp. 1725–9711.
33. Wentz, E.A.; Wills, A.J.; Kim, W.K.; Myint, S.W.; Gober, P.; Balling R.C., Jr. Factors Influencing Water Consumption in Multifamily Housing in Tempe, Arizona. *Prof. Geogr.* **2014**, *66*. [CrossRef]
34. Larson, K.L.; Polsky, C.; Gober, P.; Chang, H.; Shandas, V. Vulnerability of water systems to the effects of climate change and urbanization: A comparison of Phoenix, Arizona and Portland, Oregon (USA). *Environ. Manag.* **2013**, *52*, 179–195. [CrossRef] [PubMed]
35. Larson, K.L. An integrated theoretical approach to understanding the sociocultural basis of multidimensional environmental attitudes. *Soc. Nat. Resour.* **2010**, *23*, 898–907. [CrossRef]
36. Larson, K.L.; White, D.; Gober, P.; Harlan, S.; Wutich, A. Divergent perspectives on water resource sustainability in a public-policy-science context. *Environ. Sci. Policy* **2009**, *12*, 1012–1023. [CrossRef]
37. Seyranian, V.; Sinatra, G.M.; Polikoff, M.S. Comparing communication strategies for reducing residential water consumption. *J. Environ. Psychol.* **2015**, *41*, 81–90. [CrossRef]

38. Vano, J.A.; Nijssen, B.; Lettenmaier, D.P. Seasonal hydrologic responses to climate change in the Pacific Northwest. *Water Resour. Res.* **2015**, *51*, 1959–1976. [CrossRef]
39. Lach, D.; Ingram, H.; Rayner, S. Maintaining the status quo: How institutional norms and practices create conservative water organizations. *Texas Law Rev.* **2014**, *83*, 2027.
40. Breyer, B.; Chang, H. Urban water consumption and weather variation in the Portland, Oregon metropolitan area. *Urban Clim.* **2014**, *9*, 1–18. [CrossRef]
41. National Public Radio, Santa Fe Cuts Water Consumption by Imposing Tiered Pricing Model. Available online: http://www.npr.org/2015/05/13/406505133/santa-fe-cuts-water-consumption-by-imposing-tiered-pricing-model (accessed on 15 June 2015).
42. Wilder, M.; Scott, C.; Pablos, N.P.; Varady, R.; Garfin, G.M.; McEvoy, J. Adapting Across Boundaries: Climate Change, Social Learning, and Resilience in the U.S-Mexico Border Region. *Ann. Assoc. Am. Geogr.* **2010**, *100*, 917–928. [CrossRef]
43. Bouziotas, D.; Rozos, E.; Makropoulos, C. Water and the city: Exploring links between urban growth and water demand management. *J. Hydroinform.* **2015**, *17*, 176–192. [CrossRef]

6

Bivariate Drought Analysis using Streamflow Reconstruction with Tree Ring Indices in the Sacramento Basin, California, USA

Jaewon Kwak [1], Soojun Kim [2,†], Gilho Kim [3,*], Vijay P. Singh [4,†], Jungsool Park [1,†] and Hung Soo Kim [5,†]

1 Forecast and Control Division, Nakdong River Flood Control Office, Busan 49300, Korea; firstsword@korea.kr (J.K.); js2012park@korea.kr (J.P.)
2 Columbia Water Center, Columbia University, New York, NY 10027, USA; soojun78@gmail.com
3 Department of Hydro Science and Engineering, Korea Institute of Civil Engineering and Building Technology, Goyang-si, Gyeonggi-do 10223, Korea
4 Department of Biological and Agricultural Engineering and Zachry Department of Civil Environmental Engineering, Texas A & M University, College Station, TX 77843-2117, USA; vsingh@tamu.edu
5 Department of Civil Engineering, Inha University, Incheon 22212, Korea; sookim@inha.ac.kr
* Correspondence: kgh0518@kict.re.kr.
† These authors contributed equally to this work.

Academic Editors: Yingkui Li and Athanasios Loukas

Abstract: Long-term streamflow data are vital for analysis of hydrological droughts. Using an artificial neural network (ANN) model and nine tree-ring indices, this study reconstructed the annual streamflow of the Sacramento River for the period from 1560 to 1871. Using the reconstructed streamflow data, the copula method was used for bivariate drought analysis, deriving a hydrological drought return period plot for the Sacramento River basin. Results showed strong correlation among drought characteristics, and the drought with a 20-year return period (17.2 million acre-feet (MAF) per year) in the Sacramento River basin could be considered a critical level of drought for water shortages.

Keywords: tree ring; hydrological drought; artificial neural network; copula method

---

1. Introduction

Generally, long-term data are recommended for analyzing floods and droughts. However, although precipitation data are available from the 16th century onwards, their quality and reliability in many countries are questionable because of the methods of observation, different periods of observations, uncertainties associated with gaging sites, and temporal resolution of observations [1,2]. Many studies have, therefore, used tree-ring data as a way to acquire data for longer periods of time, up to 500 years [3].

Since Ferguson [4] correlated observed hydro-meteorological data and tree-ring data in California, many studies have utilized tree-ring data to reconstruct time series of the past. Fritts [5] did tree ring analysis and correlated the data with climate, suggesting that it can be used in water resources management [6]. Also, some studies have correlated hydro-meteorological variables with tree-ring data to reconstruct climate factors [7–10]. Some studies have reconstructed seasonal series, such as precipitation [11,12], natural hazards [13], and temperature [14], based on tree ring width data. However, most of the studies on tree rings and hydrological phenomena have focused on droughts [15–24]. This may be because annual data are normally reconstructed using tree ring data and hence it is difficult to analyze intra-annual hydrological phenomena. Long time-scale data are mainly used for droughts, because it is difficult to determine the onset or end of a drought and often the drought may last for several months or years. The long time-scale occasionally gives rise to a sample size problem [25] for drought analysis that can be overcome with long-term tree ring reconstruction [26]. Some studies have directly correlated tree rings with droughts [15,27], drought patterns or oscillations [28–30], drought index and its trend [23,31–33], return periods [34], and spatial drought characteristics [35].

Bivariate (or multivariate) analyses of drought characteristics, such as severity, duration, and arrival time, are being increasingly made [36]. These analyses have introduced multivariate drought indices, such as Multivariate Drought Index (MDI) including precipitation, runoff, evapotranspiration, and soil moisture [37]; Multivariate Standardized Drought Index (MDSI) which combines the Standardized Precipitation Index and the Standardized Soil Moisture Index [38]; and Vegetation Drought Response Index (VegDRI) which integrates climatic indicators; and satellite-derived vegetation index [39,40]. Some studies have employed conventional multivariate analysis for drought indices with PDSI [23,41] and SWSI [42,43], or bivariate frequency analysis [44,45] assuming that all variables had the same probability distribution. To overcome this restriction, the copula method has been developed [46]. For doing bivariate drought analysis by the copula method [46,47], tree ring reconstruction can be employed to our advantage.

The objective of this study, therefore, is to reconstruct the annual streamflow of the Sacramento River in California and Oregon using tree-ring width data, and use the reconstructed data for bivariate drought frequency analysis with the copula method. Selected tree-ring data were used in an artificial neural network for streamflow reconstruction and the reconstructed data were verified by comparing with actual observations. The Archimedean copula function was applied to the reconstructed streamflow data and then the return period plot of the hydrological drought in the Sacramento River basin was derived. The advantages of using the copula method have been discussed in many studies [34,36,48–55]. Since drought may last from months to years, as has happened in California, long-term reconstruction based on tree-ring data, addresses the drawback of short-term data [56].

2. Materials and Methods

2.1. Study Area and Data

Annual streamflow data for four Sacramento rivers and tree-ring data of the nearby region were employed in this study. The tree-ring data from 17 sites (Figure 1 and Table 1) in California and Oregon, which reflect standard chronologies of ring width [57], were obtained from the International Tree-Ring Databank [58]. The tree ring width data were standardized [59]. Annual streamflow is the sum of four river flows, which are the Sacramento River above Bend Bridge (SBB), the Feather River at the Lake Oroville (FTO), the Yuba River near Smartville (YRS) and the American River at Folsom (AMF), which was obtained from the California Data Exchange Center of California Department of Water Resources [60] for the period 1872 to 1977. It has a long-term mean of 18.9 MAF (million acre-feet; 1.23×10^9 m^3), median of 17.6 MAF, and maximum flow of 51.6 MAF that occurred in the year 1890. These tree ring and streamflow sites are shown in Figure 1 and Table 1.

2.2. Drought Definition Using the Run Theory

This study qualitatively defines the hydrological drought (hereafter referred to as "drought") as a significant decrease in the availability of streamflow in the river. Quantitatively, drought was defined using the run theory, which allows us to calculate drought duration, severity, and arrival time [61,62]. Thus, drought can be defined as the time when a hydro-meteorological time series x_t falls below the truncation level x_0 and that represents a hydro-meteorological event (Figure 2).

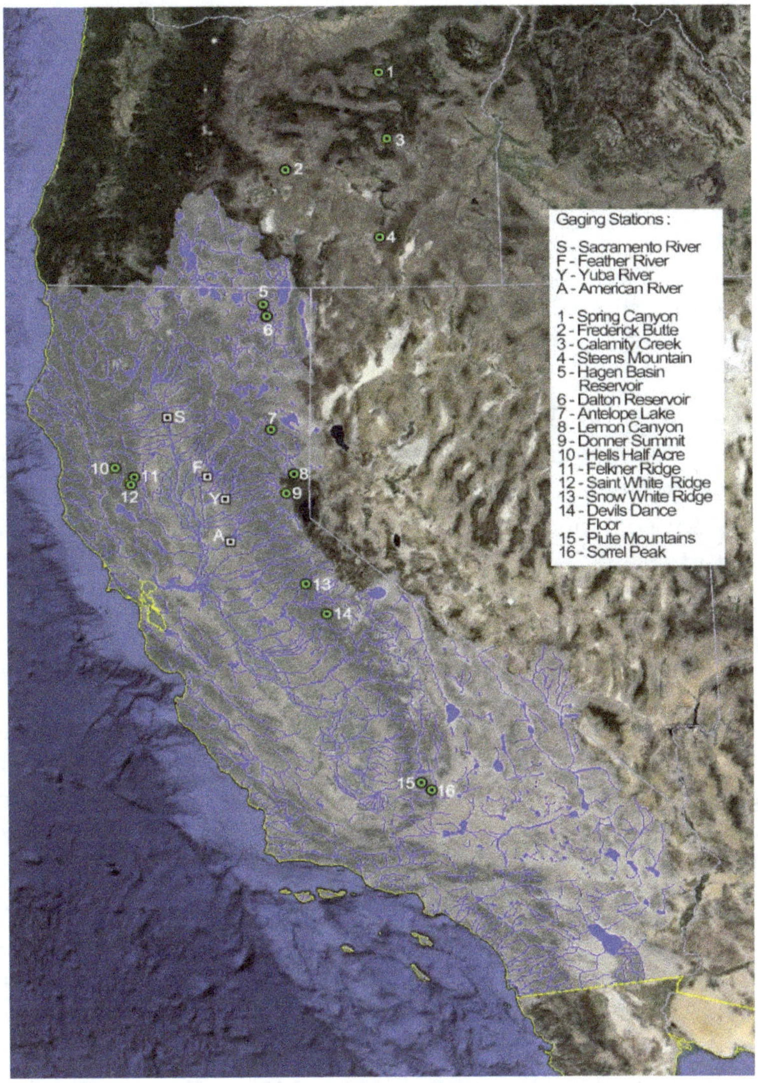

Figure 1. Study area, tree ring sites, and streamflow sites.

Table 1. Tree ring sites and streamflow observatory specification [58].

Category	Index in Figure 1	ID	Name	Site Location			Tree Ring Species
				Lat (degree)	Lon (degree)	Height (EL.m)	
Tree ring sites	7	ANTEP	Antelope Lake	40.15	−120.6	1480	*Pinus jeffreyi Balf*
	7	ANTEP	Antelope Lake	40.15	−120.6	1480	*Pinus ponderosa Douglas ex C. Lawson*
	3	CALAM	Calamity Creek	43.98	−118.8	1464	*Juniperus occidentalis Hook*
	6	DALTON	Dalton Reservoir	41.62	−120.7	1531	*Pinus ponderosa Douglas ex C. Lawson*
	14	DEVILS	Devil's Dance Floor	37.75	−119.75	2084	*Pinus jeffreyi Balf*
	9	DONNER	Donner Summit	39.32	−120.35	2265	*Pinus jeffreyi Balf*
	11	FELKN	Felkner Ridge	39.5	−122.67	1494	*Pinus lambertiana Douglas*
	2	FREDER	Frederick Butte	43.58	−120.45	1494	*Juniperus occidentalis Hook*
	5	HAGER	Hager Basin Reservoir	41.77	−120.75	1524	*Juniperus occidentalis Hook*
	10	HELLS	Hell's Half Acre	39.6	−122.95	1922	*Pinus jeffreyi Balf*
	8	LEMON	Lemon Canyon	39.57	−120.25	1859	*Pinus jeffreyi Balf*
	15	PIUTE	Piute Mountain	35.53	−118.43	1975	*Pinus jeffreyi Balf*
	13	SNOWHT	Snow White Ridge	38.13	−120.05	1731	*Pinus ponderosa Douglas ex C. Lawson*
	16	SORREL	Sorrel Peak	35.43	−118.28	2011	*Pinus jeffreyi Balf*
	1	SPRING	Spring Canyon	44.9	−118.93	1366	*Juniperus occidentalis Hook*
	4	STEENS	Steens Mountain	42.67	−118.92	1656	*Juniperus occidentalis Hook*
	13	STJOHN	St. White Mountain	39.43	−122.68	1555	*Pinus ponderosa Douglas ex C. Lawson*
Flow site	S	SBB	Sac. River, Abv bend bridge	40.29	-122.19	56.6	-
	F	FTO	Feather River, Oroville	39.52	-121.55	45.4	-
	Y	YRS	Yuba River, Smartville	39.24	-121.27	85.3	-
	A	AMF	American River, Folsom	38.68	-121.18	0	-

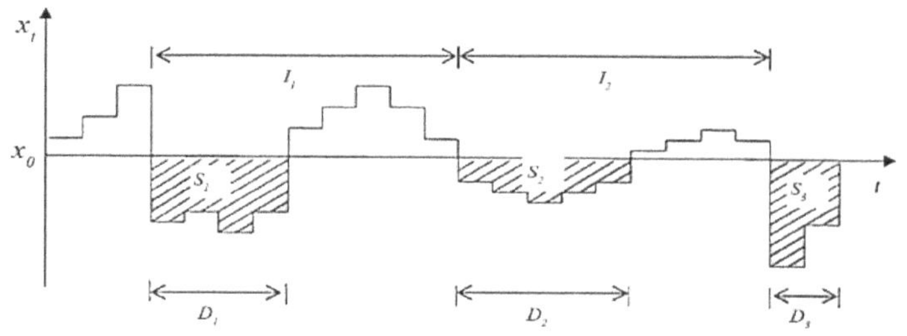

Figure 2. Drought characteristics using the run theory: $D_1, D_2, \ldots$ denote drought duration; $S_1, S_2, \ldots$ denote drought severity; $I_1, I_2, \ldots$ denote drought arrival time.

Drought events are based on the truncation level, so the selection of the level is one of the important issues for proper drought analysis. Generally, the mean value of streamflow has been widely used as the truncation level [63–68]. However, the Sacramento River streamflow shows high variability, between 5.13 and 51.65 MAF, so a median value of annual streamflow was regarded as a more reliable truncation level in this study.

2.3. Artificial Neural Network

The factors that influence annual hydro-meteorological behavior can be roughly classified into four groups: (i) atmospheric-climatic; (ii) geologic-geomorphic; (iii) soil-vegetation; and (iv) runoff-channel factors [69]. Tree-ring widths in a trunk of a tree are also influenced by atmospheric-climatic and soil-vegetation factors, such as precipitation, evapotranspiration, and soil moisture. This indicates that an appropriate modeling technique and tree rings that have a correlation with atmospheric-climatic factors can be used to reconstruct annual streamflow, and this study employed an Artificial Neural Network (ANN) model. ANN mimics the structure and functions of a biological neural system, in which neurons are connected through nodes [70]. After the perceptron was proposed to categorize information patterns [71], ANNs have been widely used to recognize nonlinear relationships between different variables. The ANN used in this study was comprised of three layers: the input layer that represents observed streamflow data, the output layer that produces simulated streamflow, and the hidden layer that is constituted by a network of neurons that are trained to recognize patterns from observations (Figure 3).

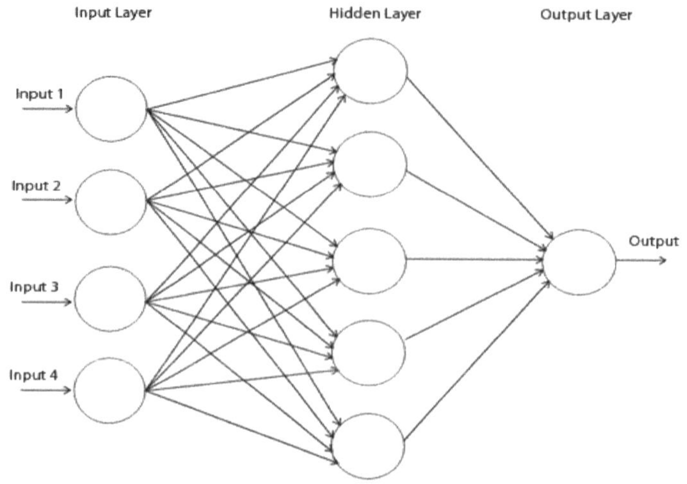

Figure 3. ANN schematization: *Input i* is the input set, *output* is the result of network delay, and each circle represents neural network [71]; each line indicates nodes between neurons that have their own connection strength.

The back-propagation algorithm was used to train the network through the adjustment of connection strength to learn about the error and optimize the neurons. It calculates the error function with respect to all the weights in the network and the gradient of error functions is fed into the optimization technique, which attempts to minimize the error of the network. Hence, the selection of back-propagation algorithm is one of the challenges when using a neural network [72]. The Levenberg–Marquardt–QNBP algorithm was selected as the back-propagation algorithm, because it is known to work well for non-linear problems, such as those related to meteorological and hydrological data [73]. Also, the number of hidden layers, which can be optimized by a trial and error method, is important for a proper network. The ANN model used in this study has six hidden layers that are optimized. One of the advantages of ANN is that it can be used as an alternative modeling technique when the data show non-linearity, which may cause error with a linear technique [74]. The tree-ring data in California and Oregon have autocorrelation and lagged-correlation characteristics [75], so ANN was employed as an alternative to reconstruct streamflow using tree-ring data. More details on ANN and the back-propagation algorithm can be found in Basheer and Hajmeer [76].

2.4. Drought Frequency Analysis Based on Copula

Unlike precipitation or flood occurrence, drought shows a different statistical behavior for a different duration [62]. Considering drought duration and severity as mutually related variables, the copula method has been employed to capture the dependence between them [47,77]. For a probability distribution $F(x_1, \ldots, x_n)$, which has n-dimensional marginal distributions $F_1(x_1), \ldots, F_n(x_n)$, the copula function C that satisfies the relationship between marginal variables can be expressed as:

$$F(x_1, \ldots, x_n) = C(F_1(x_1), \ldots, F_n(x_n)) = \psi_\alpha^{-1}(\psi_\alpha(F_1(x_1)) + \ldots + \psi_\alpha(F_n(x_n))) \quad (1)$$

$$1/T = E(L) / \{1 - F_1 - F_2 \ldots, F_n + C(F_1, F_2, \ldots, F_n)\} \quad (2)$$

where ψ_α denotes the generating function; ψ_α^{-1} is the pseudo-inverse of that function, which differs with the copula family; T denotes the return period; and *E(L)* is the interval between events. Unlike univariate frequency analysis, bivariate frequency indicates the probability that the phenomenon under study occurs if and only if a prior condition takes place. There are several types of copula functions, but the Archimedean copula family, which allows for greater flexibility and simplicity of use, is more commonly used in hydrology [78]. From the Archimedean copula family, the Clayton, Gumbel, and Frank copulas were employed and their functions are given in Table 2.

Table 2. Bivariate Archimedean copula family: C is the copula function, t denotes the drought event, α is the copula parameter, and F_1 and F_2 denote cumulative distribution function of each variable [79].

Copula Family	Copula func., $C(F_1(x_1), F_2(x_2))$	Generator func., $\psi_\alpha(t)$	Parameter (α)
Clayton	$\left(\max\left\{F_1(x_1)^{-1} + F_2(x_2)^{-1} - 1; 0\right\}\right)$	$\frac{1}{\alpha}(t^{-\alpha} - 1)$	$\alpha \in [-1, \infty]$
Frank	$\frac{-1}{\alpha}\log\left(1 + \frac{(\exp(-\alpha F_1(x_1)) - 1)(\exp(-\alpha F_2(x_2) - 1)}{\exp(-\alpha) - 1}\right)$	$-\log\left(\frac{\exp(-\alpha t) - 1}{\exp(-\alpha) - 1}\right)$	$\alpha \in [\mathcal{R}]$
Gumbel	$\exp\left(-((-\log(F_1(x_1))^\alpha + (-\log(F_1(x_1))^\alpha)^{\frac{1}{\alpha}}\right)$	$-\log(t)^\alpha$	$\alpha \in [1, \infty]$

3. Results and Discussion

3.1. Tree-Ring Data Screening

Appropriate tree rings, which have correlation with atmospheric-climatic factors, can be used as the predictor for annual streamflow. Therefore, selection of the appropriate input for the ANN model was one of the challenges in this study. Generally, a trial and error method with different input variables is employed, but it can lead to the poor performance of neural networks [80,81]. Alternatively, cross correlation coefficients were employed to select appropriate inputs in this study in the same way as Meko *et al.* [75].

Seven tree-rings indices showed correlation with streamflow (Figure 4), including Antelope Lake (Pinus Jeffreyi and Pinus Ponderosa), Felkner Ridge, Frederick Butte, Lemon Canyon, Piute Mountain, and Sorrel Peak; four tree rings had correlation with a one-year time-lag, including Dalton Reservoir, Hager Basin Reservoir and Antelope Lake (Pinus Jeffreyi and Pinus Ponderosa), as shown in Figures 4 and 5. Hence, nine tree rings were selected as predictors for the ANN model (two tree-rings were overlapped in zero-lagged and one-year lagged). Also, four tree-rings, which had one-year time-lag correlation (Antelope with Pinus Jeffreyi and Pinus Ponderosa, Dalton Reservoir, and Hager Basin Reservoir), are located on the nearby lake or reservoir. Therefore, it seems that the groundwater level or soil moisture influenced tree ring width, but there are no clues to estimate the correlation between them and further studies are thus needed. These nine tree-ring data points composed the input dataset for the ANN model.

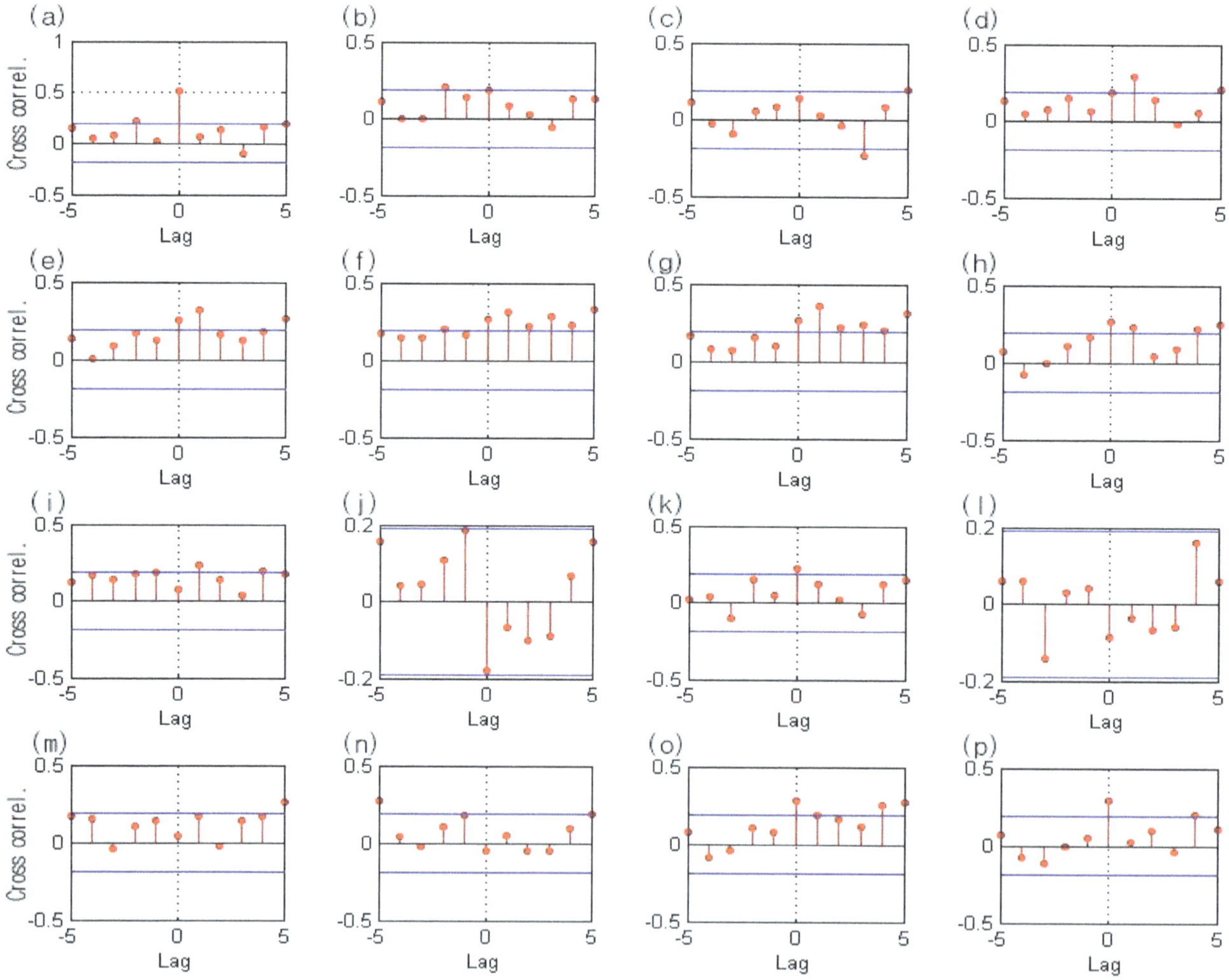

Figure 4. Cross correlation diagram between streamflow and tree ring data at: (**a**) Felkner Ridge; (**b**) Lemon Canyon; (**c**) Calamity Creek; (**d**) Dalton Reservoir; (**e**) Hager Basin Reservoir; (**f**) Antelope Lake (Pinis Jeffreyi Balf.); (**g**) Antelope Lake (Pinus Ponderosa Douglas ex C. Lawson); (**h**) Steens Mountain; (**i**) Hell's Half Acre; (**j**) Donner Summit; (**k**) Frederick Butte; (**l**) Spring Canyon; (**m**) St. White Mountain; (**n**) Devil's Dance Floor; (**o**) Sorrel Peak; and (**p**) Piute Mountain.

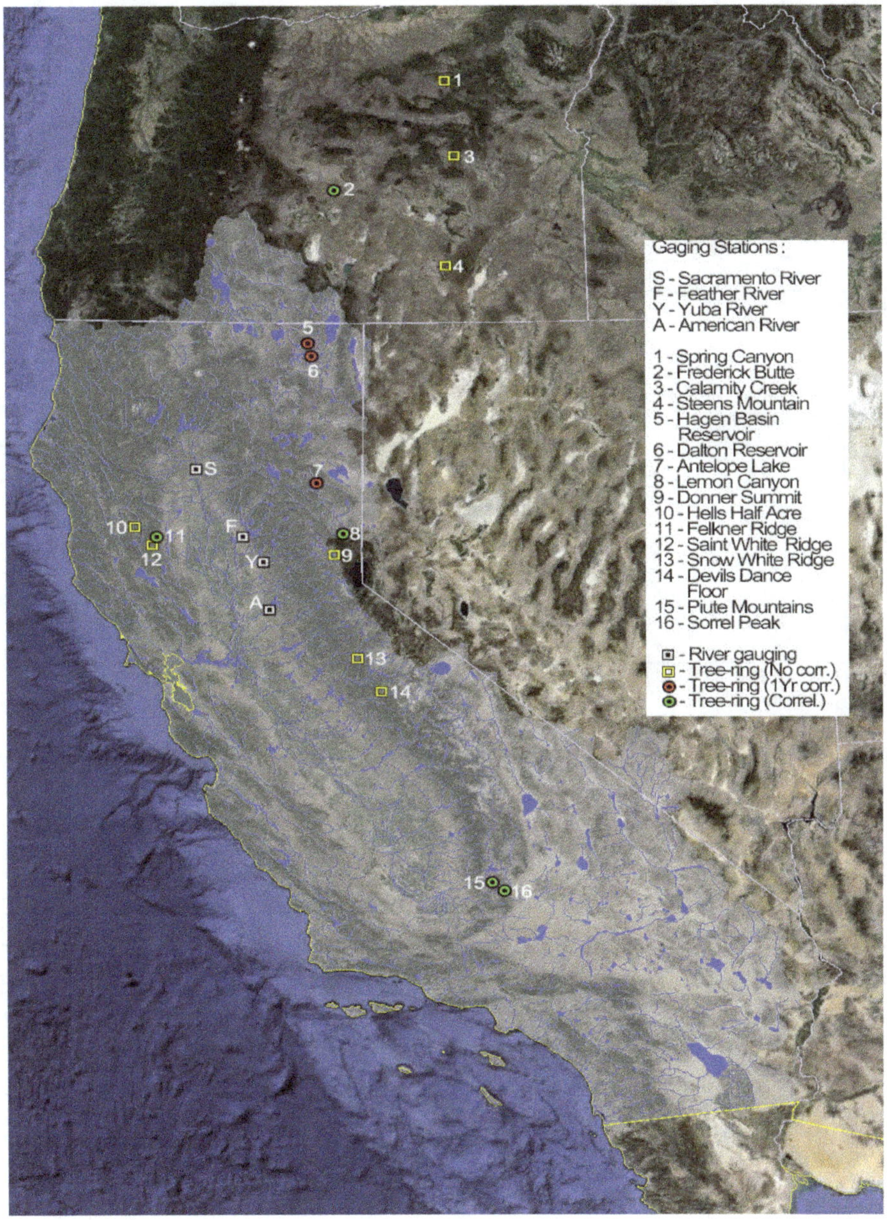

Figure 5. Study area, tree ring sites, and streamflow sites; a green circle indicates that the tree ring has correlation with streamflow, while a red circle indicates correlation with time-lag (1 year) and a yellow box indicates that there is no correlation with streamflow.

3.2. Reconstructed ANN Model Calibration and Validation

The ANN model had six hidden layers that were determined by trial and error and was established with selected predictors. The Sacramento streamflow data were divided into calibration period (1872 to 1957) and validation period (1958 to 1977). To evaluate the results of calibration and validation, R^2 and RMSE [82], and the Nash–Sutcliffe model efficiency coefficient [83] were computed. The RMSE describes a measure of average error in prediction and R^2 and Nash–Sutcliffe model efficiency coefficient have been widely used to assess the predictive performance of models [84]. The calibration and validation results with selected predictors are shown in Figure 6.

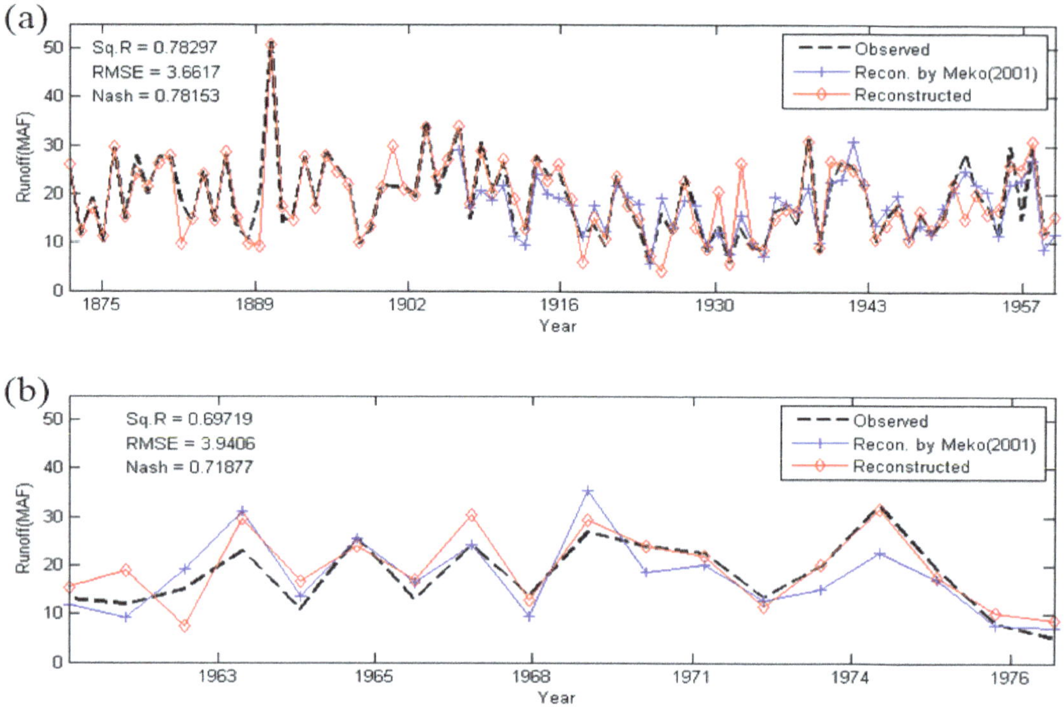

Figure 6. Calibration and validation results with tree ring: (**a**) calibration period (1872 to 1957); and (**b**) validation period (1958 to 1977).

The ANN model showed relatively high values of evaluation measures, with R^2, RMSE, and Nash–Sutcliffe efficiency of 0.78, 3.66, and 0.78 in the calibration and 0.70, 3.94, and 0.72 in the validation period, respectively. Thus, the reconstructed streamflow, based on the ANN model and selected predictor, could be used as the reconstruction model. Also, it could be used for hydro-meteorological simulations. The variability of each period was 5.74 to 51.64 MAF in the calibration period and 5.13 to 32.5 MAF in the validation period.

The reconstructed streamflow using the selected predictor (nine tree rings) is shown in Figure 7 and the basic statistics are shown in Table 3, with an average of 18.9 MAF observed and 20.4 MAF of reconstructed streamflow.

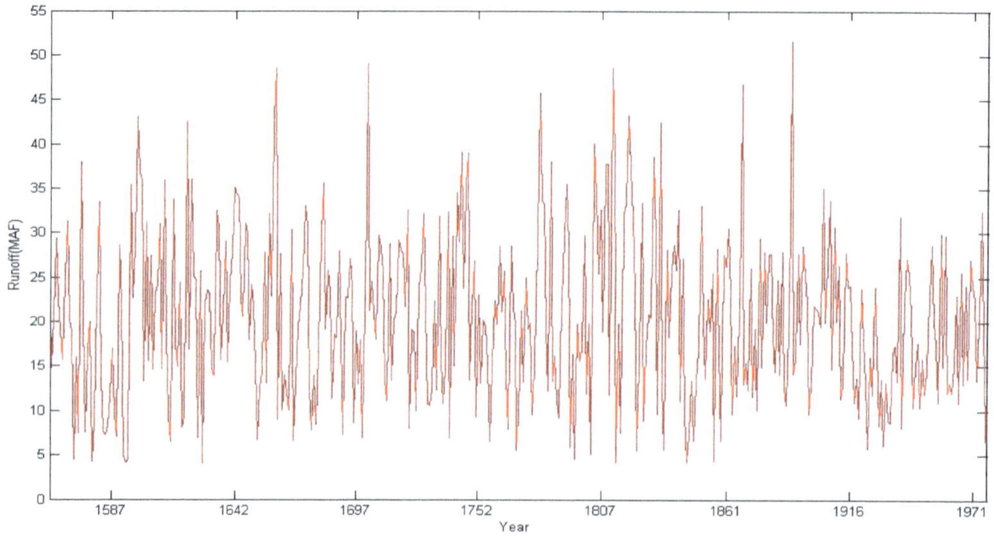

Figure 7. Reconstructed streamflow using the ANN model.

Table 3. Basic statistics of observed and reconstructed streamflow.

Period	Mean (MAF)	Median (MAF)	Standard Deviation (MAF)	Skewness
Observed (1872–1977)	18.9	17.6	7.8	0.8
Reconstructed (1560–1871)	20.4	19.8	9.6	0.6

3.3. *Bivariate Drought Analysis and Discussion*

Before drought analysis based on reconstructed streamflow, the truncation level that defines the relevant streamflow level was determined to define hydrological drought from the streamflow series. The median value of annual streamflow, which was 19.4 MAF, was employed as the truncation level to define hydrological droughts for the Sacramento River basin and results are shown in Figure 8. In total, 96 droughts occurred during the period from 1560 to 1977. Their statistical characteristics were: median drought duration of about two years, average drought severity of about 15.8 MAF, and average drought arrival time of about 2.1 years during the 15th to 20th centuries. The longest drought duration estimated was 10 years and had 75.31 MAF during 1927 to 1936 (in the observation period), and the severest drought estimated was 76.17 MAF, which had an eight-year drought duration from 1582 to 1589 (in the reconstruction period).

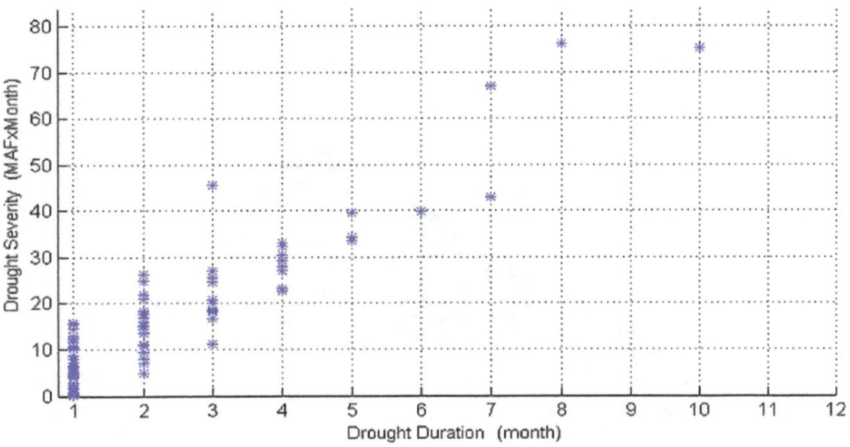

Figure 8. Truncated drought events (1560 to 1977) using observed and reconstructed streamflow.

For bivariate drought frequency analysis with the copula function, marginal distributions of drought variables (duration and severity) were derived. The drought duration was found to have an "exponential" distribution, if treated as a continuous random variable [34]. Also, the drought severity was found to have a "gamma" distribution with 95% confidential level with the PPCC (probability plot correlation coefficient; [85]) distribution goodness of fit test. Then, parameters of the Archimedean family copulas (Frank, Clayton, and Gumbel) were estimated by the method of moments according to their relationship between copula parameter and Kendall's tau [86], which has been found adequate for estimating parameters for small sample sizes [78].

The minimum quadratic distance (L^2) between the empirical and theoretical values of the K criterion, which describes the most appropriate copula [78], was calculated for each copula. As shown in Figure 9, the Frank copula, which generally fitted well throughout ($L^2 = 0.023$), was selected for bivariate drought analysis for the Sacramento River. The Frank copula parameter was estimated as 8.03, and the bivariate joint probability of drought for the Sacramento River basin was described as:

$$F(F_d, F_s) = -\frac{1}{8.03}\log\left(1 + \frac{(\exp(-8.03\,F_d(t)) - 1)(\exp(-8.03\,F_s(t) - 1)}{\exp(-8.03) - 1}\right) \quad (3)$$

where, $F_d(t)$ and $F_s(t)$ are the cumulative distribution functions of drought duration and severity. Figure 10 shows the joint CDF of the Sacramento River basin drought.

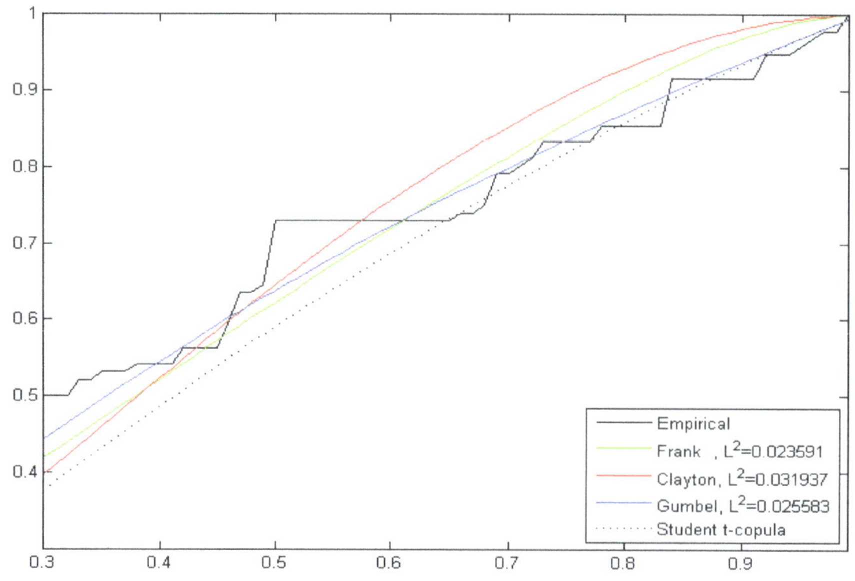

Figure 9. K criterion plot for copula families.

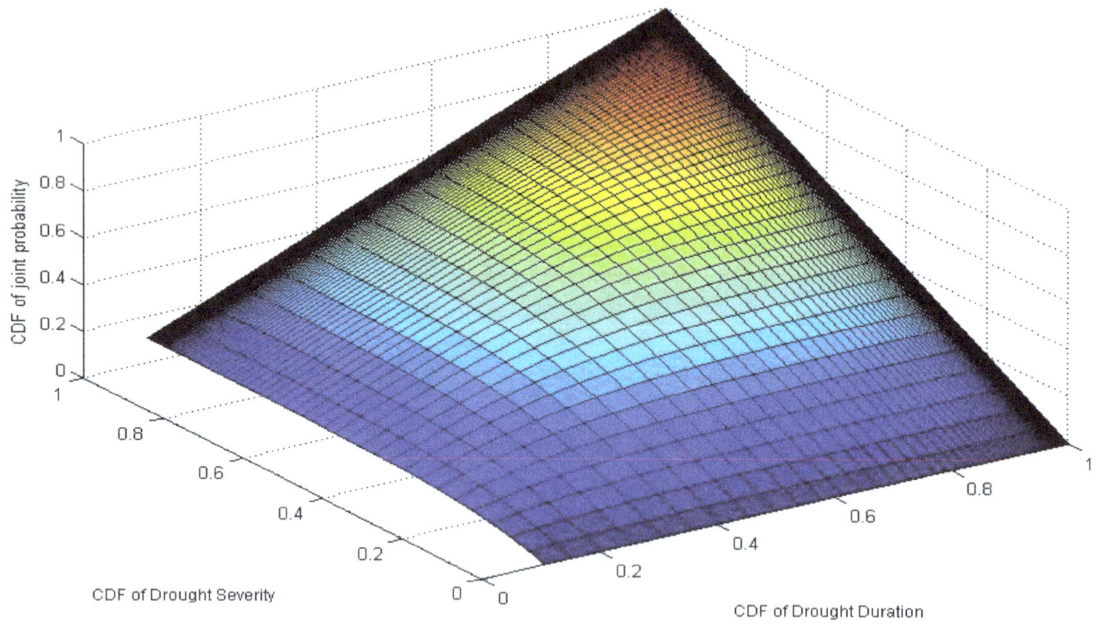

Figure 10. Joint cumulative distribution function (CDF) of the Sacramento River basin drought.

The return period is described as the average time of occurrence of events with the threshold intensity. The bivariate return period can be divided into the exceedance probabilities of both the drought duration and severity [62]. The copula-based return period with the average inter-arrival of occurrences ($E(L)$), which was 2.1 years, can be defined as:

$$T_{return\ period} = \frac{2.1}{P(D > d \,\&\, S > s)} = \frac{2.1}{1 - F_D(d) - F_S(s) + C(F_D(d), F_S(s))} \tag{4}$$

Therefore, the duration and severity of droughts can be expressed in terms of the same return period, which can be illustrated in each "return period plot", as shown in Figure 11.

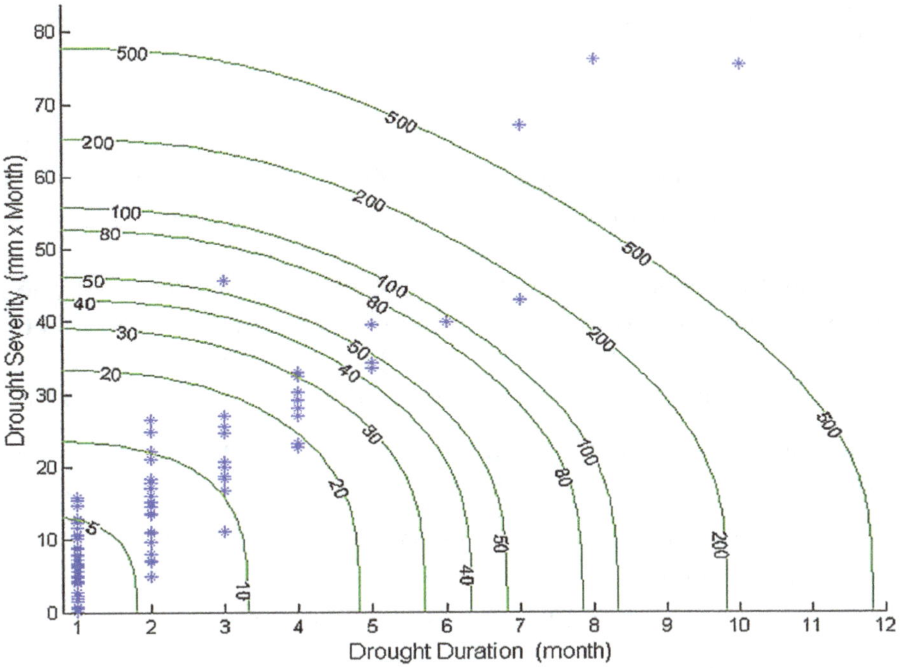

Figure 11. Return period plot for Sacramento River.

The drought event and return period plot in Figure 11 shows the hydrological drought pattern of the Sacramento River basin. Overall, the drought duration and severity seemed to have a positive correlation with each other. To identify the correlation between drought duration and severity, Pearson's linear correlation coefficient, Kendall's rank correlation coefficient (τ), and Spearman's rank correlation coefficient (ρ) were found to be 0.92, 0.73, and 0.85, respectively. Also, the severity of two-year duration drought showed higher variability than did the three-year duration drought. Strong correlation between drought characteristics and higher variability of the two-year drought was the basis for the copula method, especially higher variability of severity or larger statistical irregularity than other durations for the same return period in the univariate frequency analysis [62]. Hence, the copula method would be expected to be more reliable for drought analysis for the Sacramento River basin. Furthermore, California has a 22.4 MAF mean annual runoff and used 5.2 MAF annually to supply the southern area [87]. So, a drought that becomes over 17.2 MAF will cause the shortage of water supply and is equivalent to approximately a 20-year return period with a two-month duration (median value of drought duration) based on the return period plot in Figure 11. Therefore, any return period that causes an actual water shortage could be the appropriate critical level of drought. Also, the return period plot in Figure 11 can be used as elementary data for water resources planning. For instance, if a decision maker or agency determined a three-year design drought for a dam or reservoir to be having a 20-year return period, then its deficiency would be about 11.1 $\times$ 10^9 m^3 (27.0 MAF), and it could be the target storage volume for water resources planning.

Most of the droughts that occurred during the last five centuries did not have more than a 50-year return period, and just six droughts showed 100-year or longer return periods. These high return period events are one of the limitations of the study; for instance, the drought from 1927 to 1936 years, the longest and severest drought in historical data [88], with a 10-year duration and 75.31 MAF severity, which equated to a 9.26 $\times$ 10^9 m^3 streamflow deficiency per year, had approximately a 7500-year return period. That extreme return period was due to the Frank copula and sample size, which shows some bias in high quantile (high return period) events in Figure 9, so it could have overestimated the return period [89], and the number of drought events with high quantiles is also limited. Droughts that have high quantiles or extreme return periods depend significantly on the fitted copula function. Therefore, further studies are needed for generally well-fitted copulas throughout, and also for carefully considering the use of the return period plot in water resources planning.

4. Conclusions

This study reconstructed the past streamflow of the Sacramento River based on the ANN and tree-ring data, and bivariate drought frequency was analyzed using the Archimedean copula. Results of this study can be summarized as follows:

1. The past streamflow for the period from 1560 to 1871 is reconstructed with the ANN model and tree-ring data, which was found to be the appropriate predictor. As shown by calibration and validation results from 1872 to 1977, the R^2 and Nash values are 0.7 or higher. It is therefore concluded that the ANN model reconstructs streamflow of the Sacramento River satisfactorily.
2. Drought characteristics in the Sacramento River basin have strong correlation with each other. The Archimedean copula is found to be appropriate for bivariate drought frequency analysis.
3. It is shown that a drought with a 20-year return period or longer will cause actual water shortages in the perspective of water supply to the southern California area. Hence, it could be considered an appropriate critical level of droughts for actual water shortages.

Acknowledgments: This research was supported by a grant (13SCIPS01) from Smart Civil Infrastructure Research Program funded by Ministry of Land, Infrastructure and Transport (MOLIT) of Korea government and Korea Agency for Infrastructure Technology Advancement (KAIA) and a grant[MPSS-NH-2015-79] through the Natural Hazard Mitigation Research Group funded by Ministry of Public Safety and Security of Korean government.

Author Contributions: The research presented here was carried out in collaboration between all authors. Jaewon Kwak and Soojun Kim had the original idea for the study. Gilho Kim and Jungsool Park designed/conducted the research methods. Vijay P. Singh and Hung Soo Kim contributed to the writing of the paper. All authors discussed the structure and commented on the manuscript at all stages.

References

1. Wang, B.; Jhun, J.G.; Moon, B.K. Variability and singularity of Seoul, South Korea, rainy season (1778–2004). *J. Clim.* **2007**, *20*, 2572–2580. [CrossRef]
2. Mun, J.W.; Lee, D.Y. Tree-ring data application for drought mitigation. *J. Korean Soc. Hazard Mitig.* **2011**, *11*, 70–78. (In Korean)
3. Kim, H.S.; Hwang, S.H.; Kim, J.H. Reconstruction of River Flows Using Tree-Ring Series and Neural Network. *J. Korean Soc. Civ. Eng.* **1998**, *18*, 583–589.
4. Ferguson, C.W. Bristlecone Pine: Science and Esthetics A 7100-year tree-ring chronology aids scientists; old trees draw visitors to California mountains. *Science* **1968**, *159*, 839–846. [CrossRef] [PubMed]
5. Fritts, H.C. Tree-ring analysis: A tool for water resources research. *Eos Trans. Am. Geophys. Union* **1969**, *50*, 22–29. [CrossRef]
6. Fritts, H.C. *Tree Rings and Climate*; Academic Press Inc.: New York, NY, USA, 1976.
7. Hughes, M.K.; Xiangding, W.; Xuemei, S.; Garfin, G.M. A preliminary reconstruction of rainfall in north-central China since AD 1600 from tree-ring density and width. *Quat. Res.* **1994**, *42*, 88–99. [CrossRef]
8. Díaz, S.C.; Touchan, R.; Swetnam, T.W. A tree-ring reconstruction of past precipitation for Baja California Sur, Mexico. *Int. J. Climatol.* **2001**, *21*, 1007–1019. [CrossRef]
9. Cleaveland, M.K.; Stahle, D.W.; Therrell, M.D.; Villanueva-Diaz, J.; Burns, B.T. Tree-ring reconstructed winter precipitation and tropical teleconnections in Durango, Mexico. *Clim. Chang.* **2003**, *59*, 369–388. [CrossRef]
10. Gray, S.T.; Fastie, C.L.; Jackson, S.T.; Betancourt, J.L. Tree-ring-based reconstruction of precipitation in the Bighorn Basin, Wyoming, since 1260 AD. *J. Clim.* **2004**, *17*, 3855–3865. [CrossRef]

11. Liu, Y.; Cai, Q.; Shi, J.; Hughes, M.K.; Kutzbach, J.E.; Liu, Z.; An, Z. Seasonal precipitation in the south-central Helan Mountain region, China, reconstructed from tree-ring width for the past 224 years. *Can. J. For. Res.* **2005**, *35*, 2403–2412. [CrossRef]
12. Liu, Y.; Ma, L.; Leavitt, S.W.; Cai, Q.; Liu, W. A preliminary seasonal precipitation reconstruction from tree-ring stable carbon isotopes at Mt. Helan, China, since AD 1804. *Glob. Planet. Chang.* **2004**, *41*, 229–239. [CrossRef]
13. Schneuwly, D.M.; Stoffel, M. Tree-ring based reconstruction of the seasonal timing, major events and origin of rockfall on a case-study slope in the Swiss Alps. *Nat. Hazards Earth Syst. Sci.* **2008**, *8*, 203–211. [CrossRef]
14. Frank, D.; Esper, J. Characterization and climate response patterns of a high-elevation, multi-species tree-ring network in the European Alps. *Dendrochronologia* **2005**, *22*, 107–121. [CrossRef]
15. Cook, E.R.; Jacoby, G.C. Tree-ring-drought relationships in the Hudson Valley, New York. *Science* **1977**, *198*, 399–401. [CrossRef] [PubMed]
16. Stockton, C.W.; Meko, D.M. Drought Recurrence in the Great Plains as Reconstructed from Long-Term Tree-Ring Records. *J. Clim. Appl. Meteorol.* **1983**, *22*, 17–29. [CrossRef]
17. Graumlich, L.J. Precipitation variation in the Pacific Northwest (1675–1975) as reconstructed from tree rings. *Ann. Assoc. Am. Geogr.* **1987**, *77*, 19–29. [CrossRef]
18. Till, C.; Guiot, J. Reconstruction of precipitation in Morocco since 1100 AD Based on Cedrus atlantica tree-ring widths. *Quat. Res.* **1990**, *33*, 337–351. [CrossRef]
19. Meko, D.; Stockton, C.W.; Boggess, W.R. The Tree-ring Record of Severe Sustained Drought. *J. Am. Water Resour. Assoc.* **1995**, *31*, 789–801. [CrossRef]
20. Stahle, D.W.; Cook, E.R.; Cleaveland, M.K.; Therrell, M.D.; Meko, D.M.; Grissino-Mayer, H.D.; Luckman, B.H. Tree-ring data document 16th century mega drought over North America. *EOS Trans. Am. Geophys. Union* **2000**, *81*, 121–125. [CrossRef]
21. Raffalli-Delerce, G.; Masson-Delmotte, V.; Dupouey, J.L.; Stievenard, M.; Breda, N.; Moisselin, J.M. Reconstruction of summer droughts using tree-ring cellulose isotopes: A calibration study with living oaks from Brittany (western France). *Tellus B* **2004**, *56*, 160–174. [CrossRef]
22. Li, J.; Gou, X.; Cook, E.R.; Chen, F. Tree-ring based drought reconstruction for the central Tien Shan area in northwest China. *Geophys. Res. Lett.* **2006**, *33*. [CrossRef]
23. Li, J.; Chen, F.; Cook, E.R.; Gou, X.; Zhang, Y. Drought reconstruction for north central China from tree rings: The value of the Palmer drought severity index. *Int. J. Climatol.* **2007**, *27*, 903–909. [CrossRef]
24. Tian, Q.; Gou, X.; Zhang, Y.; Peng, J.; Wang, J.; Chen, T. Tree-ring based drought reconstruction (AD 1855–2001) for the Qilian Mountains, northwestern China. *Tree Ring Res.* **2007**, *63*, 27–36. [CrossRef]
25. Mishra, A.K.; Singh, V.P. Drought modeling—A review. *J. Hydrol.* **2011**, *403*, 157–175. [CrossRef]
26. Agüero, J.D.L.C.; Rodríguez, F.J.G. Morphometric stock structure of the Pacific sardine Sardinops sagax (Jenyns, 1842) off Baja California, Mexico. In *Morphometrics*; Springer: Berlin, Germany; Heidelberg, Germany, 2004.
27. Stockton, C.W.; Meko, D.M. A long-term history of drought occurrence in western United States as inferred from tree rings. *Weatherwise* **1975**, *28*, 244–249. [CrossRef]
28. Gray, S.T.; Betancourt, J.L.; Fastie, C.L.; Jackson, S.T. Patterns and sources of multidecadal oscillations in drought-sensitive tree-ring records from the central and southern Rocky Mountains. *Geophys. Res. Lett.* **2003**, *30*. [CrossRef]
29. Helama, S.; Meriläinen, J.; Tuomenvirta, H. Multicentennial megadrought in northern Europe coincided with a global El Niño–Southern Oscillation drought pattern during the Medieval Climate Anomaly. *Geology* **2009**, *37*, 175–178. [CrossRef]
30. Ropelewski, C.F.; Halpert, M.S. North American precipitation and temperature patterns associated with the El Niño/Southern Oscillation (ENSO). *Monthly Weather Rev.* **1986**, *114*, 2352–2362. [CrossRef]
31. Davi, N.K.; Jacoby, G.C.; D'Arrigo, R.D.; Baatarbileg, N.; Jinbao, L.; Curtis, A.E. A tree-ring-based drought index reconstruction for far-western Mongolia: 1565–2004. *Int. J. Climatol.* **2009**, *29*, 1508–1514. [CrossRef]
32. Touchan, R.; Funkhouser, G.; Hughes, M.K.; Erkan, N. Standardized precipitation index reconstructed from Turkish tree-ring widths. *Clim. Chang.* **2005**, *72*, 339–353. [CrossRef]
33. Liang, E.; Shao, X.; Liu, H.; Eckstein, D. Tree-ring based PDSI reconstruction since AD 1842 in the Ortindag Sand Land, east Inner Mongolia. *Chin. Sci. Bull.* **2007**, *52*, 2715–2721. [CrossRef]

34. Shiau, J.T. Fitting drought duration and severity with two-dimensional copulas. *Water Resour. Manag.* **2006**, *20*, 795–815. [CrossRef]
35. Chbouki, N. Spatio-Temporal Characteristics of Drought as Inferred from Tree-Ring Data in Morocco. Ph.D. Thesis, University of Arizona, Tucson, AZ, USA, 1992.
36. Shiau, J.T.; Modarres, R. Copula-based drought severity-duration-frequency analysis in Iran. *Meteorol. Appl.* **2009**, *16*, 481–489. [CrossRef]
37. Rajsekhar, D.; Singh, V.P.; Mishra, A.K. Multivariate drought index: An information theory based approach for integrated drought assessment. *J. Hydrol.* **2015**, *526*, 164–182. [CrossRef]
38. Hao, Z.; AghaKouchak, A. Multivariate standardized drought index: A parametric multi-index model. *Adv. Water Resour.* **2013**, *57*, 12–18. [CrossRef]
39. Brown, J.F.; Wardlow, B.D.; Tadesse, T.; Hayes, M.J.; Reed, B.C. The Vegetation Drought Response Index (VegDRI): A new integrated approach for monitoring drought stress in vegetation. *GISci. Remote Sens.* **2008**, *45*, 16–46. [CrossRef]
40. Tadesse, T.; Wardlow, B.D.; Brown, J.F.; Svoboda, M.D.; Hayes, M.J.; Fuchs, B.; Gutzmer, D. Assessing the vegetation condition impacts of the 2011 drought across the US Southern Great Plains using the Vegetation Drought Response Index (VegDRI). *J. Appl. Meteorol. Climatol.* **2015**, *54*, 153–169. [CrossRef]
41. Alley, W.M. The Palmer Drought Severity Index: Limitations and assumptions. *J. Clim. Appl. Meteorol.* **1984**, *23*, 1100–1109. [CrossRef]
42. Shafer, B.A.; Dezman, L.E. Development of a Surface Water Supply Index (SWSI) to assess the severity of drought conditions in snowpack runoff areas. In Proceedings of the Western Snow Conference, Reno, Nevada, April 1982; pp. 164–175.
43. Valipour, M. Use of surface water supply index to assessing of water resources management in Colorado and Oregon. *Adv. Agric.* **2013**, *3*, 631–640.
44. González, J.; Valdés, J.B. Bivariate drought recurrence analysis using tree ring reconstructions. *J. Hydrol. Eng.* **2003**, *8*, 247–258. [CrossRef]
45. Vangelis, H.; Spiliotis, M.; Tsakiris, G. Drought severity assessment based on bivariate probability analysis. *Water Resour. Manag.* **2011**, *25*, 357–371. [CrossRef]
46. Murtin, C.M.; Murtin, F. Education Inequalities among World Citizens: 1870–2000. 2006. Working Paper. Available online: http://www.eea-esem.com/files/papers/EEA-ESEM/2006/2780/EducationInequality.pdf (accessed on 21 March 2016).
47. Sklar, A. *Fonctions de Repartition 'a n Dimensions et Leura Marges*; Publication de l'Institut de Statistique de l'Université de Paris: Paris, France, 1959; pp. 229–231. (In French)
48. Shiau, J.T.; Feng, S.; Nadarajah, S. Assessment of hydrological droughts for the Yellow River, China, using copulas. *Hydrol. Process.* **2007**, *21*, 2157–2163. [CrossRef]
49. Serinaldi, F.; Bonaccorso, B.; Cancelliere, A.; Grimaldi, S. Probabilistic characterization of drought properties through Copulas. *Phys. Chem. Earth* **2009**, *34*, 596–605. [CrossRef]
50. Song, S.; Singh, V.P. Meta-elliptical copulas for drought frequency analysis of periodic hydrologic data. *Stoch Environ. Res. Risk Assess.* **2010**, *24*, 425–444. [CrossRef]
51. Mirabbasi, R.; Anagnostou, E.N.; Fakheri-Fard, A.; Dinpashoh, Y.; Eslamian, S. Analysis of meteorological drought in northwest Iran using the Joint Deficit Index. *J. Hydrol.* **2013**, *492*, 35–48. [CrossRef]
52. Chen, L.; Singh, V.P.; Guo, S.; Mishra, A.K.; Guo, J. Drought Analysis Using Copulas. *J. Hydrol. Eng.* **2012**, *18*, 797–808. [CrossRef]
53. Vergni, L.; Todisco, F.L.; Mannocchi, F. Analysis of agricultural drought characteristics through a two-dimensional copula. *Water Resour. Manag.* **2015**, *29*, 2819–2835. [CrossRef]
54. Huang, S.; Huang, Q.; Chang, J.; Chen, Y.; Xing, L.; Xie, Y. Copulas-Based Drought Evolution Characteristics and Risk Evaluation in a Typical Arid and Semi-Arid Region. *Water Resour. Manag.* **2014**, *29*, 1489–1503. [CrossRef]
55. Reddy, M.J.; Singh, V.P. Multivariate modeling of droughts using copulas and meta-heuristic methods. *Stoch. Environ. Rese. Risk Assess.* **2014**, *28*, 475–489. [CrossRef]
56. Mishra, A.; Singh, V.P.; Desai, V. Drought characterization: A probabilistic approach. *Stoch. Environ. Res. Risk Assess.* **2009**, *23*, 41–55. [CrossRef]
57. Jacoby, G.C.; Cook, E.R. Past temperature variations inferred from a 400-year tree-ring chronology from Yukon Territory, Canada. *Arct. Alp. Res.* **1981**, *13*, 409–418. [CrossRef]

58. Grissino-Mayer, H.D.; Fritts, H.C. The International Tree-Ring Data Bank: An enhanced global database serving the global scientific community. *Holocene* **1997**, *7*, 235–238. [CrossRef]
59. Cook, E.R. A time series analysis approach to tree-ring standardization (Dendrochronology, Forestry, Dendroclimatology, Autoregressive process). Ph.D. Thesis, University of Arizona, Tucson, AZ, USA, 1985.
60. California Data Exchange Center. Available online: http://cdec.water.ca.gov/index.html (accessed on 12 May 2015).
61. Yevjevich, V. An objective approach to definitions and investigations of continental hydrologic droughts. In *Hydrologic Paper*; Colorado State University: Fort Collins, CO, USA, 1967.
62. Kwak, J.; Kim, D.; Kim, S.; Singh, V.P.; Kim, H. Hydrological drought analysis in Namhan river basin, Korea. *J. Hydrol. Eng.* **2014**, *19*. [CrossRef]
63. Serinaldi, F.; Grimaldi, S. Fully nested 3-Copula: Procedure and application on hydrological data. *J. Hydrol. Eng.* **2007**, *12*, 420–430. [CrossRef]
64. Yu, K.X.; Xiong, L.; Gottschalk, L. Derivation of low flow distribution functions using copulas. *J. Hydrol.* **2014**, *508*, 273–288. [CrossRef]
65. Wong, G.; Lambert, M.F.; Metcalfe, A.V. Trivariate copulas for characterization of droughts. *ANZIAM J.* **2008**, *49*, 306–315.
66. Sadri, S.; Burn, D.H. Copula-based pooled frequency analysis of droughts in the Canadian Prairies. *J. Hydrol. Eng.* **2012**, *19*, 277–289. [CrossRef]
67. Chen, Y.D.; Zhang, Q.; Xiao, M.; Singh, V.P. Evaluation of risk of hydrological droughts by the trivariate Plackett copula in the East River basin (China). *Nat. Hazards* **2013**, *68*, 529–547. [CrossRef]
68. Saghafian, B.; Mehdikhani, H. Drought characterization using a new copula-based trivariate approach. *Nat. Hazards* **2014**, *72*, 1391–1407. [CrossRef]
69. Black, P.E. *Watershed Hydrology*; John Wiley & Sons: Hoboken, NJ, USA, 1991.
70. Hopfield, J.J. Neural networks and physical systems with emergent collective computational abilities. *Proc. Natl. Acad. Sci. USA* **1982**, *79*, 2554–2558. [CrossRef] [PubMed]
71. Rosenblatt, F. The perceptron: A probabilistic model for information storage and organization in the brain. *Psychol. Rev.* **1958**, *65*, 386–408. [CrossRef] [PubMed]
72. Battiti, R. Accelerated backpropagation learning: Two optimization methods. *Complex. Syst.* **1989**, *3*, 331–342.
73. Gill, P.E.; Murray, W.; Wright, M.H. *Practical Optimization*; Academic Press: New York, NY, USA, 1981.
74. Bourquin, J.; Schmidli, H.; van Hoogevest, P.; Leuenberger, H. Advantages of Artificial Neural Networks (ANNs) as alternative modelling technique for data sets showing non-linear relationships using data from a galenical study on a solid dosage form. *Eur. J. Pharm. Sci.* **1998**, *7*, 5–16. [CrossRef]
75. Meko, D.M.; Therrell, M.D.; Baisan, C.H.; Hughes, M.K. Sacramento river flow reconstructed to AD 869 from tree rings. *J. Am. Water Resour. Assoc.* **2001**, *37*, 1029–1039. [CrossRef]
76. Basheer, I.A.; Hajmeer, M. Artificial neural networks: Fundamentals, computing, design, and application. *J. Microbiol. Methods* **2000**, *43*, 3–31. [CrossRef]
77. AghaKouchak, A. Entropy–Copula in Hydrology and Climatology. *J. Hydrometeorol.* **2014**, *15*, 2176–2189. [CrossRef]
78. Saad, C.; El Adlouni, S.; St-Hilaire, A.; Gachon, P. A nested multivariate copula approach to hydrometeorological simulations of spring floods: The case of the Richelieu River (Québec, Canada) record flood. *Stoch. Environ. Res. Risk Assess.* **2015**, *29*, 275–294. [CrossRef]
79. Rodriguez, J.C. Measuring financial contagion: A copula approach. *J. Empir. Financ.* **2007**, *14*, 401–423. [CrossRef]
80. Haykin, S. *Neural Networks: A Comprehensive Foundation*, 2nd ed.; Prentice Hall: Upper Saddle River, NJ, USA, 1999.
81. Maier, H.R.; Dandy, G.C. Neural networks for the prediction and forecasting of water resources variables: A review of modelling issues and applications. *Environ. Model. Softw.* **2000**, *15*, 101–124. [CrossRef]
82. Hyndman, R.J.; Khandakar, Y. *Automatic Time Series for Forecasting: The Forecast Package for R*; Department of Econometrics and Business Statistics, Monash University: Melbourne, Australia, 2007.
83. Nash, J.E.; Sutcliffe, J.V. River flow forecasting through conceptual models part I—A discussion of principles. *J. Hydrol.* **1970**, *10*, 282–290. [CrossRef]

84. Moriasi, D.N.; Arnold, J.G.; Van Liew, M.W.; Bingner, R.L.; Harmel, R.D.; Veith, T.L. Model Evaluation Guidelines for Systematic Quantification of Accuracy in Watershed Simulations. *Trans. ASABE* **2007**, *50*, 885–900. [CrossRef]
85. Vogel, R.M.; Hosking, J.R.; Elphick, C.S.; Roberts, D.L.; Reed, J.M. Goodness of fit of probability distributions for sightings as species approach extinction. *Bull. Math. Biol.* **2009**, *71*, 701–719. [CrossRef] [PubMed]
86. El Adlouni, S.; Ouarda, T.B. Joint Bayesian model selection and parameter estimation of the generalized extreme value model with covariates using birth-death Markov chain Monte Carlo. *Water Resour. Res.* **2009**, *45*. [CrossRef]
87. Carle, D. *Introduction to Water in California*; University of California Press: Berkeley, CA, USA, 2004.
88. Paulson, R.W.; Chase, E.B.; Roberts, R.S.; Moody, D.W. *National Water Summary 1988–89: Hydrologic Events and Floods and Droughts (No. 2375)*; US Government Printing Office: Washington, DC, USA, 1991.
89. Poulin, A.; Huard, D.; Favre, A.C.; Pugin, S. Importance of tail dependence in bivariate frequency analysis. *J. Hydrol. Eng.* **2007**, *12*, 394–403. [CrossRef]

7

Water Resource Variability and Climate Change

Yingkui Li [1,*] **and Michael A. Urban** [2]

[1] Department of Geography, University of Tennessee, Knoxville, TN 37996, USA
[2] Department of Geography, University of Missouri, Columbia, MO 65211, USA; UrbanM@missouri.edu
* Correspondence: yli32@utk.edu.

Academic Editor: Arjen Y. Hoekstra

Abstract: A significant challenge posed by changing climates is how water cycling and surficial and subsurface water availability will be affected at global and regional scales. Such alterations are critical as they often lead to increased vulnerability in ecosystems and human society. Understanding specifically how climate change affects water resource variability in different locations is of critical importance to sustainable development in different parts of the world. The papers included in this special issue focus on three broad perspectives associated with water resource variability and climate change. Six papers employ remote sensing, meteorological station-based observational data, and tree-ring records to empirically determine how water resources have been changing over historical time periods. Eight of the contributions focus on modeling approaches to determine how known processes are likely to manifest themselves as climate shifts over time. Two others focus on human perceptions and adaptation strategies in the midst of unstable or unsettled water availability. The findings and methods presented in this collection of papers provide important contributions to the increased study and awareness of climate change on water resources.

Keywords: water resource variability; climate change; hydrological model; perception and adaptation of climate change

---

1. Introduction

Climate change and increased anthropogenic pressure on earth–atmosphere interactions affect water quantity, quality, and water-related processes, such as sediment yield, on local, regional, and global scales [1–3]. Recent decades have seen continuously increasing temperatures in most parts of the world, and changes in precipitation patterns have increased the frequency of extreme climate events such as drought and flooding [4]. The impact of changing baseline conditions coupled with increased variability can be especially complicated in regions with rapid changes in population, land development (especially urbanization), and economic disruptions. While public discussions often focus more on temperature than water availability, ecosystems and human society are highly vulnerable to water stress [5–8]. Understanding the mechanisms and geographic patterns by which anthropogenic climate change is impacting water resource variability is of critical importance to sustainable development, environmental management, and human health.

A variety of approaches have been used to examine the relationships between atmospheric variability and surficial water resources. Instrumental data collected from meteorological and hydrological gauging stations can be used to investigate altered hydrologic regimes over the timespan of several decades to a few centuries in certain areas. Relatively long-term (hundreds to thousands of years) climate and environmental records can be reconstructed using various proxies such as tree rings, sediment cores, ice cores, and landform features [9,10]. More recently, the availability of various remote sensing datasets, such as Landsat/MODIS (Moderate Resolution Imaging Spectroradiometer) imagery, ICESAT (Ice, Clouds, and Land Elevation Satellite) altimetry, GRACE (Gravity Recovery and Climate

Experiment) gravity, and LiDAR (Light Detection And Ranging) measurements have facilitated remote sensing-based approaches to quantifying water resource changes [11–13]. Computational modeling approaches, ranging from global circulation models (GCMs) to regional or watershed hydrological models, are serving to simulate and forecast the projected nature of climate variability on water resources [14,15]. Social scientists have also been investigating how groups or local communities perceive the impacts of climate change and climate vulnerability in order to implement better adaptation practices and sustainable development in coping with changing water resources of different regions [16,17].

The papers included in this special issue address three broad perspectives associated with water resource variability and climate change: (1) the *quantification* of water resource variability altered by changing climates using remote sensing assessment, meteorological station-based observational datasets, and tree-ring record reconstruction; (2) the *simulation* of such impacts on water resource variability using modeling approaches; and (3) evaluating social *perceptions and adaptation strategies* in the face of unstable water resource variability. The following section summarizes the individual contributions within each perspective.

2. Contributions

Six of the papers assess the various impacts of climate change on water resources using a variety of datasets, empirical observations, and proxies. Li et al. [18] examine surface area fluctuations occurring in 10 major lakes in the arid province of Xinjiang, China, from 2000 to 2014 using MODIS time series imagery. The authors develop a classification method to accommodate varied spectral characteristics of water pixels and derived water bodies for April, July, and September in each year for 10 major lakes (>100 km^2) in the study area. Lakes in the lowland (close to urban and agriculture areas) showed a shrinking trend, while mountain lakes have diverse changing patterns (some shrinking, some expanding), and lakes on the Tibetan Plateau exhibited significant expanding trends. By observing varied patterns of lake surface changes across the region, the authors conclude that observed lake expansion is likely driven by rising temperature, leading to accelerated melting of snow and glaciers in high mountains and on the Tibetan Plateau, and increased precipitation in this region (especially in 2010), whereas the shrinking of some lakes is likely related to anthropogenic utilization based on agricultural and industrial needs.

Ning et al. [19] analyze recent changes in water resources and grassland in the Hulun Lake region, a semi-arid region in northeastern China, using monthly GRACE and Tropical Rainfall Measuring Mission (TRMM) data. Results indicate decreasing trends in overall water storage and precipitation between 2002 and 2007, followed by increasing trends in the period from 2007 to 2012. Water storage trends are mainly correlated to precipitation and temperature patterns. As a result, a large proportion of grassland recovered to its normal state in 2008–2012, and only a small proportion of grassland (16.5% of the study area) is classified as degraded. The authors conclude that degraded grassland areas in the region are more vulnerable to climate variability and require protective strategies to prevent further degradation.

Buma et al. [20] assess observed changes in hydrological conditions of Lake Chad basin based on the total water storage (TWS) derived from GRACE, lake levels taken from satellite altimetry, and water fluxes and soil moisture obtained from the Global Land Data Assimilation System (GLDAS). The authors observe a similar pattern between TWS and lake level changes and subsurface water volume changes. The derived values for subsurface water volume changes are found to be consistent with groundwater outputs calculated from the WaterGAP Global Hydrology Model (WGHM). By utilizing recently developed remote sensing datasets, this study provides an alternative means of generating information for the management of water resources in the Lake Chad basin.

Jiang et al. [21] summarize the changing patterns, causes, and implications of surface water discharge and sediment load in Chinese rivers from 1956 to 2012 based on monthly hydrological and daily meteorological data obtained from 725 rain gauge stations across the country. Numerous patterns can be observed during this period. Streamflow discharges manifest a decreasing trend, a relatively stable state, and an increasing trend within northern, southern, and western China, respectively. Excepting the Lancang River and Yarlung Zangbo River basins, sediment loads in most Chinese river basins show gradually decreasing trends, especially after 2000. Although patterns of streamflow and sediment load are affected by the interaction of varied meso-scale climate systems—including East and South Asian monsoons and westerlies—the authors determine that water consumption for industrial and residential purposes, soil, and water conservation engineering, hydraulic engineering, and land surface changes induced by other factors are likely the main causes of observed patterns of streamflow and sediment reduction.

Wang et al. [22] investigate the impact that climate change has had on the duration of flood seasons in the Fenhe River, China, from 1957 to 2014, based on daily precipitation data from 14 meteorological stations in the basin and an analysis of the variations in the onset and retreat dates of yearly flood seasons. The results show that the observed duration of the flood season has been extended since 1975. In particular, the onset of floods has advanced 15 days, although the retreat date is relatively stable. Based on these results, the authors recommend corresponding measures to adapt to the flood season variations.

Kwak et al. [23] conduct a drought analysis using a long-term streamflow record reconstructed using tree ring indices within the Sacramento Basin, California, USA. By first identifying annual streamflow patterns of the Sacramento River from 1560 to 1871 and then analyzing the hydrological drought return period in this river basin, the authors argue that drought with a 20-year return period can be considered a critical indicator of drought for water shortages in the Sacramento River basin.

Eight of the papers aim to simulate the impact of climate change on water resource variability using various climatological and hydrological models. Pechlivanidis et al. [24] investigate the impact of changing climates have on the hydro-climatology of the Indian subcontinent by comparing current and projected future water fluxes from three RCP (Representative Concentration Pathway) scenarios (RCP2.6, RCP4.5 and RCP8.5). These results are used to depict expected changes in the annual flow cycles of three major rivers from different hydro-climatic regions, while acknowledging that conclusions can be significantly influenced by statistical uncertainty embedded in the RCP scenarios. Based on this study, the models project a gradual increase in temperature and uneven changes (ranging from −20% to +50%) in long-term average precipitation and evapotranspiration. Potential surface runoff is also expected to change anywhere from −100% to +100%. The analysis of annual cycles for the three selected regions show that the impact of climate change on discharge and evapotranspiration varies between seasons, and the magnitude of change is primarily dependent on the hydro-climatic gradient in different regions.

Li and Gao [25] simulate the impact of various precipitation change scenarios on runoff and sediment yield in a hilly-gullied watershed typical of the Loess Plateau in China using the Soil and Water Assessment Tool (SWAT). This study indicates that runoff and sediment yield both increase with increasing precipitation, while the variation in sediment yield is more sensitive to smaller rainfall events. The authors determine that under these conditions, annual runoff and sediment yield fluctuate greatly and the magnitude of the variations was especially amplified when precipitation increased by 20%. Overall escalation in runoff and sediment caused by increased precipitation is greater than corresponding decreases coincident with reduced precipitation, and runoff is the more sensitive variable compared to sediment yield.

Ligaray et al. [26] assess the hydrological response of climate change in the Chao Phraya River Basin, Thailand. Streamflow variations were simulated using a combination of SWAT and meteorological data from 2003 to 2011 for various climate sensitivity and greenhouse gas emission scenarios. Simulation results reveal that streamflow variations correspond to the changes in rainfall totals and intensity, while increased air temperature likely leads to future water shortages. The simulation also suggests that high CO_2 concentration drives plant responses that may lead to a dramatic increase in streamflow. Specifically, increased streamflow variations to 6.8%, 41.9%, and 38.4% were simulated for the three greenhouse gas emission scenarios (A1B, A2, B1) in the reference period of 2003–2011.

Mahmood et al. [27] investigate the potential impacts of climate change on the water resources of the Kunhar River basin, Pakistan under A2 and B2 climate scenarios. Using the HEC-HMS (Hydrologic Engineering Center's Hydraulic Modeling System) hydrological model, the authors simulate streamflow for the periods: 2011–2040, 2041–2070, and 2071–2099, and compare them with the baseline period (1961–1990) to explore changes in different streamflow variables. The results indicate an overall increase in mean annual flow projected under both A2 and B2 scenarios, but with a high degree of variability. Stream discharge increases mainly in summer and autumn, but decreases throughout the spring and winter months. High and median flows are predicted to increase, with peak discharges shifting from June to July, while low flow conditions are projected to decrease. The Kunhar basin will face a higher degree of variability—both more floods and droughts—by the end of the 21st century, due to the projected increase in high flow, the decrease in low flow, and greater variations in peak discharges. This study highlights key impacts of climate change on water resources to help develop suitable policies for water resource use and management in this river basin.

Hesse and Krysanova [28] simulate the impacts of climatic shifts and changing management practices on water quality and in-stream processes in the Elbe River Basin using a semi-distributed watershed model (SWIM) with implemented in-stream nutrient (N+P) turnover and algal growth processes. The set of modeled climate scenarios show a projected increase in temperature (+3 °C) and precipitation (+57 mm) on average until the end of the century, leading to varied changes in discharge (+20%), nutrient loads (NO_3-N: 5%; NH_4-N: 24%; PO_4-P: +5%), phytoplankton biomass (4%), and dissolved oxygen concentration (5%) in the Elbe River Basin. The authors utilize the model to examine the ways in which changes in climatic variables fundamentally impact the ways by which land use and nutrients are managed to reduce nutrient emissions to the river.

Liu and Chan [29] assess impacts on water quality in the Danshuei River estuarine system in northern Taiwan using a coupled three-dimensional hydrodynamic and water quality model driven by changes in climatic variables. The model is calibrated and validated using observed data and then applied to simulate water quality projections under various climate change scenarios. Results indicate that dissolved oxygen concentrations are likely to significantly decrease in the Danshuei, whereas nutrients will increase in response to expected climate changes. In particular, dissolved oxygen concentrations will be reduced to less than 2 mg/L in the main stream, failing to meet accepted water quality standards. This study suggests an appropriate strategy for effective water quality management in estuarine systems such as the Danshuei is needed to adapt to the water quality changes likely to accompany anthropogenic climate change.

Wei et al. [30] estimate flood risk that is likely to occur under the heightened hydrologic variability driven by climate change in the Tsengwen River Basin, Taiwan, using a SOBEK model (Deltares, The Netherlands). Simulated results indicate that the discharge of the Tsengwen is at increasing risk of exceeding the designed maximum streamflow at three stations from different areas of the watershed for three projected periods of 1979–2003, 2015–2039, and 2075–2099. Model results indicate that the exceedance frequency for the designed flood is 2 in 88 events in the base period (1979–2003), 6 in 82 events in the near future (2015–2039), and 10 in 81 events at the end of the century (2075–2099).

Okamoto et al. [31] turn our attention from streamflow to water fluxes driving hillslope processes. They investigate the optimal soil hydraulic parameters for simulating unsaturated flow based on a case study from the island of Miyakojima, Japan. The authors optimize the parameters for root water uptake and then examine the influence of soil hydraulic parameters on simulations of evapotranspiration. From there, they compare volumetric water content between the simulation results and those using pedotransfer estimates obtained from ROSETTA software. The resulting comparison highlights the importance of using soil hydraulic parameters based on measured data to simulate evapotranspiration and unsaturated water flow processes.

The last two papers in this special issue examine the ways by which different perceptions of climate change and adaptation strategies impact management and water resource variability. Ndamani and Watanabe [32] analyze farmer perceptions of adaptation practices using semi-structured questionnaires and focus group discussions of 100 farmer-households from four communities in the Lawra district of Ghana. The results show that adaptation is largely driven by response to dry spells and droughts (93.2%) rather than floods. Farmers in the region ranked improved crop varieties and irrigation as the most important adaptation measures, but largely lacked the capacity to implement these adaptation practices. The study also revealed that unpredictable weather, high cost of farm inputs, limited access to reliable weather information, and lack of water resources were the most critical barriers to successful adaptation. This study highlights the critical linkage between climate, hydrology, perception, and environmental management.

Shandas et al. [33] present a study of differing perspectives from the field on stressors and strategies for managing urban water scarcity in two urbanizing regions of the western US: Portland, Oregon and Phoenix, Arizona. The results show that long-term drought, population growth, and outdoor water use are the most important stressors to urban water systems, and indicate more agreement across cities than across professions in terms of effective strategies, suggesting that land-use planners and water managers remain divided in their conception of the solutions to urban water management. The authors also recommend potential pathways for coordinating the fields of land and water management to streamline strategies for urban sustainability.

3. Conclusions

This collection of papers focuses on a range of research topics influenced by the overriding hydrologic mechanisms associated with anthropogenic climate change and associated water resource variability. This includes a wide range of problems ranging from changes in surficial water levels, streamflow, sediment yields, and water quality in lakes, rivers, watersheds, and estuarine systems. The authors have brought a number of methodological tools to bear on these problems by examining various datasets and techniques, such as remote sensing, meteorological station-based observational data, tree-ring records, climate forecasts, and hydrological models used to simulate climatic impacts on streamflow, sediment yield, and water quality. Because consequent environmental problems and strategies for coping and mitigating deleterious effects must be defined in a social context, it is also important to include research examining perception, vulnerability, and adaptation. This collection of 16 papers emphasizes the importance of understanding the various interrelated facets that changing climates have on water resource variability and how focused investigations will help ground suitable strategies for mitigating and adapting to anthropogenic climate change.

Acknowledgments: The authors of this paper, who served as the guest-editors of this special issue, wish to thank the journal editors, all authors submitting papers to this special issue, and the many referees who contributed to paper revision and improvement of the 16 published papers.

References

1. Gulick, S.P.S.; Jaeger, J.M.; Mix, A.C.; Asahi, H.; Bahlburg, H.; Belanger, C.L.; Berbel, G.B.B.; Childress, L.; Cowan, E.; Drab, L.; et al. Mid-Pleistocene climate transition drives net mass loss from rapidly uplifting St. Elias Mountains, Alaska. *Proc. Natl. Acad. Sci. USA* **2015**, *112*, 15042–15047. [CrossRef] [PubMed]
2. Kouchak, A.A.; Feldman, D.; Hoerling, M.; Huxman, T.; Lund, J. Water and climate: Recognize anthropogenic drought. *Nature* **2015**, *524*, 409–411. [CrossRef] [PubMed]
3. Castle, S.L.; Thomas, B.F.; Reager, J.T.; Rodell, M.; Swenson, S.C.; Famiglietti, J.S. Groundwater depletion during drought threatens future water security of the Colorado River Basin. *Geophys. Res. Lett.* **2014**, *41*, 5904–5911. [CrossRef] [PubMed]
4. National Academies of Sciences, Engineering and Medicine. *Attribution of Extreme Weather Events in the Context of Climate Change*; The National Academies Press: Washington, DC, USA, 2016.

5. Vörösmarty, C.J.; Green, P.; Salisbury, J.; Lammers, R.B. Global water resources: Vulnerability from climate change and population growth. *Science* **2000**, *289*, 284–288. [CrossRef] [PubMed]
6. Phillips, O.L.; Aragão, L.E.O.C.; Lewis, S.L.; Fisher, J.B.; Lloyd, J.; López-González, G.; Malhi, Y.; Monteagudo, A.; Peacock, J.; Quesada, C.A.; et al. Drought sensitivity of the amazon rainforest. *Science* **2009**, *323*, 1344–1347. [CrossRef] [PubMed]
7. Piao, S.; Ciais, P.; Huang, Y.; Shen, Z.; Peng, S.; Li, J.; Zhou, L.; Liu, H.; Ma, Y.; Ding, Y.; et al. The impacts of climate change on water resources and agriculture in China. *Nature* **2010**, *467*, 43–51. [CrossRef] [PubMed]
8. Cowell, C.M.; Urban, M.A. The changing geography of the U.S. water budget: Twentieth-century patterns and twenty-first-century projections. *Ann. Assoc. Am. Geogr.* **2010**, *100*, 740–754. [CrossRef]
9. Jones, P.D.; Mann, M.E. Climate over past millennia. *Rev. Geophys.* **2004**, *42*, RG2002. [CrossRef]
10. Zhang, Q.B.; Cheng, G.; Yao, T.; Kang, X.; Huang, J. A 2326-year tree-ring record of climate variability on the northeastern Qinghai-Tibetan Plateau. *Geophys. Res. Lett.* **2003**, *30*. [CrossRef]
11. Trenberth, K.E.; Dai, A.; Rasmussen, R.M.; Parsons, D.B. The changing character of precipitation. *Bull. Am. Meteorol. Soc.* **2003**, *84*, 1205–1217. [CrossRef]
12. Song, C.; Huang, B.; Ke, L. Modeling and analysis of lake water storage changes on the Tibetan Plateau using multi-mission satellite data. *Remote Sens. Environ.* **2013**, *135*, 25–35. [CrossRef]
13. Li, Y.K.; Liao, J.J.; Guo, H.D.; Liu, Z.W.; Shen, G.Z. Patterns and potential drivers of dramatic changes in Tibetan lakes, 1972–2010. *PLoS ONE* **2014**, *9*, e111890. [CrossRef] [PubMed]
14. Seager, R.; Ting, M.; Held, I.; Kushnir, Y.; Lu, J.; Vecchi, G.; Huang, H.P.; Harnik, N.; Leetmaa, A.; Lau, N.C.; et al. Model projections of an imminent transition to a more arid climate in southwestern North America. *Science* **2007**, *316*, 1181–1184. [CrossRef] [PubMed]
15. Xu, C.Y. From GCMs to river flow: A review of downscaling methods and hydrologic modelling approaches. *Prog. Phys. Geogr.* **1999**, *23*, 229–249. [CrossRef]
16. Sheppard, S.R.J. Landscape visualisation and climate change: The potential for influencing perceptions and behavior. *Environ. Sci. Policy* **2005**, *8*, 637–654. [CrossRef]
17. Preston, B.L.; Yuen, E.J.; Westaway, R.M. Putting vulnerability to climate change on the map: A review of approaches, benefits, and risks. *Sustain. Sci.* **2011**, *6*, 177–202. [CrossRef]
18. Li, Q.T.; Lu, L.L.; Wang, C.Z.; Li, Y.K.; Sui, Y.; Guo, H.D. MODIS-Derived Spatiotemporal Changes of Major Lake Surface Areas in Arid Xinjiang, China, 2000–2014. *Water* **2015**, *7*, 5731–5751. [CrossRef]
19. Ning, S.; Ishidaira, H.; Udmale, P.; Ichikawa, Y. Remote Sensing Based Analysis of Recent Variations in Water Resources and Vegetation of a Semi-Arid Region. *Water* **2015**, *7*, 6039–6055. [CrossRef]
20. Buma, W.G.; Lee, S.I.; Seo, J.Y. Hydrological Evaluation of Lake Chad Basin using Space Borne and Hydrological Model Observation. *Water* **2016**, *8*, 205. [CrossRef]
21. Jiang, C.; Zhang, L.B.; Li, D.Q.; Li, F. Water Discharge and Sediment Load Changes in China: Change Patterns, Causes, and Implications. *Water* **2015**, *7*, 5849–5875. [CrossRef]
22. Wang, H.J.; Xiao, W.H.; Wang, J.H.; Wang, Y.C.; Huang, Y.; Hou, B.D.; Lu, C.Y. The Impact of Climate Change on the Duration and Division of Flood Season in the Fenhe River Basin, China. *Water* **2016**, *8*, 105. [CrossRef]
23. Kwak, J.; Kim, S.; Kim, G.; Singh, V.P.; Park, J.; Kim, H.S. Bivariate Drought Analysis Using Streamflow Reconstruction with Tree Ring Indices in the Sacramento Basin, California, USA. *Water* **2016**, *8*, 122. [CrossRef]
24. Pechlivanidis, I.G.; Olsson, J.; Bosshard, T.; Sharma, D.; Sharma, K.C. Multi-Basin Modelling of Future Hydrological Fluxes in the Indian Subcontinent. *Water* **2016**, *8*, 177. [CrossRef]
25. Li, T.H.; Gao, Y. Runoff and Sediment Yield Variations in Response to Precipitation Changes: A Case Study of Xichuan Watershed in the Loess Plateau, China. *Water* **2015**, *7*, 5638–5656. [CrossRef]
26. Ligaray, M.; Kim, H.; Sthiannopkao, S.; Lee, S.; Cho, K.H.; Kim, J.H. Assessment on Hydrologic Response by Climate Change in the Chao Phraya River Basin, Thailand. *Water* **2015**, *7*, 6892–6909. [CrossRef]
27. Mahmood, R.; Jia, S.F.; Babel, M.S. Potential Impacts of Climate Change on Water Resources in the Kunhar River Basin, Pakistan. *Water* **2016**, *8*, 23. [CrossRef]
28. Hesse, C.; Krysanova, V. Modeling Climate and Management Change Impacts on Water Quality and In-Stream Processes in the Elbe River Basin. *Water* **2016**, *8*, 40. [CrossRef]
29. Liu, W.C.; Chan, W.T. Assessment of Climate Change Impacts on Water Quality in a Tidal Estuarine System Using a Three-Dimensional Model. *Water* **2016**, *8*, 60. [CrossRef]
30. Wei, H.P.; Yeh, K.C.; Liou, J.J.; Chen, Y.M.; Cheng, C.T. Estimating the Risk of River Flow under Climate Change in the Tsengwen River Basin. *Water* **2016**, *8*, 81. [CrossRef]

31. Okamoto, K.; Sakai, K.; Nakamura, S.; Cho, H.; Nakandakari, T.; Ootani, S. Optimal Choice of Soil Hydraulic Parameters for Simulating the Unsaturated Flow: A Case Study on the Island of Miyakojima, Japan. *Water* **2015**, *7*, 5676–5688. [CrossRef]
32. Ndamani, F.; Watanabe, T. Farmers' Perceptions about Adaptation Practices to Climate Change and Barriers to Adaptation: A Micro-Level Study in Ghana. *Water* **2015**, *7*, 4592–4604. [CrossRef]
33. Shandas, V.; Lehman, R.; Larson, K.L.; Bunn, J.; Chang, H. Stressors and Strategies for Managing Urban Water Scarcity: Perspectives from the Field. *Water* **2015**, *7*, 6775–6787. [CrossRef]

8

Estimating the Risk of River Flow under Climate Change in the Tsengwen River Basin

Hsiao-Ping Wei [1,*], Keh-Chia Yeh [2], Jun-Jih Liou [3], Yung-Ming Chen [3] and Chao-Tzuen Cheng [3]

1 National Science and Technology Center for Disaster Reduction, New Taipei City 23143, Taiwan
2 Department of Civil Engineering in National Chiao Tung University, Hsinchu 300, Taiwan; kcyeh@cc.nctu.edu.tw
3 National Science and Technology Center for Disaster Reduction, New Taipei City 23143, Taiwan; jjliou@ncdr.nat.gov.tw (J.-J.L.); ymchen@ncdr.nat.gov.tw (Y.-M.C.); ctcheng@ncdr.nat.gov.tw (C.-T.C.)
* Correspondence: weihp@ncdr.nat.gov.tw.

Academic Editor: Yingkui Li

Abstract: This study evaluated the overflow risk of the Tsengwen River under a climate change scenario by using bias-corrected dynamic downscaled data as inputs for a SOBEK model (Deltares, the Netherlands). The results showed that the simulated river flow rate at Yufeng Bridge (upstream), Erxi Bridge (midstream), and XinZong (1) (downstream) stations are at risk of exceeding the management plan's flow rate for three projection periods (1979–2003, 2015–2039, 2075–2099). After validation with the geomorphic and hydrological data collected in this study, the frequency at which the flow rate exceeded the design flood was 2 in 88 events in the base period (1979–2003), 6 in 82 events in the near future (2015–2039), and 10 in 81 events at the end of the century (2075–2099).

Keywords: extreme typhoon events; risk analysis; SOBEK; climate change; hydrological extreme

1. Introduction

Extreme typhoon precipitation events frequently result in socioeconomic impacts and loss of human life. Increased incidences of extreme rainfall events indicate one of the common features signaling climate change worldwide. The International Panel on Climate Change [1] reported that on average, precipitation has increased globally by approximately 8%. According to Liu *et al.* [2], scientists have contended that the increase in global temperature over the past decade has prompted an increase in extreme precipitation events and a decrease in moderate and mild precipitation events. The 2010 Taiwan Climate Change Projection and Information Platform (TCCIP) Project Report II included statistical data regarding the frequency of extreme typhoon precipitation events in Taiwan from 1970 to 2009. The statistical results indicated that prior to 2000, the frequency of extreme typhoon precipitation events was approximately once every 2 years; however, this frequency increased to at least once a year after 2000 [3]. Because of river flow changes caused by extreme rainfall, discharge control structures (culverts, flap gates, weirs, and sluice gates) in river basins are at a high risk of destruction.

The main scientific tool used in long-term climate simulations is the general circulation model (GCM), the main purpose of which is to project global climate characteristics and trends. However, GCM projections (e.g., rainfall, temperature, and humidity) cannot provide adequate and effective information for simulating small areas. In past decades, scientists have developed downscaling methods to increase the spatial resolution, providing more information for correcting the error margin from the GCM simulations and presenting the influence of topographic distribution in local areas. Currently, high-resolution climate data can be obtained through high-resolution GCM, dynamical downscaling, and statistical downscaling. Although numerous recent studies have attempted to increase the spatial resolution of the output from the GCM, for example, by using statistical and dynamical downscaling, the results typically yield only certain points of information that are inadequate for resolving the climate characteristics of small areas with complex terrain such as in Taiwan.

The present study used the Tsengwen River as the study area. High-resolution dynamical downscaling data were used to simulate changes in the hourly flow rate of typhoon events. Based on the selection criteria [4], the number of extreme typhoon events selected from the base period (1979–2003), near future (2015–2039), and end of the century (2075–2099) were 88, 82, and 81, respectively. The high-resolution dynamical downscaling data were used as the input for a SOBEK river channel routing model to simulate changes in the river flow rate under climate change. Results were further compared with the design flow rate, as well as recorded river water levels of the most severe typhoon events in history, to evaluate the risk of river flooding under climate change.

2. Literature Review

The GCM is the main tool for simulating future climate conditions; however, it has a relatively low resolution (approximately 200–500 km) [5], which is inadequate for detailed assessments of land surface processes and climate change effects at local to regional scales, particularly in regions with varied topography [6–8]. Chen *et al.* [9] observed that the GCM has been widely applied in simulating future climate scenarios; however, GCM data have a relatively low spatial resolution and cannot be used for detailed discussions on climate scenarios for small areas. Present-day regional climate models (RCM) are most often used for simulating the climate of more local spatial regions. Over the past few decades, dynamical downscaling has mainly been performed using high-resolution GCM or RCM data, with a spatial resolution less than 100 km. Recently, a high-resolution atmospheric GCM model with a resolution of approximately 20 km was developed by the Meteorological Research Institute (MRI) of Japan (hereafter, MRI-AGCM) [10] to include explicitly simulated extreme weather events, such as tropical storms and meso-scale systems, in long-term climate simulations. Although the MRI-AGCM showed marked improvements in simulating extreme precipitation events, the details of local rainfall over complex terrain may still be difficult to simulate. However, a 20-km resolution remains insufficient for describing the local weather and climate characteristics in some areas of Taiwan because of the complex terrain.

In recent years, hydrologic and hydraulic models such as Hydrologic Engineering Centers River Analysis System, Mike-11, SOBEK, and the integrated flood analysis system (IFAS) have been applied to predict potential disasters by using future climate data. Linde *et al.* [11] used a SOBEK model to simulate low-probability flood-peak events in the Rhine basin. The results showed a basin-wide increase of 8%–17% in the peak discharge of the Rhine basin in 2050 for probabilities between 1/10 and 1/1250. Kimura *et al.* [12] applied the IFAS to simulate the peak discharges in Tsengwen reservoir watershed in Taiwan from extreme rainfall events (TP1–10) during three periods: the present (1979–2003), near future (2015–2039), and future (2075–2099). The peak discharges during the future climate change period were higher than those during the present climate change period. Lenderink [13] discussed the discharge of the Rhine during future climate change by investigating two periods: the present (1960–1989) and future (2070–2099). A Rhineflow method was employed to simulate discharges for the UK Met Office RCM HadRM3H [14–16]. The mean discharge in the present (1960–1989) and future (2070–2099) climate change periods increased by approximately 30% in winter and decreased by approximately 40% in summer. This model estimated the effect of climate change on river discharges. Climate data such as temperature, precipitation, and evapotranspiration were used as inputs for the hydrologic and hydraulic models of the river basin; the outputs were for typical river discharge structures [13,17].

Previous studies have rarely focused on hydrological changes in Taiwan because of the low resolution of GCM data. Taiwan can currently generate its own high-resolution data for future climate scenarios, which were employed in the present study for hydraulic and hydrologic routing to project future flow rates under climate change. This study directly compared the river flow and water level determined through hydraulic and hydrologic routing. In addition, the risk of flood protection facilities under climate change was evaluated.

3. Research Methodology

This study was aimed at quantifying the effects of climate change in the Tsengwen catchment area. A flowchart of the research process employed in this study is shown in Figure 1. The first stage focused on introducing related climate change data. For the second stage, the major focus was developing a hydrodynamic model, including its calibration and validation. In the final stage, river discharge changes and river bank overtopping-frequency results were evaluated.

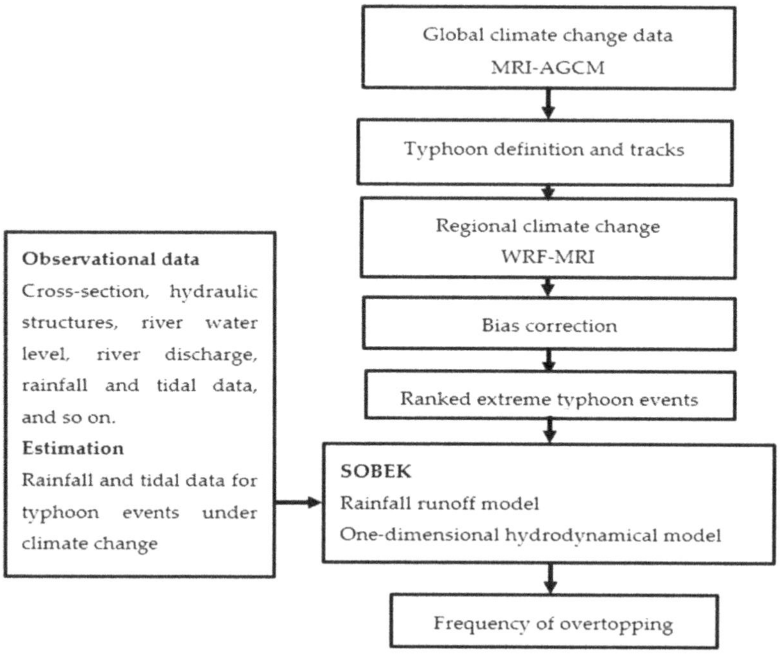

Figure 1. Conceptual scheme of the evaluation of the effects of climate change on river flow and water level.

3.1. SOBEK Model

The SOBEK model, which is in the SOBEK modeling suite developed by Deltares (formerly WL Delft Hydraulics), the Netherlands, integrates the commercial hydrologic and hydrodynamic programs of urban drainage systems along with river and regional drainages. The present study used the SOBEK channel-flow (CF) module along with the rainfall-runoff (RR) module for river channel simulations. The estimated RR volume was calculated as the lateral inflow (node) that converges in the main stream when calculating the unsteady flow of the river channel [18].

3.1.1. Rainfall Runoff

The SOBEK model incorporates the Sacramento RR model for simulating the process of rainfall forming runoff, including evaporation, infiltration, subsurface runoff, and underground water. The concept is to convert effective rainfall at the surface through a unit hydrograph into surface runoff, and to then add soil surface moisture, intermediate flow, and ground water discharge (base flow) to obtain the total runoff [19]. The Sacramento model defines a mathematical equation that accounts for each process in the transformation of rainfall into outflow toward a river. The concept of the Sacramento model and its parameters are shown in Figure 2. According to the Sacramento model, the soil column is divided into two soil zones: upper and lower [18]. The model has 17 parameters, the values of which must be specified [18]. Table 1 lists all the Sacramento parameters [19] and parameter ranges used in this study [20].

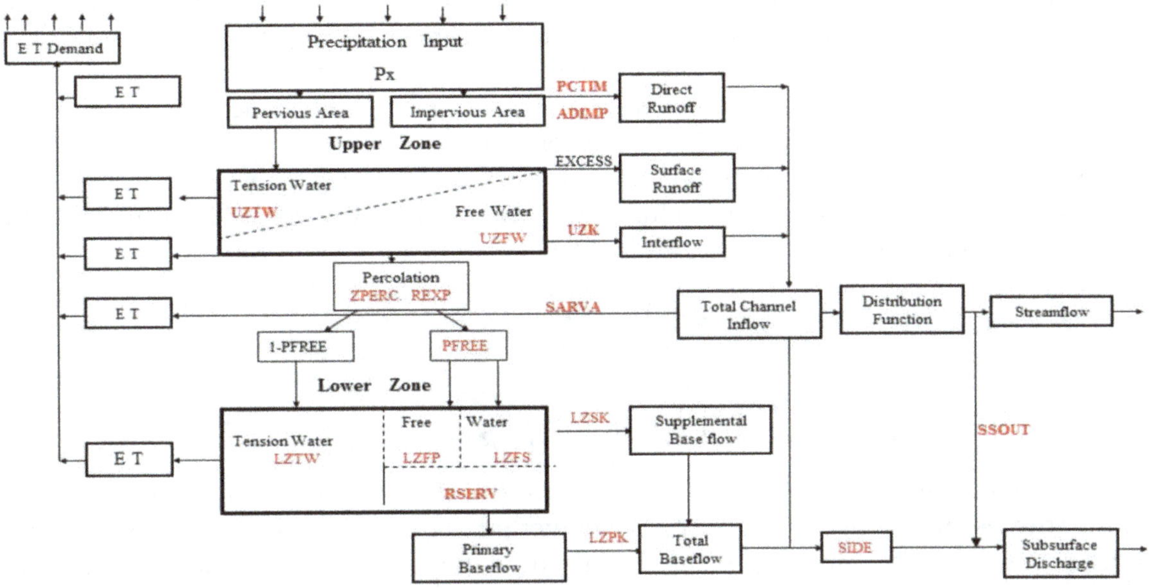

Figure 2. Conceptualization of the Sacramento model and parameters.

Table 1. Sacramento model parameters and their allowable ranges.

Parameters	Description	Allowable Range
UZTWM	Capacity of the upper tension water zone (mm)	250–300
UZFWM	Capacity of the upper free water zone (mm)	240–300
UZK	Upper zone lateral drainage rate (fraction of contents per day)	0.2
PCTIM	Permanent impervious fraction of the segment contiguous with stream channels	0.02
ADIMP	Additional impervious fraction when all tension water requirements are met	0.3–0.5
SARVA	Fraction of the segment covered by streams, lakes, and riparian vegetation	0.01
ZPERC	Proportional increase in the percolation under saturated to dry conditions in the lower zone	10–20
REXP	Exponent in the percolation equation, for determining the rate at which percolation demand changes from dry to wet conditions	1.5–2.5
LZTW	Capacity of the lower zone tension water storage (mm)	210–330
LZFPM	Capacity of the lower zone primary free water storage (mm)	230–450
LZFSM	Capacity of the lower zone supplemental free water storage (mm)	200–340
LZPK	Drainage rate of the lower zone primary free water storage (fraction of contents per day)	0.004–0.04
LZSK	Drainage rate of the lower zone supplemental free water storage (fraction of contents per day)	0.06–0.14
PFREE	Fraction of percolated water that drains directly to the lower zone free water storage	0.2
RSERV	Fraction of the lower zone free water storage that is unavailable for transpiration purposes	0.3
SIDE	Ratio of the unobserved to observed base flow	0
SSOUT	Fixed rate of discharge lost during the total CF (mm/t)	0

3.1.2. River Hydraulics

River flood routing is based on the dynamic wave transfer theory for one-dimensional (1D) varied flow; that is, de Saint Venant's gradually varied flow equation for describing water flow in rivers. This study used the nonlinear implicit difference method for calculating the depth and flow rate for each period. Water depth and flow rate at each cross-section point where main and branch streams converge were determined on the basis of conditions that the main and branch streams have the same water level, and inflow equals outflow. Equations of continuity (1) and motion (2) were considered for flood routing on the basis of de St. Venant's 1D gradually varied flow equation, which is the dynamic wave model. River simulations included the simulation of bridges, reservoirs, and cross-river structures such as weirs, culverts, orifices, and pump stations.

$$\frac{\partial A_f}{\partial t} + \frac{\partial Q}{\partial x} = q_{lat} \tag{1}$$

$$\frac{\partial Q}{\partial t} + \frac{\partial}{\partial x}\left(\frac{Q^2}{A_f}\right) + gA_f\frac{\partial h}{\partial x} + \frac{gQ\,|Q|}{C^2RA_f} - w_f\frac{\tau_{wind}}{\rho_w} = 0 \tag{2}$$

where Q is the discharge (m^3/s), h is the water depth (m), R is the hydraulic radius (m), q_{lat} is the lateral discharge per unit length (m^2/s), A_f is the wetted area (m^2), w_f is the flow width (m), τ_{wind} is the wind shear stress ($N = m^2$), ρ_w is the density of water (kg/m^3), t denotes time (s), x refers to distance (m), and g denotes acceleration due to gravity (m/s^2) ($\approx$9.81).

When the SOBEK CF module processes the equation of motion, the influence of wind shear is considered, and wind force and direction are set to a fixed value or time sequence. Moreover, the SOBEK model can account for the influence of wind on the water level, which is not considered in the urban drainage and flood model. When considering the lateral inflow of a unit length of a river, including culverts, pumps, and weirs, the flow rate can be computed using the stage–discharge relationship of the hydraulic structures.

3.2. Indicators for Model Error Analysis

To validate the simulation model, the simulated and observed water level values were compared, and three statistical indices, namely, the coefficient of efficiency (CE), error of peak water level (EL_P), and error of the time to peak (E_{TP}), were calculated. The three indices are computed as follows:

$$CE = 1 - \frac{\sum_{i=1}^{n}\left(L_{obs} - L_{est}\right)^2}{\sum_{i=1}^{n}\left(L_{obs} - \overline{L}_{obs}\right)^2} \tag{3}$$

$$EL_P = \frac{L_{est} - L_{obs}}{L_{obs}} \tag{4}$$

$$E_{TP} = T_{est} - T_{obs} \tag{5}$$

where L_{est} denotes the estimated flood discharge (cm), L_{obs} represents the observed flood discharge (cm), and $\overline{L}_{obs}$ is the mean value of the observed flood discharge (cm); L_{est} and L_{obs} are the observed and estimated peak water levels of the flood, respectively; and T_{est} and T_{obs} denote the estimated and observed time to peak discharges, respectively.

3.3. Study Area

We selected the Tsengwen River basin as the study area, which covers an area of approximately 1176.7 km^2. The Tsengwen river basin is complex; the mountains are over 3000 m high, and the valley is narrower than 20 km. The average annual rainfall received by the drainage basin is 2643 mm. The Tsengwen River basin comprises the Tsengwen, Nanhua, and Wu Shantou Reservoirs. The Tsengwen Reservoir is located upstream of the Tsengwen Creek, and is the largest reservoir in

Taiwan and the major source of water supply for downstream irrigation systems in Chiayi and Tainan Counties. The Tsengwen Reservoir has a large net water storage capacity (approximately 0.5 billion m^3). The mean annual inflow to the reservoir is approximately 1.1 billion m^3 [21]. The Tsengwen River Basin includes the Tsengwen River main stream, Cailiao River, Guantian River, and Houjue Creek. The location of the Tsengwen River Basin is shown in Figure 3.

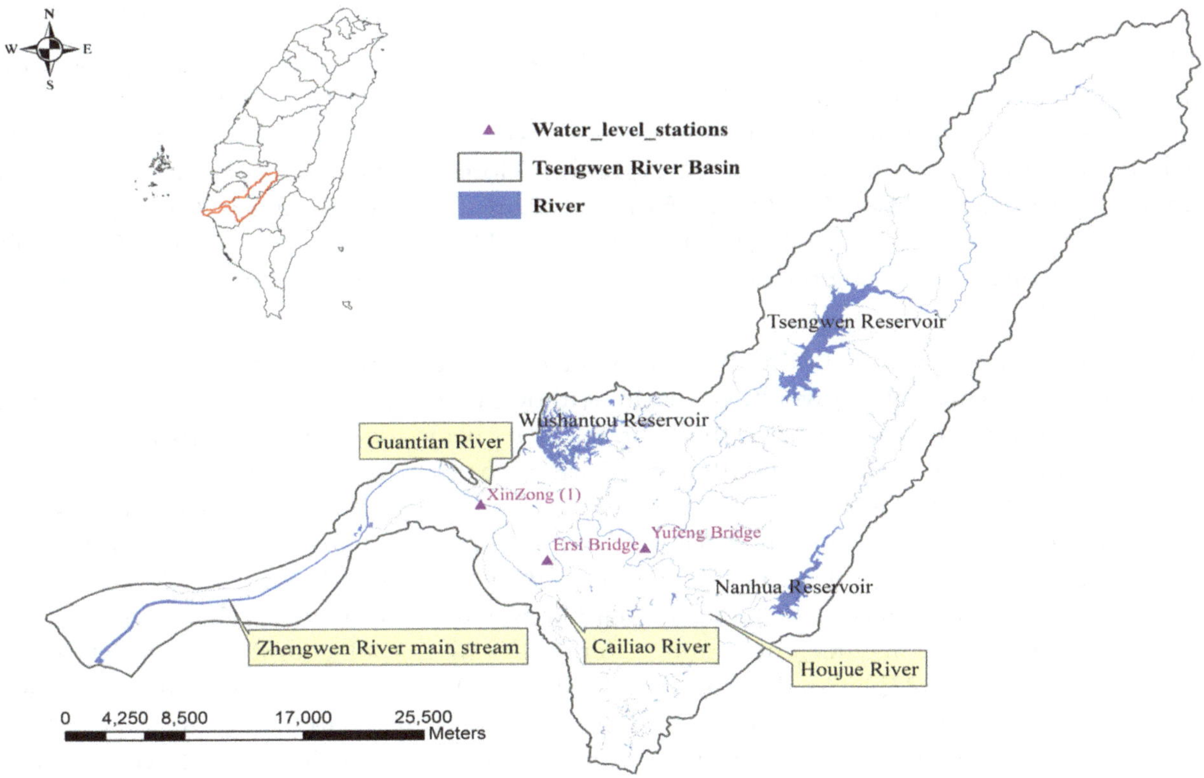

Figure 3. Tsengwen River Basin.

3.4. Hydrologic and Geomorphic Data

Hydrologic and geomorphic data must be collected before simulating the river flow rate and water depth for different scenarios. River cross-sections, hydraulic structures, rainfall in future climate, land use, river flow, and water level data are the basic data.

Rainfall data include observations from historical typhoon events and simulated rainfall data of extreme typhoon events under future climate change. The water level data include those of gauging stations along the river and tidal stations near the estuary (120°06′43″ E, 23°01′25″ N). Hourly water level data from current gauging stations were collected to validate hydraulic routing. Tide levels at the estuary were considered downstream boundary conditions in the model. The finite volume coastal ocean model (FVCOM) was employed to project changes in the astronomical tide at the estuary under a future climate change scenario. Chen and Liu [22] provided a detailed description of the FVCOM structure and parameters. River cross-sectional data of 2010 were provided by the projects of the WRA's Sixth River Management Office and Water Resources Planning Institute, and these include data of the cross-sections of the Tsengwen main stream, Cailiao River, Guantian River, and Houku Creek. The reservoir data include data of the Tsengwen, Nanhua, and Wushantou Reservoirs. The SOBEK model is based on reservoir operations [23–25], in which settings for the reservoir include reservoir area, volume, and contributing area, and settings for the dam include spillway, water gate, power plant discharge, and emergency spillway. Discharge functions were set according to reservoir operation rules.

This study used the atmospheric general circulation model MRI-AGCM; the climate during three periods over a total of 75 years was simulated: base period (1979–2003), near future (2015–2039), and end of the century (2075–2099), developed by Japan Meteorological Agency (JMA) and Meteorological Research Institute (MRI); and ECHAM5, the climate model developed by the German research institute MPI, for climate projections. Simulation results were used as the initial field and boundary conditions for dynamical downscaling in the WRF modeling system, which was developed by the U.S. National Center for Atmospheric Research (please refer to the report of TCCIP (2010) for details).

This study adopted the high-resolution MRI-AGCM (20 km) to define typhoon events. MRI-AGCM revises the definition of typhoon provided by Vitart *et al.* [4], and uses the conditions of 850-hpa vorticity, sea-level pressure, presence or absence of warm-core structure near the typhoon center, and maximum local thickness to detect typhoons. Moreover, the wind speed at the bottom layer of typhoons must reach at least 17 m/s for 1.5 days or more. The process of selecting typhoon events can be divided into two steps: screening typhoon events and tracing typhoon routes [4]. The number of typhoon events determined using the aforementioned definition for the three periods and MRI-AGCM are 88, 82, and 81.

We ranked extreme typhoon rainfall events from each of the three 25-year periods based on the total rainfall over 24 h in the Tsengwen River basin. Although the TCCIP (2/3) reported that projections must be revised, this study bias-corrected the rainfall data by using the cumulative distribution function model [26] for the extreme typhoon rainfall events during the three periods. Figure 4 shows the average rainfall of the TOP1–20 events during the base period, near future, and end of the century. Moreover, we observed that the rainfall of the typhoon events at the end of the century (2075–2099) is higher than that of the base period and near future. Table 2 shows the statistical values for the TOP1–20 extreme events during the three periods. The Central Weather Bureau of Taiwan defines 24-h accumulated rainfall of 250 mm as extremely torrential rain. After observing rainfall characteristics that resulted in floods in Taiwan, Yu *et al.* [27] defined 3-h accumulated rainfall of 130 mm as short-duration disastrous rain. Table 1 shows that the TOP1–2 events in the base period, TOP1–5 in the future and the TOP1–12 events at the end of the century are extremely torrential rain events. The TOP2 event in the near future and the TOP1-3, TOP5, and TOP6 events at the end of the century are short-duration disastrous rainfall events.

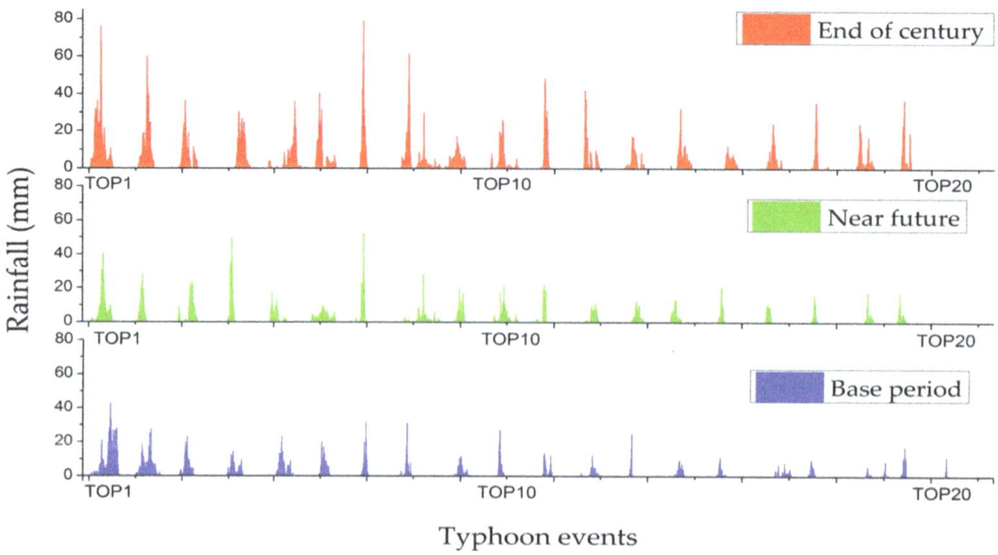

Figure 4. Rainfall hyetograph for the TOP1-20 extreme typhoon rainfall events during the three periods.

Table 2. Precipitation analysis of extreme events.

Typhoon Events	Base Period (1979–2003)				Near Future (2015–2039)				End of This Century (2075–2099)			
No.	(1)	(2)	(3)	(4)	(1)	(2)	(3)	(4)	(1)	(2)	(3)	(4)
Top1	851.4	108.6	491.3	120	548.2	111.7	420.3	90	1027.8	191.7	722.1	48
Top2	505.9	63	268.4	90	370	132.3	367.8	78	738.2	160.6	572.5	66
Top3	298.6	57.1	247.3	90	304.9	68.8	296.3	66	551.1	200.4	549.5	36
Top4	295.7	58.7	223.1	60	344.3	64.2	295.8	60	534.6	84.8	433.2	132
Top5	248.8	55.7	222	66	288.3	117.0	280.9	67	677.0	154.3	430.1	48
Top6	194.8	73.5	192.6	78	227	39.6	197.2	90	407.5	135.9	404.9	66
Top7	174.8	75.7	164.4	42	197.7	47.0	183.3	42	503.0	94.2	394.7	102
Top8	149.1	68	147.6	42	171.9	55.5	162.7	72	475.7	91.2	366.6	48
Top9	242.5	36.3	141.8	48	167.5	40.3	158.2	72	484.8	97.2	359.9	42
Top10	132.4	117.1	131.8	96	218.1	28.2	149.9	48	344.9	63.9	283.4	48
Top11	153.4	31.0	123.0	54	156.9	33.7	145.9	48	338.3	92.0	268.2	54
Top12	112.9	106.0	112.9	48	295.2	27.1	145.7	150	328.0	82.2	250.8	42
Top13	104.3	89.6	98.7	72	147.9	62.7	143.3	72	258.0	61.9	219.6	66
Top14	95.6	95.6	95.6	108	225.4	32.7	133.4	42	259.7	79.0	203.6	42
Top15	92.1	90.6	92.1	84	122.2	56.7	121.8	66	201.1	94.6	195.4	54
Top16	88.6	88.6	88.6	24	120.8	28.7	120.2	84	239.0	43.6	191.8	72
Top17	85.7	77.7	85.7	30	112.2	37.3	106.8	60	306.8	42.7	186.1	60
Top18	91.8	52.7	85.3	54	110.9	37.3	106.8	78	290.1	57.3	173.7	30
Top19	75.2	75.2	75.2	48	98.5	42.6	98.1	30	190.8	28.6	147.1	48
Top20	67.5	34.8	37.0	138	76.5	23.6	74.8	42	68.7	21.5	66.8	78
Typhoon Morakot	1007.5	144.3	636.2	72	–	–	–	–	–	–	–	–

(1) Total precipitation (mm); (2) Maximum of 3 h; (3) Maximum of 24 h; (4) Total duration (h).

In 2009, typhoon Morakot induced long-duration continuous rainfall, and the total rainfall received was approximately equal to the rainfall received during the most extreme typhoon event at the end of the century. However, the maximum rainfall induced by typhoon Morakot after continuously raining for 3 and 24 h was lower than the rainfall induced by the top few extreme typhoon events at the end of the century. In other words, rainfall distribution during extreme typhoon events during the future climate will be high over a short period.

Because the weather research and forecasting (WRF) climate data are grid data, this study collected the rainfall data from WRF grid points of the WRA's rainfall stations nearby, and used the data as input for the SOBEK model. The location of rainfall stations and WRF grid points are shown in Figure 5. Because of the historical rainfall data length and data acquisition constraints, this study selected 11 rainfall stations: MUZHA, TSOCHEN, BEILIAO, CHIKULAOS, BEILIAOS, SHANHUA, YUTEN, NANXI, WANGYEGONG, ZHENGWEN, and BIAOHU.

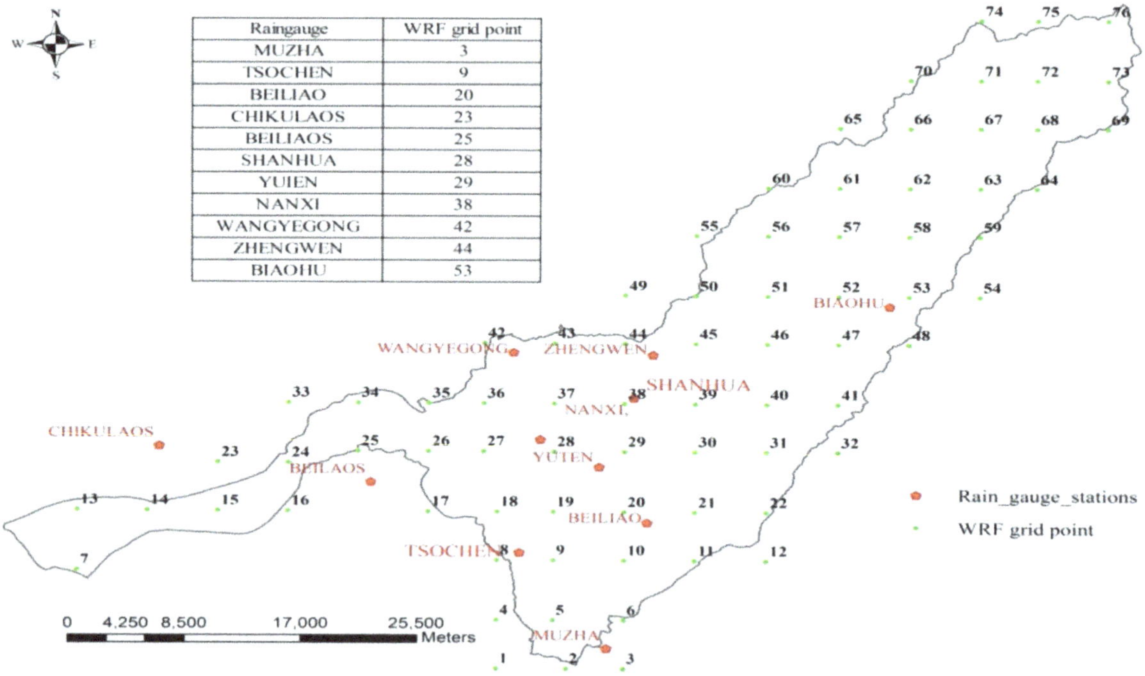

Figure 5. Map of the Tsengwen River Basin rainfall stations and WRF grid points.

4. Case Analysis

This study used 88 extreme typhoon rainfall events for the base period, 81 for the near future, and 82 for the end of the century periods. The data were used as inputs for the SOBEK routing model, which is used for simulating changes in the river flow rate during future climate change.

4.1. River Hydraulic Structure Impact Assessment

Common hydraulic structures in rivers include weirs, piers, dams, embankments, and groundsills. When a river channel requires hydraulic structures, which can be for various purposes (flood disaster prevention or hydraulic design), the flow rate and flood stage of the river channel must first be estimated to protect the hydraulic structures as well as the lives and assets of residents.

Conventional river flood prevention plans incorporate the concept of a return period when considering risk [28]; the design standard of river flood prevention facilities in Taiwan considers return periods of 50, 100, or 200 years. Hydrological data used for return period analysis are obtained through statistical analysis of historical data (20–60 years). During flood prevention facility planning and designing, the flow rate is projected on the basis of the hydrological data of the return period and geomorphic data of the river channel, along with a safety factor to reduce the uncertainty. The projected flood stage is calculated using a hydraulic model test or 1D hydraulic model based on the river's physical characteristics. Table 3 shows the design flow rate and flood stage at XinZong (1), Erxi Bridge, and Yufeng Bridge, as well as the highest water level observed in the past.

Table 3. Design discharge and water level.

Gauge Station	Return Period (Years)	Design Discharge (cm)	Design Stage (m)	* Historical Maximum Stage (m)
XinZong (1)		9890	15.71	18.36
Erxi Bridge	100	8740	21.37	23.56
Yufeng Bridge		6900	46.06	46.98

* occurred during typhoon Morakot.

4.2. Model Calibration and Validation

This study used the data of the rainfall for Typhoon Kalmaegi (2008) and Typhoon Morakot (2009) to calibrate the SOBEK model parameters and rainfall of 0610 torrential rains to validate the model parameters. Figures 6–8 compare the water levels measured at the XinZong Bridge No. 1 station in the Tsengwen River basin by using the SOBEK model. The figures show that the SOBEK simulations match the measured water levels.

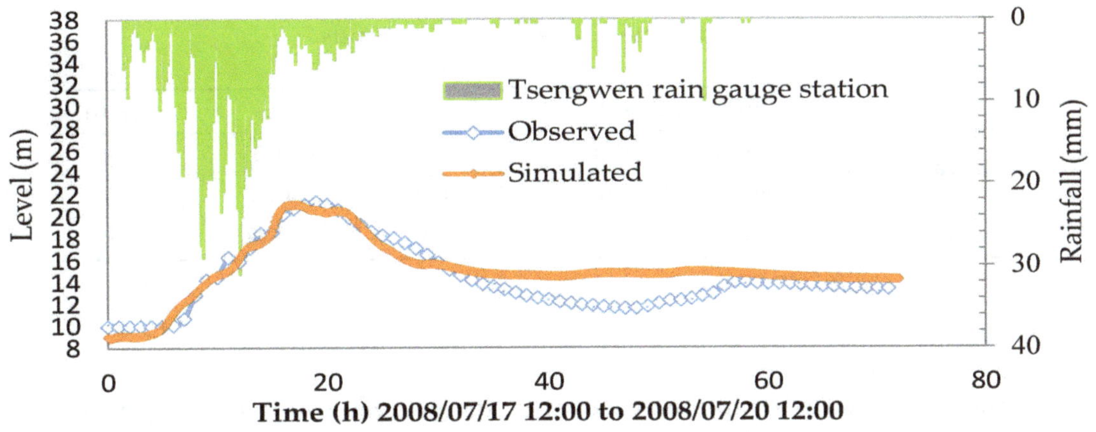

Figure 6. Comparison between the estimated and observed results of the water level at the XinZong (1) water level station during Typhoon Kalmaegi.

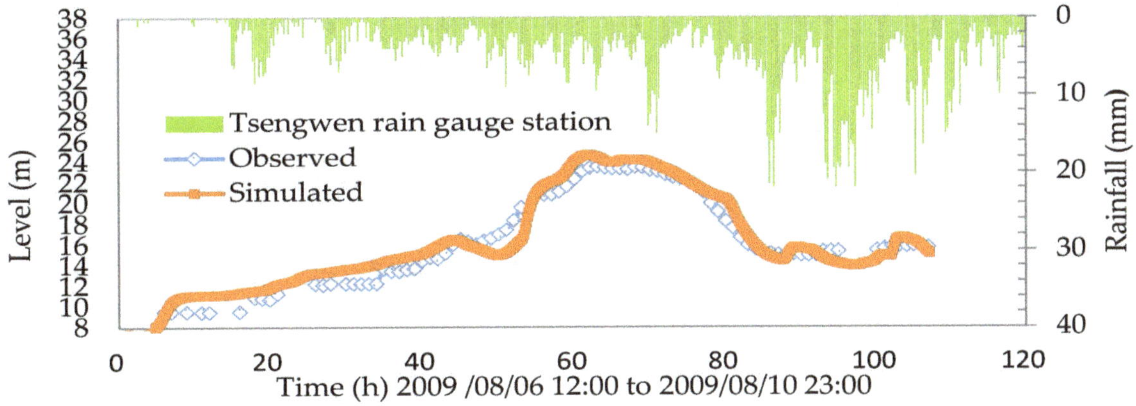

Figure 7. Comparison between the estimated and observed results of the water level at the XinZong (1) water level station during Typhoon Morakot.

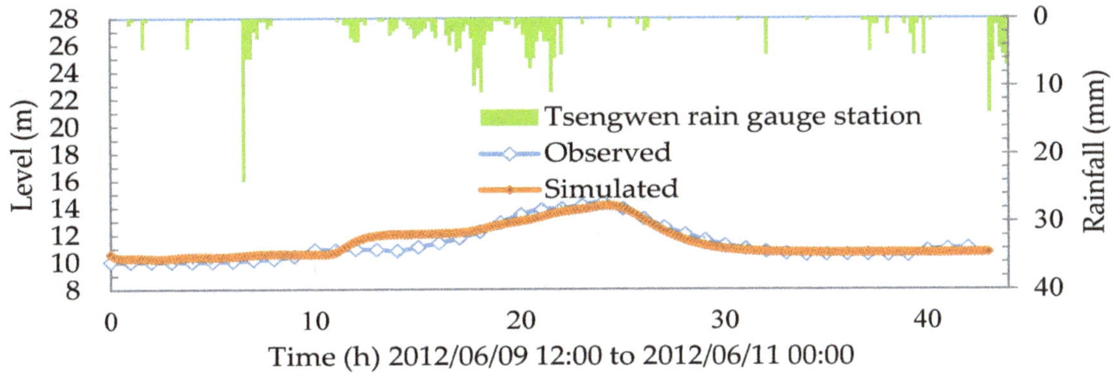

Figure 8. Comparison between the estimated and observed results of the water level at the XinZong (1) water level station for 0610 extreme rain.

To discuss the performance of the model, this study used CE, EL_P, and E_{TP} as a basis for the model validation. A CE approximating 1 indicates that the routing model has a higher goodness of fit, an EL_P greater than 0 indicates that the peak water level projected by the model is higher than the observed peak water level, an EL_P less than 0 indicates that the peak water level projected by the model is lower than the observed peak water level, and a lower E_{TP} indicates that the model is more accurate when projecting the time to peak. The results in Table 4 show that the simulated water level approximated the observed water level.

Table 4. Calibrated and verified results.

Item	Typhoon Events	CE	EL_P	E_{TP} (h)
Calibrated	Kalmaegi(2008)	0.8	−1.31	−2
	Morakot(2009)	0.9	4.75	−1
Verified	0610 Extreme rain (2009)	0.9	−1.66	0

4.3. Simulation Results

Figures 9 and 10 show the simulated discharge and water level hydrograph for the TOP1–20 extreme typhoon events during the base period, near future, and end of the century at the XinZong (1), Erxi Bridge, and Yufeng Bridge. According to these figures, the peak discharge and water level for the end of the century is higher than those for the base period and near future.

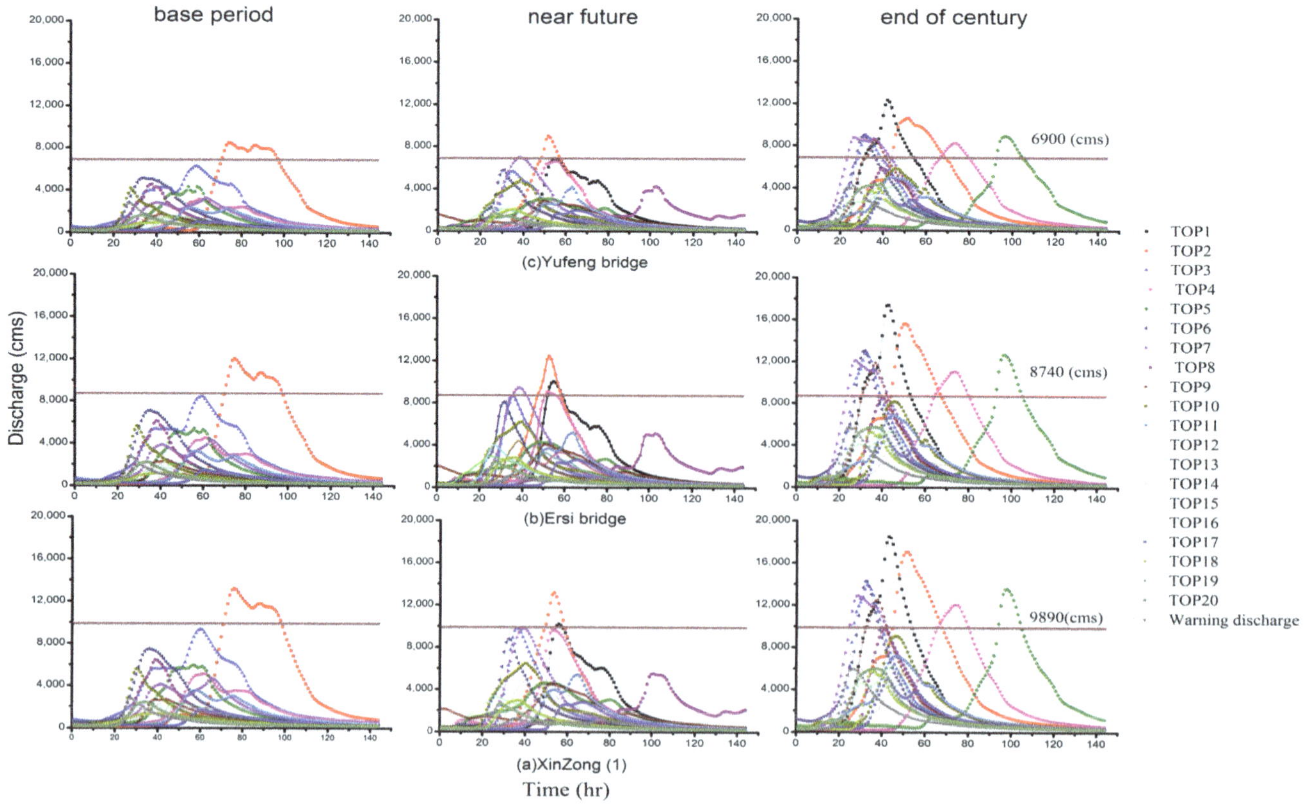

Figure 9. Discharge hydrographs for the TOP1-20 extreme typhoon events during the three periods at (**a**) XinZong (1); (**b**) Erxi Bridge; and (**c**) Yufeng Bridge.

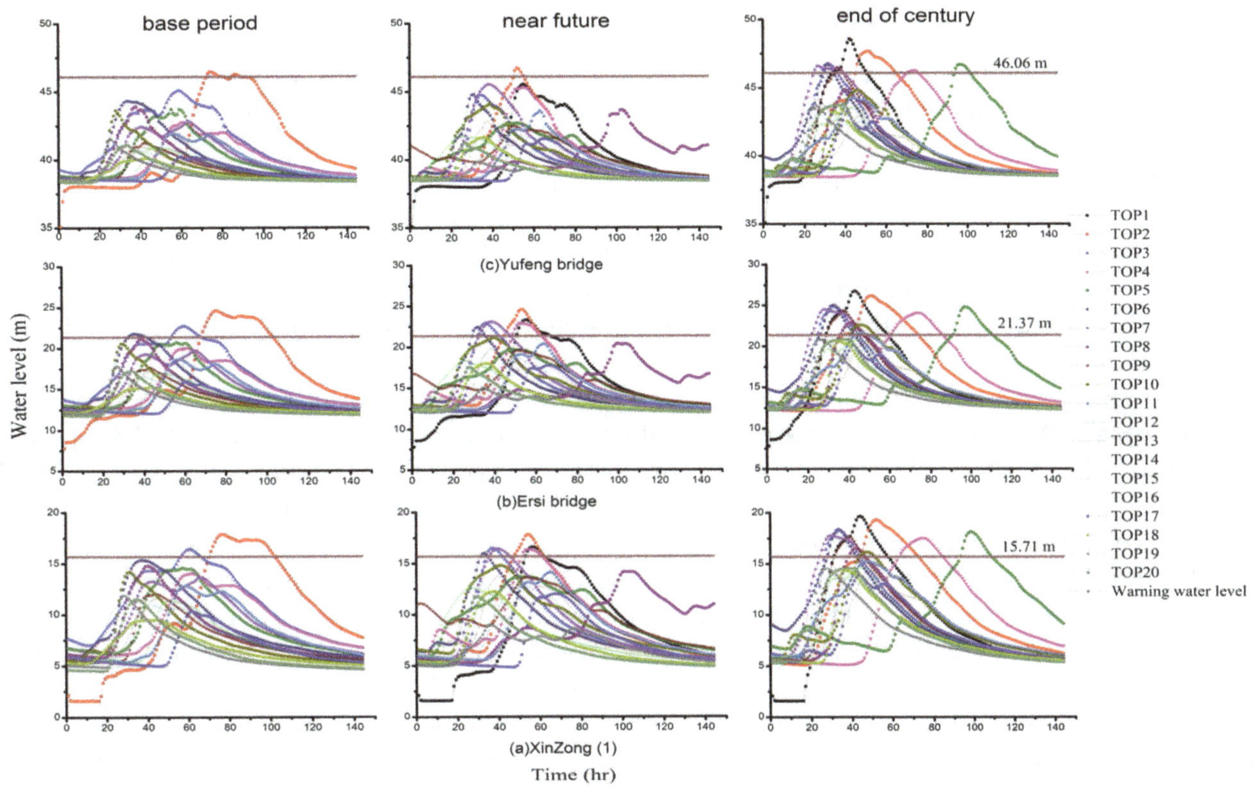

Figure 10. Water level hydrographs for the TOP1–20 extreme typhoon events during three periods at (**a**) XinZong (1); (**b**) Erxi Bridge; and (**c**) Yufeng Bridge.

The design flood stages at XinZong (1), Erxi Bridge, and Yufeng Bridge gauging stations are 15.71, 21.37, and 46.06 m, respectively. Table 5 shows the water levels that exceeded the design values for extreme typhoon events in the base period, near future, and end of the century at XinZong (1), Erxi Bridge, and Yufeng Bridge. The simulated water levels at these three gauging stations exceeded the design values for the water levels in the three periods. In the base period, the peak flows at XinZong 1, Erxi Bridge, and Yufeng Bridge exceeded the management plan flow rate in 2 of 88, 3 of 88, and 1 of 88 events, respectively. For the near future, the corresponding peak flow rates exceeded the design discharge in 6 of 82, 6 of 82, and 1 of 82 events, and at the end of the century, the corresponding flow rates exceeded the flow rate in 10 of 81, 12 of 81, and 8 of 81 events. At the end of the century, extreme peak flow events were forecasted to increase in both frequency and intensity. The simulation results show that the upstream area of the Tsengwen River is already at risk of flooding at the end of the century.

Table 5. Water levels exceeding the design stage during extreme typhoon events.

Water Level Station	Design Water Level (m)		
	Base Period (88)	**Future (82)**	**End of Century (81)**
XinZong (1)	2	6	10
Erxi Bridge	3	6	12
Yufeng Bridge	1	1	8

The highest water levels measured at Erxi Bridge during the TOP1 and TOP2 extreme typhoon events at the end of the century were 24.55 and 24.32 m, respectively, which are higher than the highest water level of 23.56 m during typhoon Morakot. This simulation result indicates that a severe flood could reoccur under climate change.

5. Conclusions and Recommendations

This study used dynamic downscaling data produced by the TCCIP project for river flow rate simulation, and the results highlight the risk of overflow in the Tsengwen River in the future under a climate change scenario.

In 2009, Typhoon Morakot induced continuous rainfall over a long period, and the total rainfall received was lower than that received during the most extreme typhoon events forecasted for the end of the century. Furthermore, the maximum total rainfall received in 3 and 24 h during the top extreme typhoon events at the end of the century was higher than that received at those times during typhoon Morakot, indicating that extreme typhoon events under future climate change will induce strong rainfall over a short period.

Based on the flow rate simulation results, the flow rate at Yufeng Bridge (upstream,) Erxi Bridge (midstream), and XinZong. (1) (downstream) will potentially exceed the management plan at the end of the century. At XinZong (1), the number of times that the flow rate exceeded the management plan rate was 2 in 88 events in the base period, 6 in 82 events in the near future, and 10 in 81 events at the end of the century; that for the end of the century was 5-fold higher than that of the near future and 3-fold higher than that of the base period. At the end of the century, extreme peak flow events will increase in frequency and intensity. Simulation results show that the peak flow rate at the end of the century will be higher than that during Typhoon Morakot. Therefore, a severe flood could reoccur in the future.

In this study, the river cross-section was assumed to be the same when simulating flow rates for future climate change. In future, we will consider the influence of erosion and land use change on the river cross-section when carrying out simulations for future climate change.

Acknowledgments: The authors thank Akio Kitoh of the Meteorological Research Institute, Japan, for providing the MRI data. The authors also thank the Taiwan Central Weather Bureau and Water Resources Agency for providing the observational data. This project was funded by the National Science Council of Taiwan. The authors thank Yuan-Fong Su and Wei-Bo Chen for allowing use of their data in our research.

Author Contributions: Hsiao-Ping Wei executed the model simulations and discussed the results with Keh-Chia Yeh, Jun-Jih Liou, Yung-Ming Chen, and Chao-Tzuen Cheng.

References

1. Intergovernmental Panel on Climate Change (IPCC). *Climate Change 2007: The Physical Science Basis. Contribution of Working Group I to the Fourth Assessment Report of the Intergovermental. Panel on Climate Change*; Cambridge University Press: Cambridge, UK, 2007.
2. Liu, K.F.; Li, H.C.; Hsu, Y.C. Debris flow hazard assessment with numerical Simulation. *Nat. Hazards* **2009**, *49*, 137–161. [CrossRef]
3. National Science and Technology Center for Disaster Reduction (NCDR): Introduction of Taiwan Climate Change Projection and Information Platform Project (TCCIP). *NCDR Rep.* **2010**, *7*, 2010–2054. (In Chinese)
4. Vitart, F.; Anderson, J.L.; Stern, W.F. Simulation of the interannual variability of tropical storm frequency in an ensemble of GCM integrations. *J. Clim.* **1997**, *10*, 745–760. [CrossRef]
5. Lenderink, G.; Buishand, A.; Deursen, W.V. Estimates of future discharges of the river Rhine using two scenario methodologies: Direct *versus* delta approach. *Hydrol. Earth Syst. Sci.* **2007**, *11*, 1145–1159. [CrossRef]
6. Wilby, R.L.; Wigley, T.M.L.; Conway, D.; Jones, P.D.; Hewitson, B.C.; Main, J.; Wilks, D.S. Statistical downscaling of general circulation model output: A Comparison of Methods. *Water. Resour. Res.* **1998**, *34*, 2995–3008. [CrossRef]
7. Wilby, R.L.; Charles, S.P.; Zorita, E.; Timbal, B.; Whetton, P.; Mearns, L.O. *IPCC Task Group on Data and Scenario Support for Impact and Climate Analysis (TGICA). Guidelines for Use of Climate Scenarios Developed from Statistical Downscaling Methods*; IPCC Data Distribution Centre: Geneva, Switzerland, 2004.
8. Tabor, K.; Williams, J.W. Globally downscaled climate projections for assessing the conservation impacts of climate change. *Ecol. Appl.* **2010**, *20*, 554–565. [CrossRef] [PubMed]

9. Chen, H.; Xu, C.Y.; Guo, S. Comparison and evaluation of multiple GCMs, statistical downscaling and hydrological models in the study of climate change impacts on runoff. *J. Hydrol.* **2012**, *434–435*, 36–45. [CrossRef]
10. Mizuta, R.; Yoshimura, H.; Murakami, H.; Matsueda, M.; Endo, H.; Ose, T.; Kamiguchi, K.; Hosaka, M.; Sugi, M.; Yukimoto, S.; *et al.* Climate simulations using the improved MRI-AGCM with 20-km grid. *J. Meteor. Soc.* **2012**, *90A*. [CrossRef]
11. Linde, A.H.T.; Aerts, J.C.J.H.; Bakker, A.M.R.; Kwadijk, J.C.J. Simulating low probability peak discharges for the Rhine basin using resampled climate modeling data. *Water. Resour. Res.* **2010**, *46*, 1–19. [CrossRef]
12. Kimura, N.; Chiang, S.; Wei, H.P.; Su, Y.F.; Chu, J.L.; Cheng, C.T.; Liou, J.J.; Chen, Y.M.; Lin, L.Y. Tsengwen reservoir watershed hydrological flood simulation under global climate change using the 20 km mesh meteorological research institute atmospheric general circulation model (MRI-AGCM). *Terr. Atmos. Ocean. Sci.* **2014**, *25*, 449–461. [CrossRef]
13. Lenderink, G.; Ulden, A.V.B.; van den Hurk, B.; Keller, F. A study on combining global and regional climate model results for generating climate scenarios of temperature and precipitation for the Netherlands. *Clim. Dyn.* **2007**, *29*, 157–176. [CrossRef]
14. Jones, R.; Murphy, J.; Hassell, D.; Taylor, R. Ensemble mean changes in a simulation of the European climate of 2071–2100: Using the New Hadley Centre Regional Climate Modelling System HadAM3H/HadRM3H. Available online: http://prudence.dmi.dk/public/publications/hadley_200208.pdf (accessed on 6 October 2005).
15. Jones, C.G.; Willen, U.; Ullerstig, A.; Hansson, U. The Rossby Centre Regional Atmospheric Climate Model (RCA). Part I: Model Climatology and Performance for the Present Climate over Europe. *Ambio* **2004**, *33*, 199–210. [CrossRef] [PubMed]
16. Giorgi, F.; Bi, X.; Pal, J. Means, trends and interannual variability in a regional climate change experiment over Europe. PartI: Present-Day Climate (1961–1990). *Clim. Dyn.* **2004**, *22*, 733–756. [CrossRef]
17. Linde, A.H.T.; Aerts, J.C.J.H.; Kwadijk, J.C.J. Effectiveness of flood management strategies on peak discharges in the Rhine basin. *J. Flood Risk Manag.* **2010**, *3*, 248–269. [CrossRef]
18. *Delft Hydraulics SOBEK User Manuals*; Deltares: Delft, The Netherlands, 2007.
19. Burnash, R.J.C. The NWS River Forecast System—Catchment model. In *Computer Models of Watershed Hydrology*; Water Resources Publications: Littleton, CO, USA, 1995.
20. Republic of China Ministry of Economic Affairs. *Water Resources Agency (WRA) Calibration of Flood Forecasting Model and Review and Establishment of Warning Stages*; Water Resources Agency report (2/2). Republic of China Ministry of Economic Affairs: Taipei, Taiwan, 2014. (In Chinese)
21. Republic of China Ministry of Economic Affairs. *Water Resources Agency (WRA) Operation Directions for Nanhua Reservoir*; (2011 updated version). Republic of China Ministry of Economic Affairs: Taipei, Taiwan, 2004. (In Chinese)
22. Chen, W.B.; Liu, W.C. Modeling Flood Inundation Induced by River Flow and Storm Surges over a River Basin. *Water* **2014**, *6*, 3182–3199. [CrossRef]
23. Republic of China Ministry of Economic Affairs. *Water Resources Agency (WRA) Operation Directions for Tsengwen Reservoir*; (2014 updated version). Republic of China Ministry of Economic Affairs: Taipei, Taiwan, 2002. (In Chinese)
24. Republic of China Ministry of Economic Affairs. *Water Resources Agency (WRA) Operation Directions for Wusanto Reservoir*; (2011 updated version). Republic of China Ministry of Economic Affairs: Taipei, Taiwan, 2008. (In Chinese)
25. Tang, Y.; Reed, P.M.; Wagener, T. How effective and efficient are multiobjective evolutionary algorithms at hydrologic model calibration? *Hydrol. Earth Syst.* **2006**, *10*, 289–307. [CrossRef]
26. Su, Y.F.; Cheng, C.T.; Liu, J.J.; Chen, Y.M. Bias correction of MRI-WRF dynamic downscaling datasets. *Terr. Atmos. Ocean. Sci.* **2015**. Submitted.
27. Yu, Y.C.; Lee, T.J.; Kung, C.Y. Disaster warning and scenario analysis in Typhoons and Heavy Rainfall events. *NCDR Rep.* **2014**, *2014*, 1–15. (In Chinese)
28. Johnson, F.; Sharma, A. Accounting for interannual variability: A Comparison of Options for Water Resources Climate Change Impact Assessments. *Water Resour. Res.* **2011**, *47*. [CrossRef]

9

Multi-Basin Modelling of Future Hydrological Fluxes in the Indian Subcontinent

Ilias G. Pechlivanidis [1,*], Jonas Olsson [1], Thomas Bosshard [1], Devesh Sharma [2] and K.C. Sharma [2]

[1] Swedish Meteorological and Hydrological Institute, Norrköping 60176, Sweden; jonas.olsson@smhi.se (J.O.); thomas.bosshard@smhi.se (T.B.)

[2] Department of Environmental Science, Central University of Rajasthan, Kishangarh, Dist-Ajmer, Rajasthan 305817, India; devesh.water@gmail.com (D.S.); kcsharma@curaj.ac.in (K.C.S.)

* Correspondence: ilias.pechlivanidis@smhi.se.

Academic Editor: Yingkui Li

Abstract: The impact of climate change on the hydro-climatology of the Indian subcontinent is investigated by comparing statistics of current and projected future fluxes resulting from three RCP scenarios (RCP2.6, RCP4.5, and RCP8.5). Climate projections from the CORDEX-South Asia framework have been bias-corrected using the Distribution-Based Scaling (DBS) method and used to force the HYPE hydrological model to generate projections of evapotranspiration, runoff, soil moisture deficit, snow depth, and applied irrigation water to soil. We also assess the changes in the annual cycles in three major rivers located in different hydro-climatic regions. Results show that conclusions can be influenced by uncertainty in the RCP scenarios. Future scenarios project a gradual increase in temperature (up to 7 °C on average), whilst changes (both increase and decrease) in the long-term average precipitation and evapotranspiration are more severe at the end of the century. The potential change (increase and decrease) in runoff could reach 100% depending on the region and time horizon. Analysis of annual cycles for three selected regions showed that changes in discharge and evapotranspiration due to climate change vary between seasons, whereas the magnitude of change is dependent on the region's hydro-climatic gradient. Irrigation needs and the snow depth in the Himalayas are also affected.

Keywords: multi-basin modelling; HYPE; climate change impacts; India; CORDEX; DBS

---

1. Introduction

Climate change impacts can be particularly complex in regions which are additionally subject to other environmental and socio-economic changes, *i.e.*, population growth, urbanization, land use change, and change in industrial and hydropower sectors [1]. India is a developing country with nearly two-thirds of the population depending directly on the climate- and water-sensitive sectors. The country already faces high risks of water shortages due to population growth, urbanization, and increasing demands in the agricultural, industrial, and hydropower sectors; hence, India offers a unique opportunity to examine the impacts of climate change, which in some areas have already been observed [2–4]. The region is characterised by a strong hydro-climatic gradient due to the monsoon (tropical climatic regions in the south; temperate and alpine regions in the Himalayan north, where elevated areas receive sustained winter snowfall) and the geographic features; hence, posing extraordinary scientific challenges to understand, quantify, and predict future availability of water resources. Of particular interest are the Northern Indian Himalayan plains given the sensitivity of snow and glacier melt processes to climate variability and change [5–7]. Arid and semi-arid regions might also experience changes in their hydrological cycle [8,9].

Assessment of future climate change impacts on water resources commonly involves climate variables (*i.e.* precipitation, temperature) from global circulation models (GCMs) in combination with hydrological models [10,11]. GCMs demonstrate significant skill at the continental and hemispheric spatial scales and incorporate a large proportion of the complexity of the global system; however, they are inherently unable to represent local basin-scale features and dynamics [12]. To narrow the gap between GCMs' abilities and hydrological needs, regional climate models (RCMs) have been developed to downscale the GCM output and, thus, provide high-resolution meteorological inputs to hydrological models. To improve the confidence in regional trends of hydro-climatic key variables and increase robustness in hydrological long term predictions, the World Climate Research Programme (WCRP) has recently launched a framework, called COordinated Regional climate Downscaling EXperiment (CORDEX), to generate and evaluate fine-scale ensembles of regional climate projections for all continents globally [13]. CORDEX has several domains that are defined as regions for which the regional downscaling is taking place. In particular, the efforts in the South Asia (SA) domain aim to translate regionally-downscaled climate data into meaningful sustainable development information in the monsoon South Asia area [14]. CORDEX-SA was initiated in 2012 and the RCM outputs have only recently become available.

While RCMs transfer the large-scale information from GCMs to scales which are closer to the basin scale (10–50 km), the output often shows large bias in the magnitude and spatial distribution of precipitation and, to a lesser extent, temperature [15]. RCM data are, therefore, not considered to be directly useful for assessing hydrological impacts at the regional and/or local scale [16]. A way to tackle the problem of RCM misrepresentation is to bias correct the RCM data to make them reproduce historical observed statistics to the degree possible [17]. Different approaches to bias correction have been made, with various complexity [18]. Simpler methods include shifting long-term annual or seasonal means to agree with observations whereas more advanced methods include adjustment of the full frequency distribution. A distribution-based approach is attractive not least for precipitation, for which both bias and future change are generally found to depend on the intensity level [19]. Bias correction often includes an implicit downscaling component, in that higher-resolution reference observations are used when fitting the RCM mapping functions. Bias correction generally preserves the variability described by different climatic conditions generated by RCM projections [20]; however, the RCM may perform differently depending on the season or governing atmospheric circulation. For instance, a typically wet weather regime (e.g., pattern or season) can have a different precipitation distribution in time and space than a dry regime.

Projected hydrologic information is prone to considerable uncertainty/errors at various steps of the modelling chain, *i.e.*, climate projection, bias correction and downscaling techniques, and hydrological simulation [21–23]. These errors can propagate in a very complex way (e.g., magnitude of error could vary both in space and time) which could be misinformative for management decisions [24,25]. A major source of uncertainty, among others, concerns the future emission scenarios, described by the representative concentration pathways (RCP), which further results in different climate projections. [26] showed that towards the end of the 21st century, the emission scenarios (here RCPs) are the dominant source of uncertainty in climate projections. The spatiotemporal variability of water fluxes differs between RCPs, particularly in areas with unique weather systems, *i.e.*, monsoon [27,28]. However, the choice of GCM and RCM may also have a large impact on the results and generally an ensemble of projections—encompassing different GCMs, RCMs, and emission scenarios—is recommended in hydrological climate change impact assessments [29,30].

Conventionally, hydrologic impacts are investigated on small (~0.1–10^2 km^2) or medium-sized basins (~10^2–10^3 km^2); however, current needs require assessment on larger areas and river basins, which requires the use of large scale hydrological models [31,32]. This type of modelling has the potential to encompass many river basins, cross-regional, and international boundaries and represents a number of different geophysical and climatic zones [33]. In addition, according to [34], large scale modelling can balance "depth with breadth", enhance process understanding, increase robustness of generalizations, facilitate catchment classification and regionalization schemes, and support better understanding of prediction uncertainty.

This paper contributes to ongoing efforts on assessing the potential impacts of climate change on water availability in the Indian subcontinent. In particular, we aim to answer the following questions: i) what is the quantified impact of climate change on India's water resources? ii) how is the uncertainty due to RCP propagated in hydrological impact modelling? and iii) how does the potential impact vary in different climatic regions (*i.e.*, tropical, humid subtropical, and montane)? Although previous investigations have, at least to some degree, addressed similar questions, our contribution is associated with three novel features in that we apply: 1) three recently-generated high-resolution CORDEX-SA projections (RCP2.6, RCP4.5, and RCP8.5), 2) the DBS (distribution-based scaling; [35]) method to correct biases in the climate projections, and 3) the large-scale multi-basin HYPE (HYdrological Predictions for the Environment; [36]) hydrological model, to quantify climatic as well as anthropogenic impacts of climate change on hydrology and water availability over the entire Indian subcontinent. Section 2 introduces the study area, whereas the hydrological model and methodology are presented in Section 3. Section 4 presents the results of climate change impacts, followed by a discussion in Section 5 and conclusions in Section 6.

2. Study Area and Data

2.1. Study Area

India is the seventh-largest country by area and the second-most populated country with over 1.2 billion people. The country covers an area of about 3.3 million km^2 and some of its river basins extend into several neighboring countries (*i.e.*, China, Nepal, Pakistan, and Bangladesh; see Figure 1). Major rivers of Himalayan origin that are mainly located in India include the Ganga and the Brahmaputra, both of which drain into the Bay of Bengal. Major peninsular rivers, whose steeper gradients prevent them from flooding, include the Godavari, the Mahanadi, and the Krishna, which also drain into the Bay of Bengal; and the Narmada and the Tapi, which drain into the Arabian Sea. Coastal features include the marshy Rann of Kutch of Western India and the alluvial Sundarbans delta of Eastern India; the latter is shared with Bangladesh.

The spatiotemporal variation in climate is perhaps greater than in any other area of similar size in the world. The climate is strongly influenced by the Himalayas and the Thar Desert in the northwest, both of which drive the summer and winter monsoons [37]. Four seasons can be distinguished: winter (January–February), pre-monsoon (March–May), monsoon (June–September), and post-monsoon (October–December). In terms of spatial variability, the rainfall pattern roughly reflects the different climate regimes of the country, which vary from humid in the northeast (precipitation is 2068 mm/year and occurs about 180 days/year), to arid in Rajasthan (precipitation is 313 mm/year and occurs about 20 days/year). Moreover, India is characterized by strong temperature variations in different seasons ranging from a mean temperature of about 10 °C in winter to about 32 °C in pre-monsoon season.

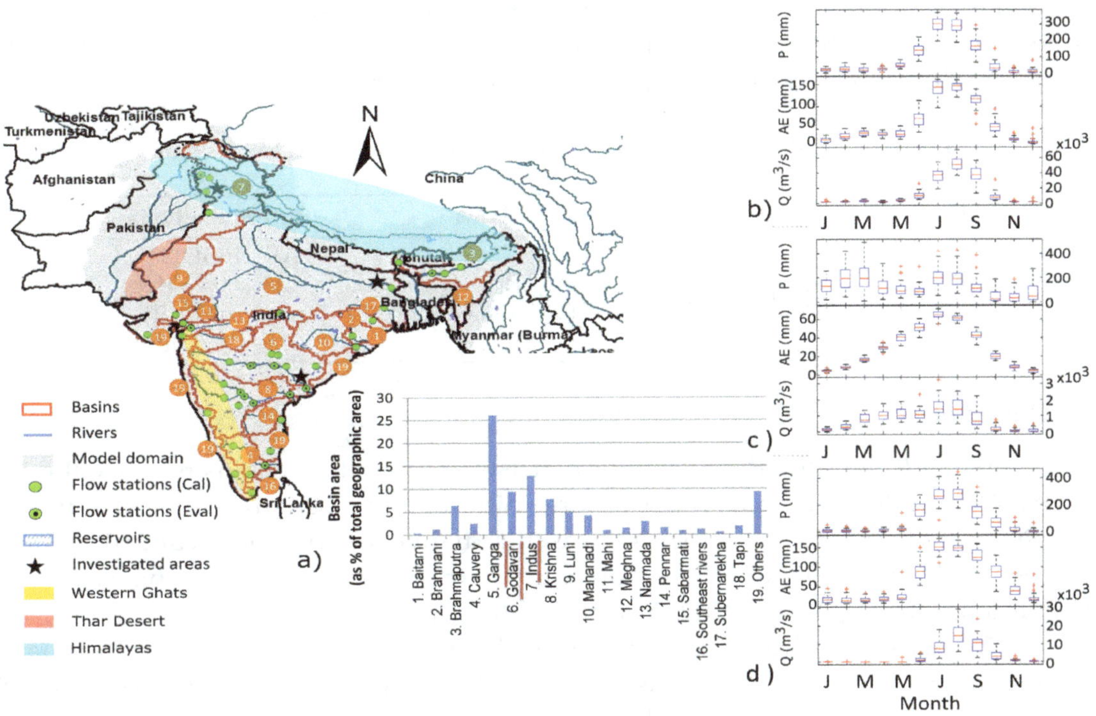

Figure 1. (**a**) The Indian subcontinent (model domain)—the numbers correspond to the basins in which regional analysis is conducted. Mean annual hydro-climatic cycles (precipitation (P), actual evapotranspiration (AE), and discharge (Q)) during the period 1976–2005 for the rivers: (**b**) Ganga (at Farakka station), (**c**) Indus (Chenab at Akhnoor station), and (**d**) Godavari (at Polavaram station). These results are shown from investigation areas in (**a**) with a star.

The monsoon season is very important for water resources (and in turn their use for, e.g., power generation and agriculture) in the country since 75% of the annual rainfall (877 out of 1182 mm) is received in this period [38]. In particular, India's mean monthly rainfall during July (286.5 mm) is highest and constitutes about 24% of the annual total. The contribution in August is slightly lower (~21%) and in June and September ~14% [39]. The contribution of pre-monsoon and post-monsoon rainfall to the annual total is roughly the same (11%). Higher variation is observed during the end of post-monsoon and winter (*i.e.*, November–February). To provide a better understanding of the system behavior, Figure 1b–d shows the annual cycles of the hydro-climatic fluxes in three different climatic regions, *i.e.*, humid subtropical (Ganga), montane (Indus), and tropical (Godavari).

2.2. Spatial Input Data

Data availability is usually a severe constraint in the analysis of large-scale domains. To overcome such a problem, we use global datasets to extract the information required for hydrological applications (see Table 1 in [40]).

2.3. Meteorological Reference Data

Daily precipitation inputs for the period 1971–2005 are obtained from the Asian Precipitation—Highly-Resolved Observational Data Integration Towards Evaluation of Water Resources (APHRODITE) project [41,42] at 0.25° resolution (Table 1). Similarly, AphroTEMP [43] provides daily temperature inputs for the same period at 0.5° resolution. APHRODITE and AphroTEMP (in the following jointly denoted APHRODITE) are the only long-term continental-scale gridded datasets that are based on a dense network of daily data for Asia including the Himalayas. Therefore, the datasets have contributed to studies including among others water resources, climate change analysis, and statistical downscaling [28]. In this study, as a reference period we have chosen the 30-year period 1976–2005. The APHRODITE dataset fully covers this reference period and the period does not overlap with the future climate projections that start in 2006.

Table 1. The three CORDEX-SA climate projections used in this study.

RCP	GCM	RCM	Reference Data
2.6 4.5 8.5	EC-EARTH	RCA4 (0.44 × 0.44 deg)	APHRODITE (0.25 × 0.25 deg) AphroTEMP (0.5 × 0.5 deg)

2.4. Climate Projections

Our ensemble of three climate projections consists of modelling chains that use the same GCM (EC-EARTH; [44]) and RCM (RCA4; [45]), but three different representative concentration pathways, RCPs (see). RCPs are numbered after their increased radiative forcing until year 2100 (+2.6, +4.5, and +8.5 W/m^2, respectively; [46]). Note that more climate projections are becoming available over the South Asian domain through the CORDEX initiative (e.g., [47]).

A total of 129 years of hydrological simulations have been conducted for each climate scenario (1971–2099). However, the analysis is based on three 30-year periods: reference period (1976–2005), mid-century period (2021–2050), and end-century period (2070–2099). Note that in here we only analyze three projections from the CORDEX-SA ensemble; the CORDEX experiment is ongoing and more projections are being generated but these were the only ones available at the time of the study.

3. Methodology

3.1. India-HYPE: Description, Setup, and Calibration

The Hydrological Predictions for the Environment, HYPE, model [36] is a semi-distributed rainfall-runoff model capable of describing the hydrological processes at the basin scale. The model represents processes for snow and ice accumulation and melting, evapotranspiration, soil moisture, discharge generation, groundwater recharge, and routing through rivers and lakes. HYPE simulates the water flow paths in soil, which is divided into three layers with a fluctuating groundwater table. Parameters are linked to physiographical characteristics in the landscape, such as hydrological response units (HRUs) linked to soil type and depths and vegetation. Elevation is used to get temperature variations within a sub-basin for estimating the snow cover dynamics.

Lakes are defined as classes with specified areas and receive the local runoff and the river flow from upstream sub-basins. Precipitation falls directly on lake surfaces and lake water evaporates at the potential rate until the lake is dry. Each lake has a defined depth below an outflow threshold. The outflow from lakes is determined by a general rating curve, unless a specific one is given or if the lake is regulated. Lakes and man-made reservoirs are not separated in the simulation. A simple regulation rule can be used, in which the outflow is constant or follows a seasonal function for water levels above the threshold. A rating curve for the spillways can be used when the reservoir is full.

Irrigation in HYPE is simulated based on crop water demands calculated either with the FAO-56 crop coefficient method [48] or relative to a reference flooding level for submerged crops (e.g., rice). The demands are withdrawn from rivers, lakes, reservoirs, and/or groundwater within and/or external to the sub-basin where the demands originated. The demands are constrained by the water availability at these sources. After subtraction of conveyance losses, the withdrawn water is applied as additional infiltration to the irrigated soils from which the demands originated (in here named as applied irrigation water to soil (AIW)).

The HYPE model is set up for the entire Indian subcontinent (4.9 million km^2) divided into 6010 sub-basins, *i.e.*, with an average size of 810 km^2, and is referred to as India-HYPE. The model runs at a daily time step using APHRODITE as input data, but due to lack of daily discharge observations it was calibrated and evaluated (both in space and time) against monthly observations from 42 stations in the GRDC (Global Runoff Data Centre) database. For the Indian subcontinent, GRDC data are limited to monthly discharge for chosen river basins in the period 1971–1979. More discharge data are held in the Indian government agencies, but are not released to the public domain due to confidentiality; this generally sets a constraining factor for a model setup. Many of the parameters in the model are coupled to soil type or land use, while others are assumed to be general to a larger region. This approach fosters the potential of parameter transferability within reasonably homogeneous regions. In applications of HYPE, we generally consider the parameter identifiability and their regionalization to ungauged regions to be acceptable if the model performs adequately in the gauged basins over the entire model domain.

The HYPE model was spatiotemporally calibrated and evaluated in a multi-basin approach by considering the median performance in selected stations; 30 stations were selected for model calibration and 12 "blind" stations for spatial evaluation. The years 1969–1970 were used as a model warm-up period, the next five years for model calibration (1971–1975) and the final four years for temporal performance evaluation (1976–1979). The model's predictability was tested using various performance measures (*i.e.*, objective functions and flow signatures) and additional data sources (*i.e.*, remote sensing potential evapotranspiration records); see details in [40]. Here, we assess the model's predictability based on the Nash-Sutcliffe Efficiency, NSE [49] and relative error, RE (defined as the difference between the mean modelled value and the mean observed value divided by the mean observed value). The former has been widely applied in hydrology as a benchmark measure of fit. It can also be interpreted as a classic skill score, where skill is interpreted as the comparative ability with regards to a baseline model (here taken to be the "mean of the observations"; *i.e.*, NSE < 0 indicates that the mean of the observed time series provides, on average, a better prediction than the model). NSE ranges between 1 (perfect fit) and $-\infty$ whereas RE ranges between $-\infty$ and $+\infty$, with the "ideal" value being 0.

3.2. *Bias Correction of RCM Data*

The RCM projections (mean daily precipitation and temperature) were bias corrected against the APHRODITE dataset using the distribution based scaling, DBS, statistical method [35]. In brief, DBS aims to map the quantile distributions of precipitation and temperature in the RCM data to those of the reference data. For precipitation, a two-step procedure is applied: 1) correction of the wet-day frequency by applying a wet-day cut-off threshold (in case of wet frequency bias) or by adding wet-days to pre-existing wet-spells (if dry frequency bias), and 2) quantile-mapping of the precipitation data using a double-gamma distribution to accurately represent both normal and extreme precipitation intensity ranges. For temperature, a quantile-mapping correction based on a Gaussian distribution is used. The temperature correction model is dependent on the wet/dry state of the corresponding precipitation. This means that DBS takes into account different biases on wet and dry days (see details in [35]). The bias-corrected projections were used to force the hydrological model for the assessment of climate change impacts on water resources. DBS was used for bias-correcting GCM projections over India (Mumbai region) by [50].

3.3. *Climate Change Impact Assessment*

For the present climate, water availability in space and time was simulated using the reference APHRODITE dataset as input. The same framework is then used to project the impact of climate change on the water resources with the assumption that the land use shall not change over time.

3.3.1. Long-Term Averages

Firstly, we assess the impact of climate change on the hydro-climatic variables, *i.e.*, long-term means of precipitation (P, in mm), temperature (T, in C), actual evapotranspiration (AE, in mm), runoff

(R, in mm), soil moisture deficit (SMD, in mm), snow depth (SD, in cm), and applied irrigation water to soil (AIW, in Mm^3). The daily series for each 30-year period (reference, mid- and end-century) is used to extract the statistics. The relative future change in the long-term average (%) between two periods (mid- or end-century *versus* reference period) due to climate change is estimated for each sub-basin. Positive (negative) change indicates increase (decrease) from the average value in the reference period. Note that for T we calculate absolute differences between future and reference periods; hence, we can express properly also changes in the sign of T.

The spatial variability of change at the basin scale is further summarized (here presented as a boxplot) allowing comparison of the overall change between basins; note that this analysis is only presented for P and R.

3.3.2. Annual Cycles

To complement the above assessment, we investigate the changes in the annual cycles in different locations and governing climatic conditions. We, therefore, compare the annual cycles of P, E, and discharge (Q, in m^3/s) for the Ganga (at Farakka station), Indus (Chenab at Akhnoor station), and Godavari (at Polavaram station) basins (see annual cycles of the present climate in Figure 1b–d). The climate in these basins is characterized as humid subtropical, montane, and tropical, respectively; hence, the analysis can capture the sensitivity to the dominant climatic gradients in the subcontinent.

4. Results

4.1. Model Evaluation

Here, we investigate the model's reliability (which involves measures sensitive to high flows, timing, variability, and volume), and present performances for the benchmark objective functions, *i.e.*, NSE and RE. The NSE and RE for all calibration and evaluation stations and periods are presented in Table 2. Overall, the India-HYPE model achieved an acceptable performance for our purpose and is, therefore, considered adequate to describe the dominant hydrological processes in the region (median NSE and RE for the calibration period is 0.76 and −5.26%, respectively, while for the evaluation period the values are 0.68 and +8.01%, respectively). However, as expected, the performance decreased when the model is validated both in space and time (NSE and RE equal 0.40 and 16.81%, respectively); the model is not "trained" to capture the flow conditions at independent areas and time periods. A more detailed analysis during the evaluation period showed that the model could not fully capture the variability (e.g., standard deviation) of the observed data in the validation stations. The ratio of the standard deviation of modelled over the standard deviation of observed data decreases during the evaluation period at the evaluation stations from 0.78 to 0.58 which consequently affects the NSE values; see the discussion on the decomposition of the NSE and the importance of its decomposed terms (dealing with timing, variability, and volume) on the overall NSE in [51]. However other flow characteristics, *i.e.* timing and volume, are better represented than variability during the evaluation period. Further analysis and discussion on model performance and consistency can be found in [40].

Table 2. Median model performance for calibration and evaluation stations and periods.

Space	Time/Periods	NSE	RE (%)
Cal. (30 stations)	1971–1975	0.76	−5.26
	1976–1979	0.63	−5.43
Eval. (12 stations)	1971–1975	0.68	8.01
	1976–1979	0.40	16.81

4.2. Bias Correction

The effect of bias-correction using the DBS methodology is illustrated here for the case of mean P (Figure 2). The highest P amounts are measured along the mountain ridges, e.g., on the western side of the Western Ghats, Southwest India, and at the foothills of the Himalaya in the northern part of the study region. Those wet regions are strongly associated with atmospheric flows during the monsoon season. Dry regions are located in the northwestern part and on the eastern side of the Western Ghats. The projections employed in this study all share the same historical run, thus Figure 2 is representative for all three projections. The historical projection shows large positive and negative biases in mean P, with sharp gradients between regions of positive and negative biases. The regions with sharp gradients coincide very well with regions of complex topography (e.g., in the Himalaya or on the lee and luv side of the Western Ghats). DBS is able to correct for most of the bias although some remaining dry bias is apparent in regions where the historical projection strongly underestimates mean precipitation. This is related to a dry frequency bias, *i.e.*, too few wet days in the projection. DBS is not able to fully correct for the pronounced deficits in rain days, but can only correct moderate dry frequency bias by adding wet-days to existing wet-spells. Such a limitation of bias-correction methods in case of dry frequency biases is a general limitation of current quantile-mapping bias-correction methods [35,52].

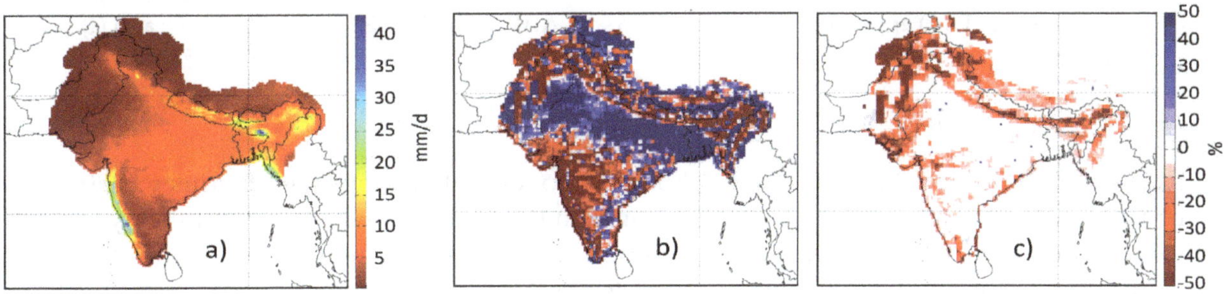

Figure 2. Observed precipitation and RCM bias in the period 1976–2005: (**a**) mean daily precipitation in APHRODITE; (**b**) relative bias in uncorrected historical GCM-RCM projection, and (**c**) relative bias in bias-corrected historical GCM-RCM projection.

Concerning T, in most parts of India it was underestimated by 1–2 C in the historical GCM-RCM projection (not shown). After the DBS bias-correction, only a negligible bias remained and the spatial pattern became essentially identical to the observations (see Figure 3c).

4.3. Reference Data Analysis

To infer a quantitative understanding of the magnitude of the climate change impact on the hydro-climatic components (P was analyzed in Section 4.2), it is necessary to estimate the long-term averages of India-HYPE driven by APHRODITE for the entire reference period (Figure 3). Long-term average R is controlled by the spatial patterns of P and AE; consequently high R is generated at the southwestern rivers and at the Himalayan mountain range. AE, which in this model is controlled by water availability and T, shows similar spatial patterns to P but lower than the latter by almost 25%; hence, indicating that most precipitated water is evaporated back to the atmosphere. Note, also, the high temperatures which are almost homogeneously distributed over the region (average T is about 25 °C); however, the T gradient is very strong at the Himalayan mountain range (average T varies between −5 and 2 °C). The AIW varies between 100 and 900 Mm^3 per year with high spatial variability in the region (note that the spatial pattern of this variable is subject to the GMIA irrigation map). The SMD is between 0 and 140 mm/year, with the Himalayas exhibiting very low values and the Thar Desert exhibiting high values; note the presence of wetlands covering this northwest region. Finally, SD can reach 100 cm in the Himalayas with large local variability.

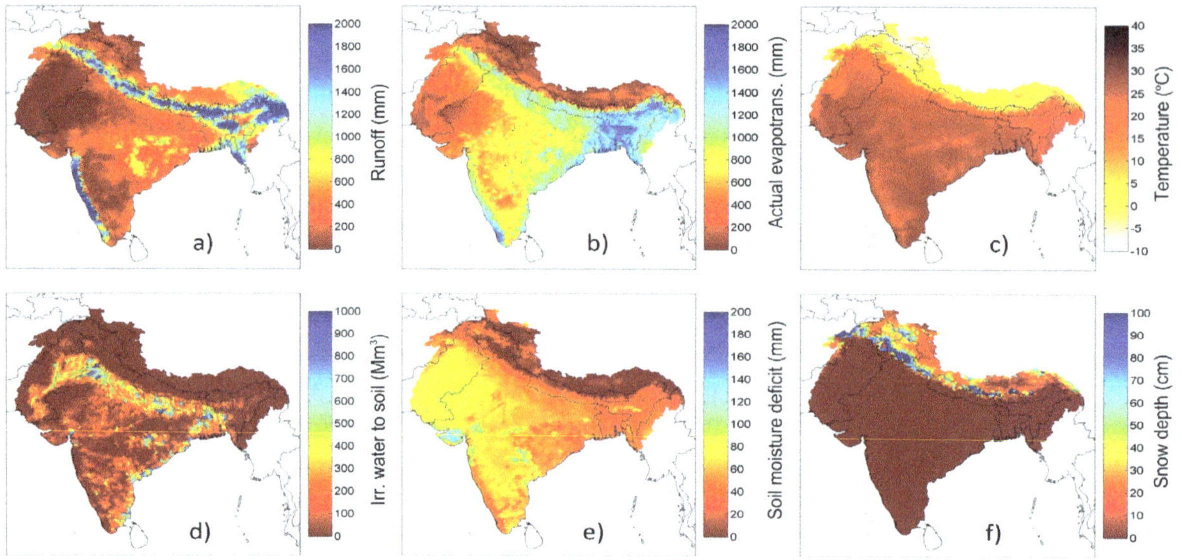

Figure 3. Annual averages (period 1976–2005) for the variables: (**a**) runoff, R; (**b**) actual evapotranspiration, AE; (**c**) temperature, T; (**d**) applied irrigation water to soil, AIW; (**e**) soil moisture deficit, SMD; and (**f**) snow depth, SD.

4.4. Climate Change Impacts

As outlined in Section 3.3, we analyze the effects of climate change for various variables in two different scenario periods. We firstly present changes in the variables which control the hydro-climatic conditions and continue with the applied irrigation water to soil which is more "end-user related". We then investigate the changes in the annual cycles of three regions.

4.4.1. Long-Term Averages

All projections show an increase of T for both scenario periods and in the whole modelling domain (Figure 4). In RCP2.6, the increase in T is limited to ranges below +1.5 °C until the end of the 21st century. For the same scenario period and RCP4.5, the increase in T is projected to fall into the range of +1.0 to +2.0 °C over most of the domain, but is higher—up to +3.0 °C—in the mountainous parts in the north. The same spatial pattern is seen in RCP8.5 in which T is projected to increase between 1.5 and 6.0 °C with the highest increase also occurring in the mountainous regions. This elevation dependency could be due to the snow albedo effect which leads to higher increases in T when the snow cover duration decreases [53].

Regarding P, the projections show strongly increasing mean P over most of the region (Figure 5). This also includes areas where present day P is already very high (e.g. along the Western Ghats and the southern side of the Himalayan mountain range). The signal is stronger towards the end of the 21st century than in mid-century and the higher the RCP is, with the exception of RCP2.6 for which there is no intensification of the signal from the mid- to the end-century period. The clear dependency of the change signal on the period and the RCP is true for all variables discussed below. The dry regions in the northwestern and central parts, including e.g., the Indus, Luni, Sabarmati, and parts of the Ganga River, are projected to receive less P. Additionally, the spatial variability of the changes within the major river basins (see boxplots in Figure 5) is generally increasing towards the end of the century. The above-mentioned river basins include sub-basins with a very heterogeneous pattern of P changes, ranging from decreases to increases for all projections except for the RCP8.5 end-century evaluation. Those basins are located in a zone where slight changes in the spatial patterns of P decrease or increase might lead to a pronounced effect on the basin-wide qualitative changes (*i.e.*, decrease or increase of mean P).

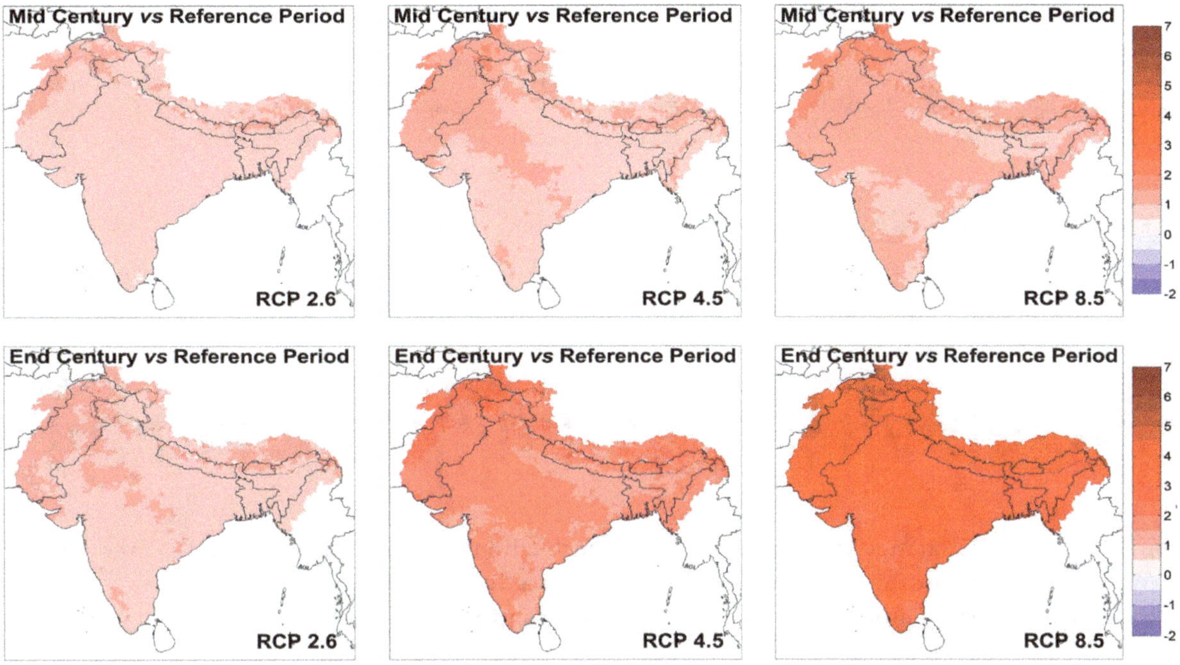

Figure 4. Change in temperature T for each climate projection (columns 1–3) and period (top and bottom row for the mid- and end-century, respectively).

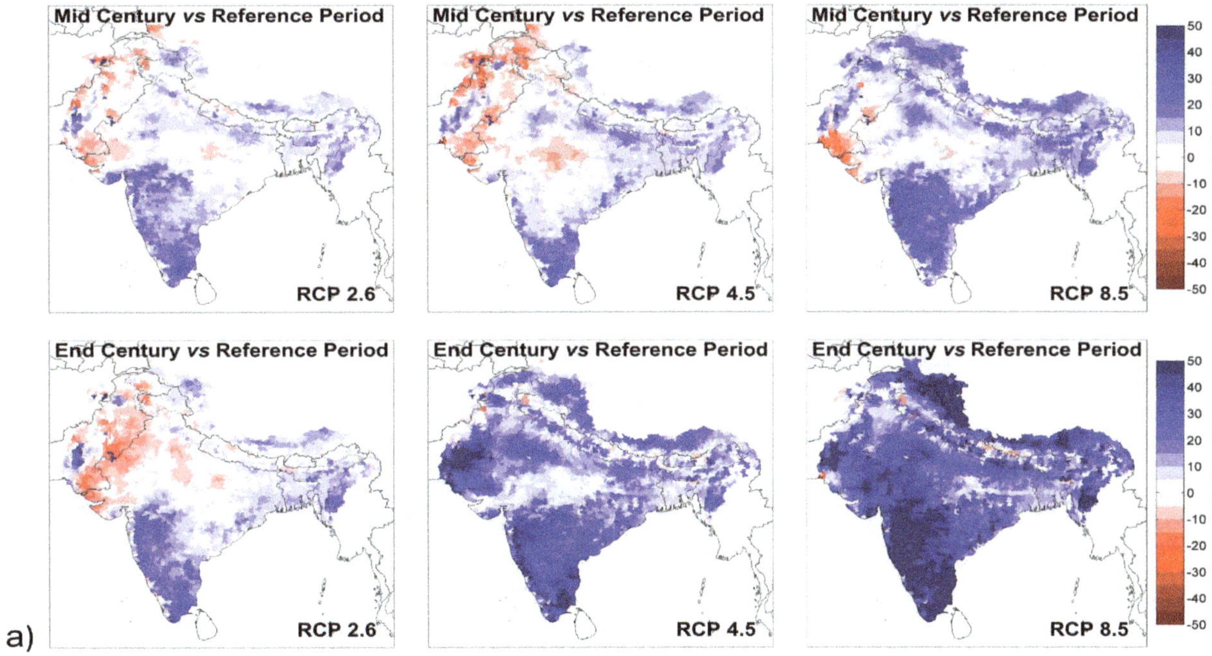

Figure 5. *Cont.*

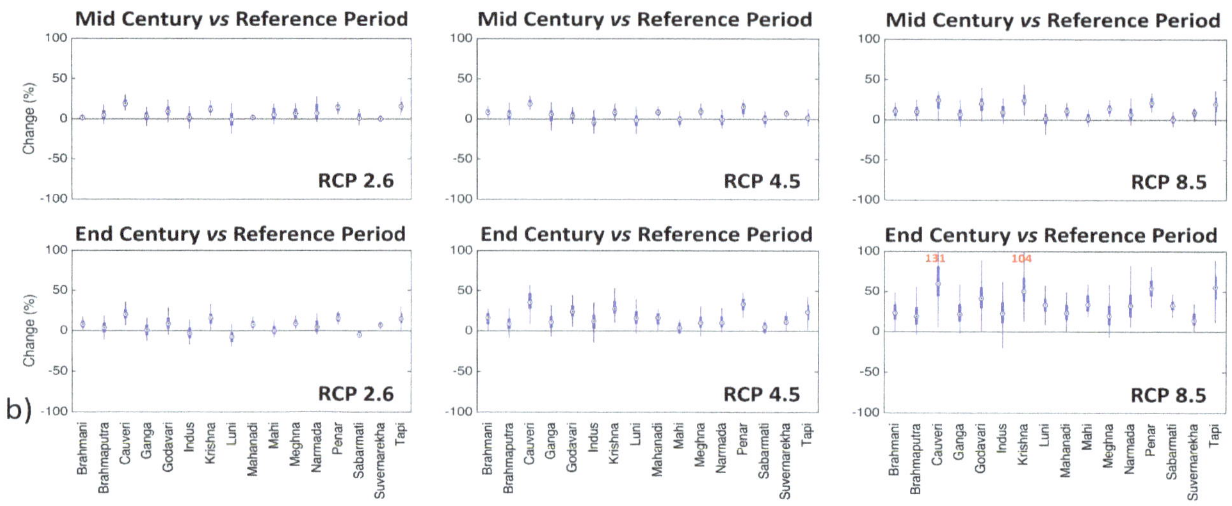

Figure 5. (**a**) Relative change in precipitation P for each climate projection (columns 1–3) and period (top and bottom row for the mid- and end-century, respectively); and (**b**) variability of the change at the basin scale for each climate projection (columns 1–3) and period (rows 1–2).

AE generally either remains essentially unchanged or increases. In areas with a pronounced decrease in P, such as in the northwest, a slight decrease of AE may, however, occur (see Figure 6). The increase of evaporation in the south is due to decrease in SMD (see Figure 8). Over the Himalayan mountain range, strong increases are projected in all RCPs and both scenario periods. In that area, the increase of potential evapotranspiration (not shown) due to increased temperature and a shortened period of snow cover (see Figure 8 for snow depth changes as an indicator of snow cover changes) both contribute to higher AE.

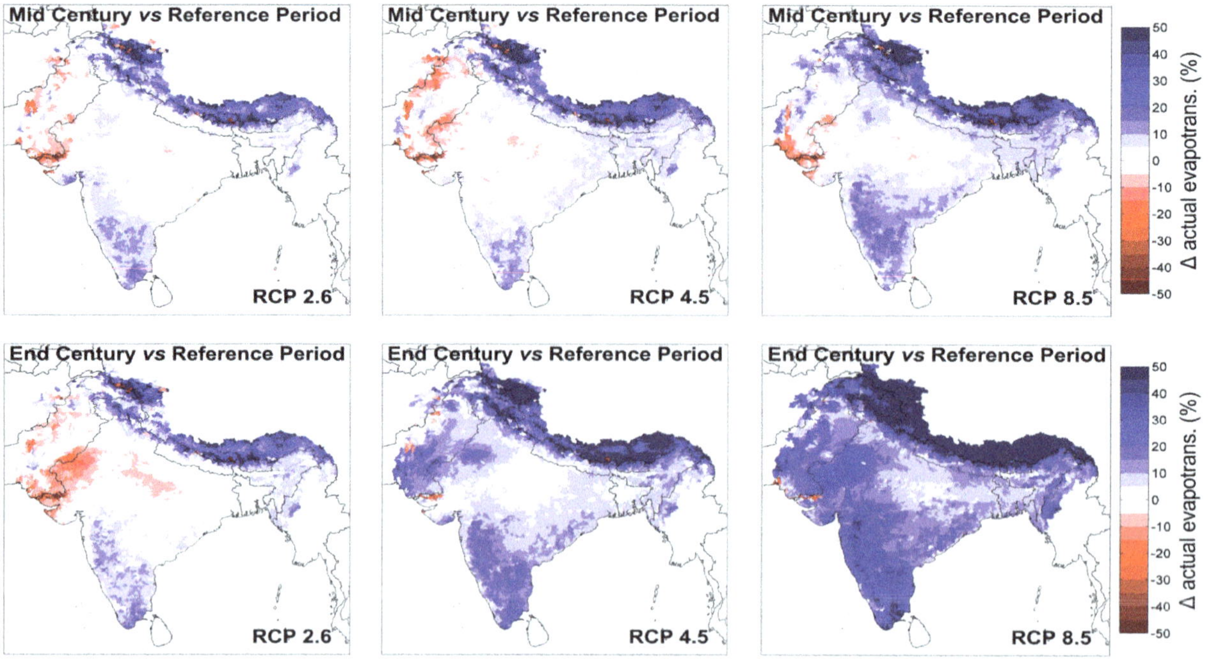

Figure 6. Same as Figure 5a for actual evapotranspiration AE.

The pattern of R changes follows the one of P but the magnitudes of the relative changes are larger than for P (Figure 7). Consequently, the spatial variability within the major river basins is also much larger than for P, shown by the wider boxplots in the Figure 7. Thus, the basins, which are susceptible to small spatial changes of areas with increasing and decreasing P, are even more sensitive to spatial changes when it comes to R.

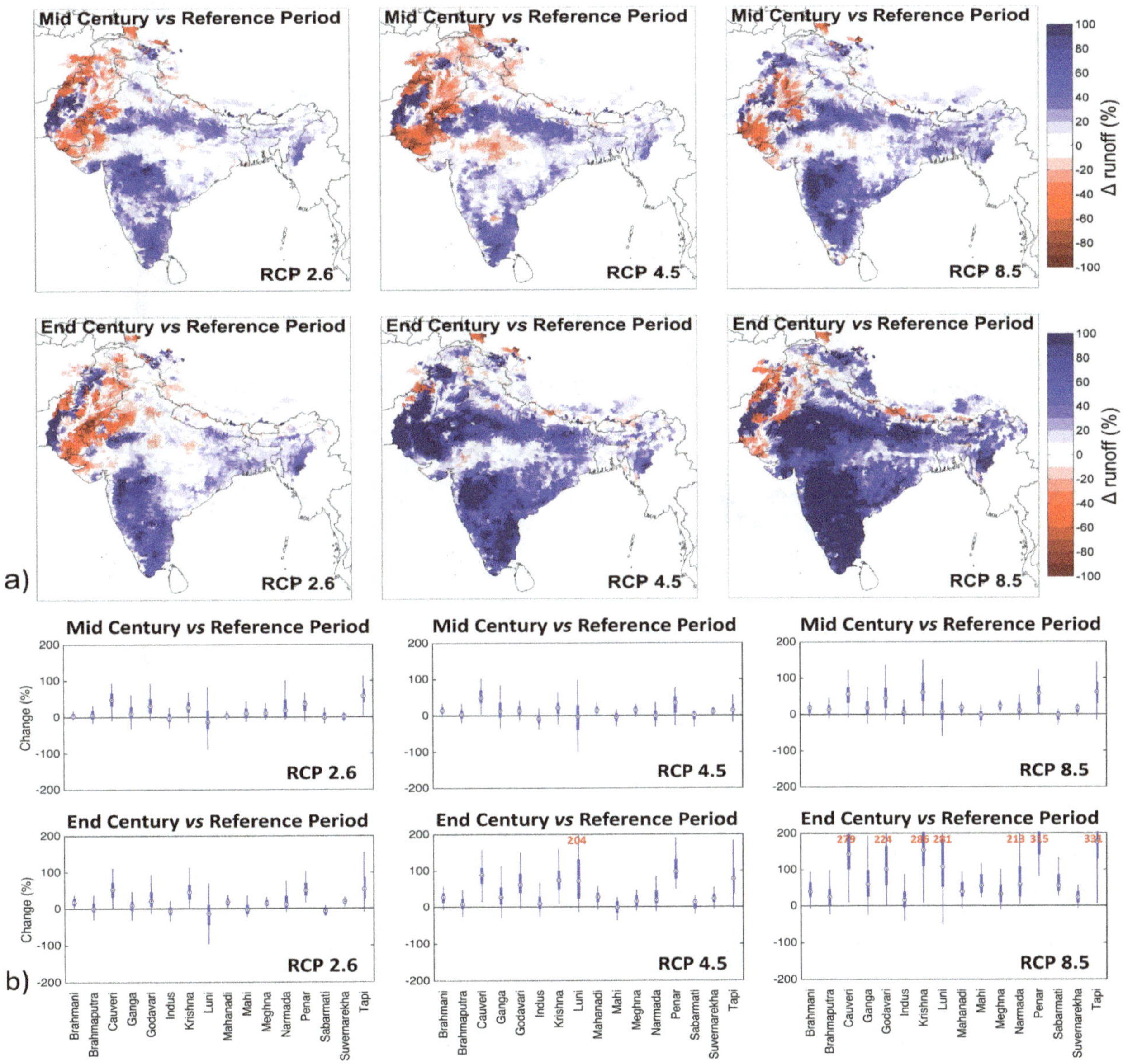

Figure 7. (**a**) Same as Figure 5a for runoff R. Maximum values for the regions in which change is greater than 200% are presented in red and (**b**) variability of the change at the basin scale for each climate projection (columns 1–3) and period (rows 1–2).

Within the model domain, snow only occurs in the Himalayas (Figure 3). For a large fraction of that area, SD is projected to decrease strongly, a change which is already apparent in the mid-century period (Figure 8). Only in some high-elevated areas is T cold enough to make SD increase due to increased mean P.

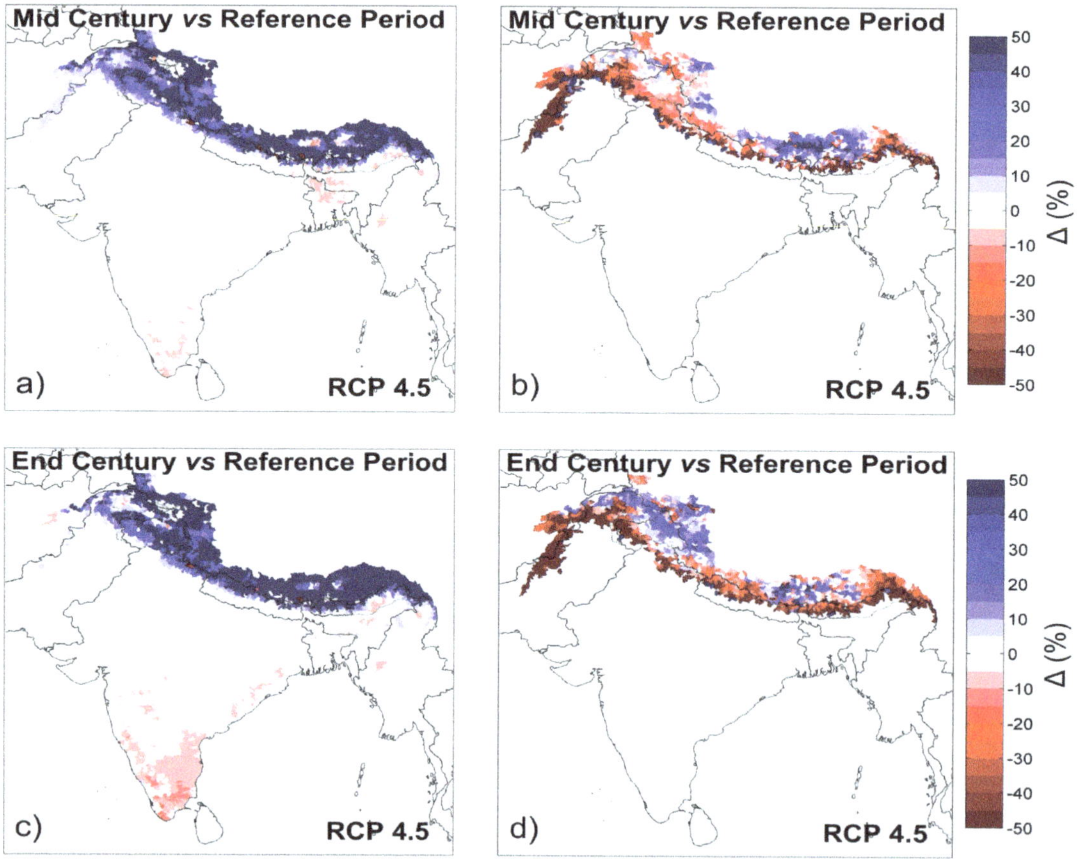

Figure 8. Relative change in soil moisture deficit SMD (column 1) and snow depth SD (column 2) for climate projection RCP4.5 until mid-century (row 1) and end-century (row 2). Only RCP4.5 is shown; the spatial pattern of changes is very similar for the other RCPs. (**a**) SMD change for mid-century; (**b**) SD change for mid-century; (**c**) SMD change for end-century, and (**d**) SD change for end-century. See the results for all RCPs and future periods in Figures S1 and S2 in the Supplement.

The changes in irrigated water to soil (Figure 9) should be interpreted in a way that they represent the changes in future demand for irrigation. Wherever irrigation is projected to increase, a higher demand for irrigation is implied. In that sense, the results indicate that the demand for irrigation will increase in the Northwestern and central regions in all RCPs and both scenario periods. Towards the end of the century in RCP8.5, the region with increased irrigation demand is enlarged and encompasses the whole Indus and Ganga river basins. Decreases in irrigation needs are projected for the southern part of the model domain.

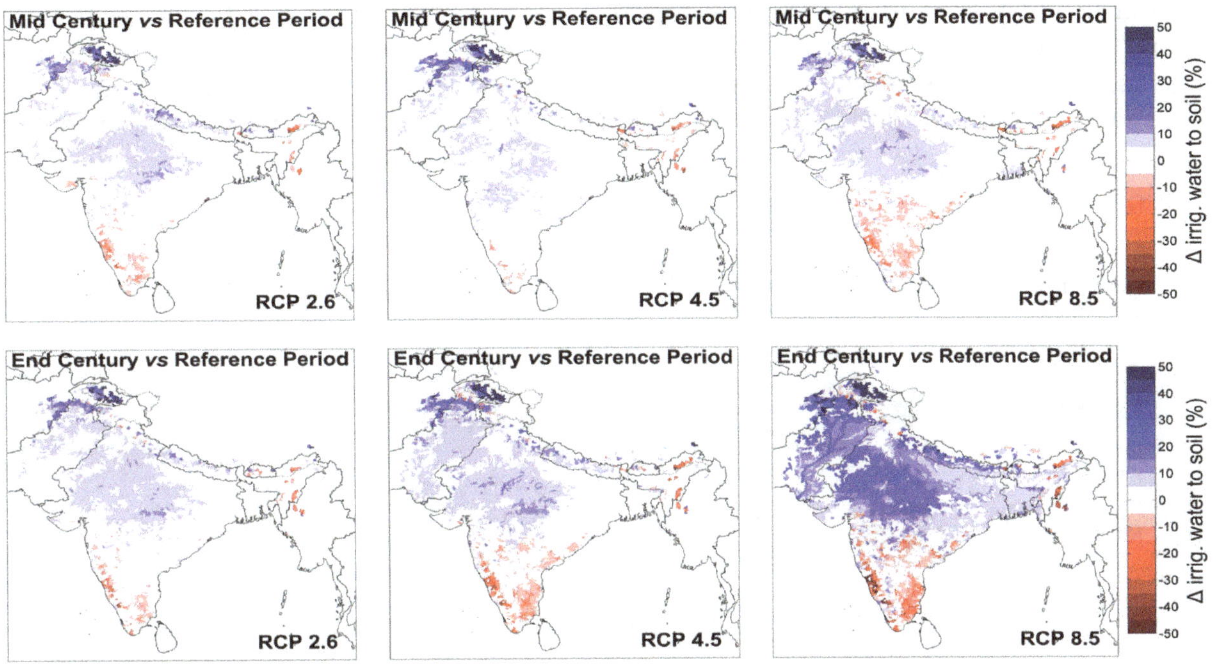

Figure 9. Same as Figure 5a for applied irrigation water to soil AIW.

4.4.2. Annual Cycles

Whereas in Section 4.4.1 we presented spatially detailed maps of changes in annual mean quantities, we show here the results of changes in the annual cycle at three selected gauges (see stars in Figure 1) in order to provide more details about the seasonal dynamics.

The Ganga River has clear Q seasonality with peak Q in the monsoon season and very little Q in the remaining months (Figure 1b). P is projected to increase during the monsoon period, and little change happens in the rest of the year (Figure 10a). This trend is stronger for the end-century than for the mid-century changes in RCP4.5 and RCP8.5, while RCP2.6's changes remain on a low level even at the end of the century. For the end-century period, RCP8.5 also shows a longer season of increasing P, lasting until the start of the post-monsoon season in October. The seasonal pattern of changes in AE closely follows the one of P, but the magnitudes of the changes are smaller. Thus, the increases in P partly become runoff effective, which is seen as increasing Q in the monsoon season. The largest increases in Q towards the end of the century are projected to range from 5000 to 20,000 m^3/s depending on the chosen RCP, corresponding to a 12 to 50 % increase compared to the reference period.

The gauge in the Indus catchment lies close to the Himalayas. In present day conditions, the seasonal cycle of P shows a two-peak pattern with one peak in March and the second one in July/August (Figure 1c). The three projections do not agree on the change signal (Figure 10b). For the end-century period, a slight tendency towards increasing P in summer and decreasing P in March is noticeable. This would mean that the first P peak weakens, whereas the second one in summer gets stronger. AE increases throughout the year, with the largest increases being projected for the summer. There is no clear link between the P and the AE change signal, indicating that the soil moisture is not the limiting factor for AE in that area. The change signal in Q is associated with more uncertainty and small projected changes switch sign several times in the course of the annual cycle. This is also reflected in the boxplots in Figure 7 where the Indus River shows a large spatial variability of runoff changes around the no-change level. Small changes in some sub-basins might lead to a switch of the sign of change in the response on the aggregated basin level. Overall, it seems that the climate change signal of discharge is small.

Under the current conditions, the Godavari River has a clear seasonal discharge regime with a peak in the monsoon season and low discharge in the rest of the year (Figure 1d). Compared to the Ganga River, the seasonality is even more pronounced due to the lack of snow-melt from the Himalayas during the pre- and post-monsoon season. Additionally, the seasonal pattern of changes in all variables looks similar to the ones in the Ganga River; however, the magnitudes are larger (Figure 10c). For total Q in the monsoon and post-monsoon period, the changes amount to +10% to +80%.

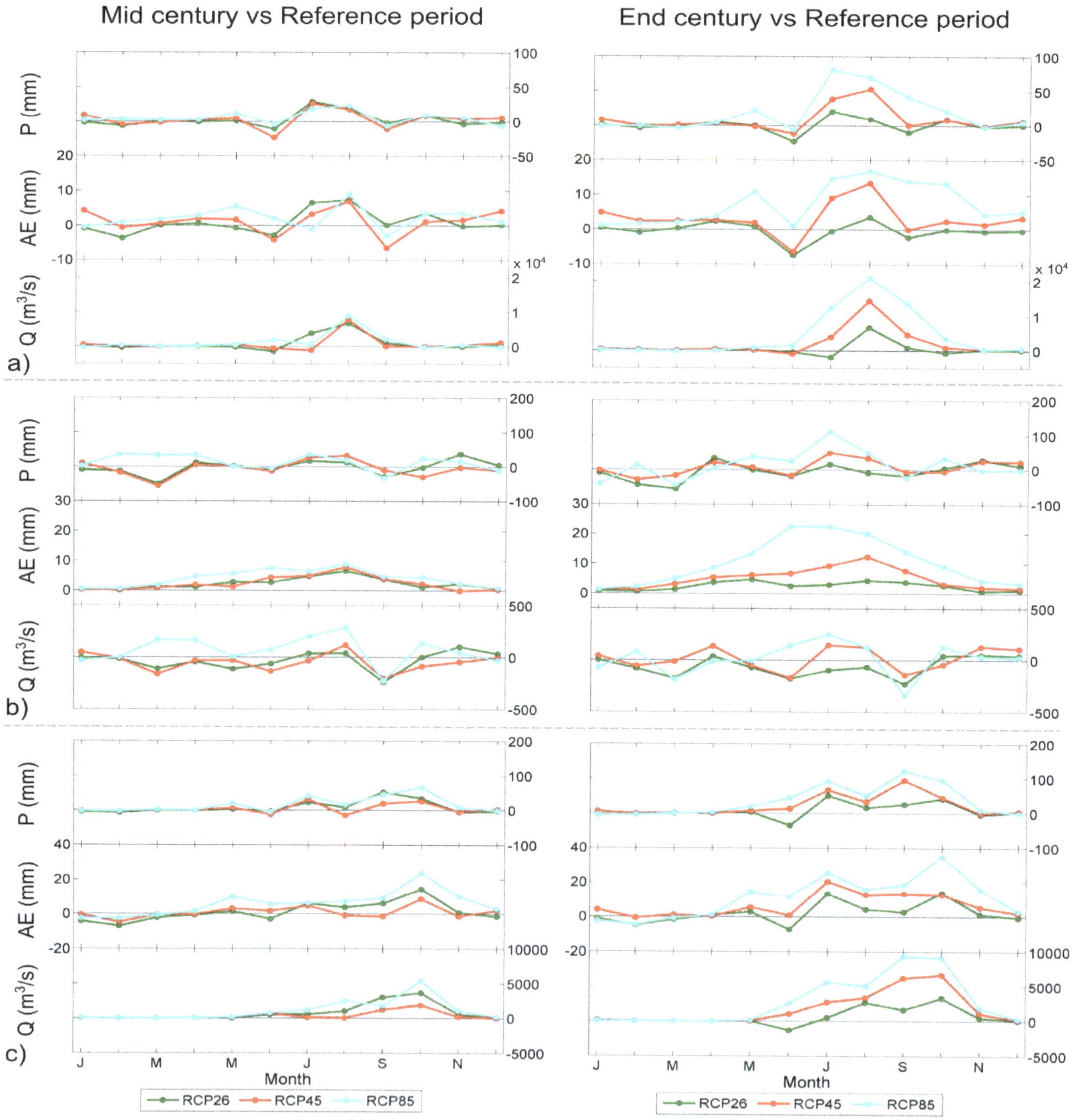

Figure 10. Absolute changes in the annual cycles of precipitation P, actual evapotranspiration AE and discharge Q for each climate scenario and period (column 1 and 2 for the mid- and end-century, respectively) for the rivers: (**a**) Ganga (at Farakka station); (**b**) Indus (Chenab at Akhnoor station), and (**c**) Godavari (at Polavaram station).

5. Discussion

5.1. Enhancing Understanding of Future Climate Change Impacts

Our study contributes to the previously reported assessments of future climatic patterns (which in most cases are driven by projections from different GCMs and RCMs) in India and their impact on water resources. Overall our results support previous findings that climate projections reveal an increase in monsoon precipitation in the mid-century [54,55] and a possible extension of the monsoon period [56–58]. Our results over Northern and Western India and the southeastern coastline differ from previous projections based on the CGCM3.1-PRECIS model chain and the A1B emission scenario [31,59]. According to [59], in those regions precipitation is expected to decrease both in the mid- and end-century. Moreover, analysis (not shown here) on the number of wet spells was consistent with the general trends of previous finding, showing a higher frequency of wet spells from the 2060s and beyond, mostly in northern and coastal regions [14,60]. [61] define the river basins of Kutch, Saurashtra, Luni, and Indus in the northwest and Pennar in the southeast as physically water-scarce, most of the rest of India as economically water-scarce, and only Brahmaputra, Meghna, Brahmani, Mahanadi, and some smaller basins along the east coast as non-water-scarce (Figure 1). Our results support this regional classification, however attention is needed in results for R in the northwestern basins due to their pronounced spatial variability (although overall results point at decreasing runoff). The overall decreasing future trend of water availability in the northwestern basins with a high spatial variability agrees with the results in [31]. In our results river Pennar will clearly become less water-scarce by the end of the century, which is in contrast to [31] who found a reduced water yield in this basin on this time horizon.

A key component for assessing water scarcity is the evapotranspiration. Similarly to P, AE in our results exhibit a pronounced spatial variability in the northwest and overall an increase is projected for the Indian subcontinent. This can be contrasted with recent findings of observed decreasing trends in temperature (and consequently evaporation and water scarcity) over Northeastern India [62], the Godavari basin [63], as well as many regions world-wide [64,65]. For India, this finding has been attributed to mainly a reduction of wind speeds ('stilling') and an increase of atmospheric humidity [63]. These impacts cannot be simulated with the purely T-based relationship for AE used here.

In the Himalayan region, we found a generally consistent increase in runoff for different projections and future periods, which agrees with the general trend in the literature [5,6,66–68]; however this increase varies significantly in space (following the topographic pattern). In addition, it is generally expected that the changes in the Ganga and Meghna rivers will be larger than in the Brahmaputra river, probably due to the impact of melted water from snow and glaciers [2]. This conclusion is consistent with our results. Change in runoff in the Krishna river is ambiguous with studies showing both a future increase [32,69], as here, and a decrease [31].

5.2. Limitations of This Study

Reliability of the input data used to drive large-scale multi-basin hydrological models, particularly those derived from global datasets, has been questioned [70]. Although a preliminary comparison of the applied climatic and physiographic data against national data did not show significant discrepancies (e.g., when comparing the APHRODITE data against P and T observations from the Indian Meteorological Department), some inconsistencies were generally observed at the local scale.

Here, we have only focused on the uncertainty from the RCP scenario. This was motivated by previous findings indicating that towards the end of the 21st century, generally the emission scenario (here, RCP) is the dominant source of uncertainty in climate projections (e.g., [26]). However, it is clear that also the other sources may substantially contribute to the uncertainty, e.g., the choice of GCMs, RCMs, bias-correction method, hydrological model structure, and parameterization can have a substantial impact [71,72].

Climate change impact results are subject to the impact model's predictability and consistency, here assessed by the model's performance in the present climate. The India-HYPE model can adequately represent the long-term average fluxes and their seasonal variation (see details in [40]); however, it is expected that a larger number of discharge gauges (representing different hydro-climatic systems) and/or temporally extended time series (in terms of length and resolution) and/or additional variables (e.g., evapotranspiration, snow cover area *etc.*) would provide additional information to drive the parameterization of the model and guide towards model structural improvements.

Finally, our study is limited by the assumed stationarity in the investigated hydrological systems. It is recognized that non-stationarity exists as a characteristic of the natural world due to various environmental changes (land use and other man-made alterations) [73].

6. Conclusions

We have explored the potential impact of climate change on the hydrology and water resources of the Indian subcontinent, based on an RCP-ensemble of regional climate projections from the CORDEX-SA framework. Climate projections with GCM EC-EARTH and RCM RCA4 based on three RCP scenarios (RCP2.6, RCP4.5, and RCP8.5) are bias-corrected and introduced into the HYPE hydrological model to assess average changes in various hydro-climatic fluxes for two future periods (mid- and end-century). Additionally, the intra-annual variability of key fluxes (precipitation, evapotranspiration, and discharge) in three river systems.

Overall, the distribution of change in runoff varies considerably with both hydro-climatic region and climate projection. In particular, the following future changes were indicated:

- Temperature will increase in the entire subcontinent, with the highest increase in the mountainous regions.
- An increase in long-term average precipitation and evapotranspiration in wet regions; however, less precipitation and evapotranspiration are expected at the dry regions.
- Average snow depth in the Himalayan region will be reduced; this is consistent in all projections and time horizons.
- A general increase in the need for irrigation; however, the need is reduced in the south.
- Large relative changes in runoff, and large spatial variability at the basin scale, particularly towards the end of the century.
- Changed seasonality in discharge, with more pronounced changes in the tropical and subtropical zones than in the mountainous regions.

Overall, the conclusions support previous findings in most parts of India (including the Himalayas), with respect to future trends in runoff. This study, however, indicates larger changes under the severe emission scenario RCP8.5. Regionally, some differences are found in comparison with existing results, e.g., for precipitation and evapotranspiration in northern and western India and the southeastern coastline and for runoff changes in the Krishna river. These differences highlight the need for further studies focusing on climate change impacts on hydrology in India. In addition to pure climate impact modelling, as performed here, we believe more efforts are needed to better understand and quantify the sensitivity of the river systems to climate changes, as well as other changes (e.g., land-use and population). This work should preferably include multiple scenarios of all considered changes used in combination with multiple hydrological models. Work in this direction is ongoing and will be reported elsewhere.

Supplementary Materials:
Figure S1: Relative change in soil moisture deficit (SMD) for each climate projection (columns 1–3) and period (top and bottom row for mid- and end-century respectively), Figure S2: Relative change in snow depth (SD) for each climate projection (columns 1–3) and period (top and bottom row for mid- and end-century, respectively).

Acknowledgments: This work was funded mainly by the Swedish International Development Cooperation Agency (SIDA; grant no. AKT-2012-022) with additional support from the Swedish Research Council Formas (grant No. 2010-121) and the Swedish Secretariat for Environmental Earth System Sciences (SSEESS; grant No. KVA/2015/111/75). We would also like to acknowledge David Gustafsson, Kean Foster, Kristina Isberg, Jörgen Rosberg, Jafet Andersson and Sian de Koster for assistance with background material for this study and manuscript checks.

Author Contributions: Ilias G. Pechlivanidis led the project-related activities on setting up the hydrological model, conducting the analyses, and writing the manuscript. Jonas Olsson assisted on the manuscript writing and supervised the efforts on bias-correction of the climate projections. Thomas Bosshard led the activities on bias-correcting the climate projections and assisted on writing the manuscript. Devesh Sharma and K.C. Sharma provided local knowledge during the hydrological model setup and made improvements to the manuscript.

References

1. Buytaert, W.; De Bièvre, B. Water for cities: The impact of climate change and demographic growth in the tropical Andes. *Water Resour. Res.* **2012**, *48*. [CrossRef]
2. Mirza, M.; Warrick, R.; Ericksen, N. The implications of climate change on floods of the Ganges, Brahmaputra and Meghna rivers in Bangladesh. *Clim. Chang.* **2003**, *57*, 287–318. [CrossRef]
3. Neupane, R.P.; Yao, J.; White, J.D. Estimating the effects of climate change on the intensification of monsoonal-driven stream discharge in a Himalayan watershed. *Hydrol. Process.* **2014**, *28*, 6236–6250. [CrossRef]
4. Singh, P.; Kumar, N. Impact assessment of climate change on the hydrological response of a snow and glacier melt runoff dominated Himalayan river. *J. Hydrol.* **1997**, *193*, 316–350. [CrossRef]
5. Pervez, M.S.; Henebry, G.M. Assessing the impacts of climate and land use and land cover change on the freshwater availability in the Brahmaputra River basin. *J. Hydrol. Reg. Stud.* **2014**, *3*, 285–311. [CrossRef]
6. Singh, P.; Bengtsson, L. Impact of warmer climate on melt and evaporation for the rainfed, snowfed and glacierfed basins in the Himalayan region. *J. Hydrol.* **2005**, *300*, 140–154. [CrossRef]
7. Singh, P.; Arora, M.; Goel, N.K. Effect of climate change on runoff of a glacierized Himalayan basin. *Hydrol. Process.* **2006**, *20*, 1979–1992. [CrossRef]
8. Gosain, A.; Rao, S.; Basuray, D. Climate change impact assessment on hydrology of Indian river basins. *Curr. Sci.* **2006**, *90*, 346–353.
9. Pechlivanidis, I.G.; Olsson, J.; Sharma, D.; Bosshard, T.; Sharma, K.C. Assessment of the climate change impacts on the water resources of the Luni region, India. *Glob. NEST J.* **2015**, *17*, 29–40.
10. Graham, L.P.; Hagemann, S.; Jaun, S.; Beniston, M. On interpreting hydrological change from regional climate models. *Clim. Chang.* **2007**, *81*, 97–122. [CrossRef]
11. Pechlivanidis, I.G.; Jackson, B.; McIntyre, N.; Wheater, H.S. Catchment scale hydrological modelling: A review of model types, calibration approaches and uncertainty analysis methods in the context of recent developments in technology and applications. *Glob. NEST J.* **2011**, *13*, 193–214.
12. Fowler, H.J.; Blenkinsop, S.; Tebaldi, C. Linking climate change modelling to impacts studies: Recent advances in downscaling techniques for hydrological. *Int. J. Climatol.* **2007**, *27*, 1547–1578. [CrossRef]
13. Giorgi, F.; Jones, C.; Asrar, G.R. Addressing climate information needs at the regional level: The CORDEX framework. *WMO Bull.* **2009**, *58*, 175–183.
14. Chaturvedi, R.; Joshi, J.; Jayaraman, M.; Bala, G.; Ravindranath, N. Multi-model climate change projections for India under representative concentration pathways. *Curr. Sci.* **2012**, *103*, 1–12.
15. Foley, A. Uncertainty in regional climate modelling: A review. *Prog. Phys. Geogr.* **2010**, *34*, 647–670. [CrossRef]
16. Haylock, M.R.; Cawley, G.C.; Harpham, C.; Wilby, R.L.; Goodess, C.M. Downscaling heavy precipitation over the United Kingdom: A comparison of dynamical and statistical methods and their future scenarios. *Int. J. Climatol.* **2006**, *26*, 1397–1415. [CrossRef]
17. Wetterhall, F.; Pappenberger, F.; He, Y.; Freer, J.; Cloke, H.L. Conditioning model output statistics of regional climate model precipitation on circulation patterns. *Nonlinear Process. Geophys.* **2012**, *19*, 623–633. [CrossRef]
18. Teutschbein, C.; Seibert, J. Bias correction of regional climate model simulations for hydrological climate-change impact studies: Review and evaluation of different methods. *J. Hydrol.* **2012**, *456–457*, 12–29. [CrossRef]

19. Déqué, M. Frequency of precipitation and temperature extremes over France in an anthropogenic scenario: Model results and statistical correction according to observed values. *Glob. Planet. Change* **2007**, *57*, 16–26. [CrossRef]
20. Lenderink, G.; Buishand, A.; van Deursen, W. Estimates of future discharges of the river Rhine using two scenario methodologies: Direct *versus* delta approach. *Hydrol. Earth Syst. Sci.* **2007**, *11*, 1145–1159. [CrossRef]
21. Chen, H.; Xu, C.-Y.; Guo, S. Comparison and evaluation of multiple GCMs, statistical downscaling and hydrological models in the study of climate change impacts on runoff. *J. Hydrol.* **2012**, *434–435*, 36–45. [CrossRef]
22. Minville, M.; Brissette, F.; Leconte, R. Uncertainty of the impact of climate change on the hydrology of a nordic watershed. *J. Hydrol.* **2008**, *358*, 70–83. [CrossRef]
23. Preston, B.L.; Jones, R.N. Evaluating sources of uncertainty in Australian runoff projections. *Adv. Water Resour.* **2008**, *31*, 758–775. [CrossRef]
24. Chiew, F.H.S.; Teng, J.; Vaze, J.; Kirono, D.G.C. Influence of global climate model selection on runoff impact assessment. *J. Hydrol.* **2009**, *379*, 172–180. [CrossRef]
25. Jones, R.N.; Chiew, F.H.S.; Boughton, W.C.; Zhang, L. Estimating the sensitivity of mean annual runoff to climate change using selected hydrological models. *Adv. Water Resour.* **2006**, *29*, 1419–1429. [CrossRef]
26. Hawkins, E.; Sutton, R. The potential to narrow uncertainty in regional climate predictions. *Bull. Am. Meteorol. Soc.* **2009**, *90*, 1095–1107. [CrossRef]
27. Bueh, C.; Cubasch, U.; Hagemann, S. Impacts of global warming on changes in the East Asian monsoon and the related river discharge in a global time-slice experiment. *Clim. Res.* **2003**, *24*, 47–57. [CrossRef]
28. Lucas-Picher, P.; Christensen, J.H.; Saeed, F.; Kumar, P.; Asharaf, S.; Ahrens, B.; Wiltshire, A.J.; Jacob, D.; Hagemann, S. Can regional climate models represent the Indian monsoon? *J. Hydrometeorol.* **2011**, *12*, 849–868. [CrossRef]
29. Aich, V.; Liersch, S.; Vetter, T.; Huang, S.; Tecklenburg, J.; Hoffmann, P.; Koch, H.; Fournet, S.; Krysanova, V.; Müller, E.N.; Hattermann, F.F. Comparing impacts of climate change on streamflow in four large African river basins. *Hydrol. Earth Syst. Sci.* **2014**, *18*, 1305–1321. [CrossRef]
30. Huang, S.; Krysanova, V.; Hattermann, F. Projections of climate change impacts on floods and droughts in Germany using an ensemble of climate change scenarios. *Reg. Environ. Chang.* **2014**, *15*, 461–473. [CrossRef]
31. Gosain, A.; Rao, S.; Arora, A. Climate change impact assessment of water resources of India. *Curr. Sci.* **2011**, *101*, 356–371.
32. Raje, D.; Priya, P.; Krishnan, R. Macroscale hydrological modelling approach for study of large scale hydrologic impacts under climate change in Indian river basins. *Hydrol. Process.* **2013**, *28*, 1874–1889. [CrossRef]
33. Blöschl, G.; Sivapalan, M.; Wagener, T.; Viglione, A.; Savenije, H. *Runoff Prediction in Ungauged Basins. Synthesis across Processes, Places and Scales*; Cambridge University Press: Cambridge, UK, 2013.
34. Gupta, H.V.; Perrin, C.; Blöschl, G.; Montanari, A.; Kumar, R.; Clark, M.; Andréassian, V. Large-sample hydrology: A need to balance depth with breadth. *Hydrol. Earth Syst. Sci.* **2014**, *18*, 463–477. [CrossRef]
35. Yang, W.; Andréasson, J.; Graham, P.L.; Olsson, J.; Rosberg, J.; Wetterhall, F. Distribution-based scaling to improve usability of regional climate model projections for hydrological climate change impacts studies. *Hydrol. Res.* **2010**, *41*, 211–229. [CrossRef]
36. Lindström, G.; Pers, C.; Rosberg, J.; Strömqvist, J.; Arheimer, B. Development and testing of the HYPE (Hydrological Predictions for the Environment) water quality model for different spatial scales. *Hydrol. Res.* **2010**, *41*, 295–319. [CrossRef]
37. Attri, S.D.; Tyagi, A. *Climate Profile of India*; Government of India Ministry of Earth Sciences: New Delhi, India, 2010; p. 129.
38. Mall, R.K.; Singh, R.; Gupta, A.; Srinivasan, G.; Rathore, L.S. Impact of Climate Change on Indian Agriculture: A Review. *Clim. Change* **2006**, *78*, 445–478. [CrossRef]
39. Li, L.; Xu, C.-Y.; Zhang, Z.; Jain, S.K. Validation of a new meteorological forcing data in analysis of spatial and temporal variability of precipitation in India. *Stoch. Environ. Res. Risk Assess.* **2013**, *28*, 239–252. [CrossRef]
40. Pechlivanidis, I.G.; Arheimer, B. Large-scale hydrological modelling by using modified PUB recommendations: the India-HYPE case. *Hydrol. Earth Syst. Sci.* **2015**, *19*, 4559–4579. [CrossRef]

41. Yatagai, A.; Kamiguchi, K.; Arakawa, O.; Hamada, A.; Yasutomi, N.; Kitoh, A. APHRODITE: Constructing a long-term daily gridded precipitation dataset for Asia based on a dense network of rain gauges. *Bull. Am. Meteorol. Soc.* **2012**, *93*, 1401–1415. [CrossRef]
42. Yatagai, A.; Arakawa, O.; Kamiguchi, K. A 44-year daily gridded precipitation dataset for Asia based on a dense network of rain gauges. *Sola* **2009**, *5*, 137–140. [CrossRef]
43. Yasutomi, N.; Hamada, A.; Yatagai, A. Development of a long-term daily gridded temperature dataset and its application to rain/snow discrimination of daily precipitation. *Glob. Environ. Res.* **2011**, *3*, 165–172.
44. Hazeleger, W.; Wang, X.; Severijns, C.; Ştefǎnescu, S.; Bintanja, R.; Sterl, A.; Wyser, K.; Semmler, T.; Yang, S.; van den Hurk, B.; van Noije, T.; van der Linden, E.; van der Wiel, K. EC-Earth V2.2: Description and validation of a new seamless earth system prediction model. *Clim. Dyn.* **2012**, *39*, 2611–2629. [CrossRef]
45. Samuelsson, P.; Collin, G.J.; Willén, U.; Ullerstig, A.; Gollvik, S.; Hansson, U.; Jansson, C.; Kjellström, E.; Nikulin, G.; Wyser, K. The Rossby Centre Regional Climate model RCA3: Model description and performance. *Tellus* **2011**, *63A*, 4–23. [CrossRef]
46. Moss, R.H.; Edmonds, J.A.; Hibbard, K.A.; Manning, M.R.; Rose, S.K.; van Vuuren, D.; Carter, T.R.; Emori, S.; Kainuma, M.; Kram, T.; *et al.* The next generation of scenarios for climate change research and assessment. *Nature* **2010**, *463*, 747–756. [CrossRef] [PubMed]
47. Ghimire, S.; Choudhary, A.; Dimri, A.P. Assessment of the performance of CORDEX-South Asia experiments for monsoonal precipitation over the Himalayan region during present climate: Part I. *Clim. Dyn.* **2015**. [CrossRef]
48. Allen, R.G.; Pereira, L.S.; Raes, D.; Smith, M. *Crop Evapotranspiration, Guidelines for Computing Crop Water Requirements-FAO Irrigation and Drainage Paper 56*; Food and Agriculture Organization of the United Nations (FAO): Rome, Italy, 1998.
49. Nash, J.E.; Sutcliffe, J.V. River flow forecasting through conceptual models. *J. Hydrol.* **1970**, *10*, 282–290. [CrossRef]
50. Rana, A.; Foster, K.; Bosshard, T.; Olsson, J.; Bengtsson, L. Impact of climate change on rainfall over Mumbai using Distribution-based Scaling of Global Climate Model projections. *J. Hydrol. Reg. Stud.* **2014**, *1*, 107–128. [CrossRef]
51. Gupta, H.V.; Kling, H.; Yilmaz, K.K.; Martinez, G.F. Decomposition of the mean squared error and NSE performance criteria: Implications for improving hydrological modelling. *J. Hydrol.* **2009**, *377*, 80–91. [CrossRef]
52. Themeßl, M.J.; Gobiet, A.; Heinrich, G. Empirical-statistical downscaling and error correction of regional climate models and its impact on the climate change signal. *Clim. Change* **2012**, *112*, 449–468. [CrossRef]
53. Kotlarski, S.; Bosshard, T.; Lüthi, D.; Pall, P.; Schär, C. Elevation gradients of European climate change in the regional climate model COSMO-CLM. *Clim. Change* **2012**, *112*, 189–215. [CrossRef]
54. Cherchi, A.; Alessandri, A.; Masina, S.; Navarra, A. Effects of increased CO_2 levels on monsoons. *Clim. Dyn.* **2010**, *37*, 83–101. [CrossRef]
55. Kripalani, R.H.; Oh, J.H.; Kulkarni, A.; Sabade, S.S.; Chaudhari, H.S. South Asian summer monsoon precipitation variability: Coupled climate model simulations and projections under IPCC AR4. *Theor. Appl. Climatol.* **2007**, *90*, 133–159. [CrossRef]
56. Goswami, B.N.; Venugopal, V.; Sengupta, D.; Madhusoodanan, M.S.; Xavier, P.K. Increasing trend of extreme rain events over India in a warming environment. *Science* **2006**, *314*, 1442–1445. [CrossRef] [PubMed]
57. Kumar, K.K.; Patwardhan, S.K.; Kulkarni, A.; Kamala, K.; Rao, K.K.; Jones, R. Future projection of Indian summer monsoon variability under climate change scenario: An assessment from CMIP5 climate models. *Curr. Sci.* **2011**, *101*, 312–326.
58. Sharmila, S.; Joseph, S.; Sahai, A.K.; Abhilash, S.; Chattopadhyay, R. Future projection of Indian summer monsoon variability under climate change scenario: An assessment from CMIP5 climate models. *Glob. Planet. Chang.* **2014**, *124*, 62–78. [CrossRef]
59. Salvi, K.; Kannan, S.; Ghosh, S. High-resolution multisite daily rainfall projections in India with statistical downscaling for climate change impacts assessment. *J. Geophys. Res. Atmos.* **2013**, *118*, 3557–3578. [CrossRef]
60. Ojha, R.; Kumar, D.N.; Sharma, A.; Mehrotra, R. Assessing severe drought and wet events over India in a future climate using a nested bias-correction approach. *J. Hydrol. Eng.* **2013**, *18*, 760–772. [CrossRef]

61. Amarasinghe, U.A.; Sharma, B.R.; Aloysius, N.; Scott, C.; Smakhtin, V.; de Fraiture, C. *Spatial Variation in Water Supply and Demand across River Basins of India*; Research Report 83; International Water Management Institute: Colombo, Sri Lanka, 2004. [CrossRef]
62. Jhajharia, D.; Shrivastava, S.K.; Sarkar, D.; Sarkar, S. Temporal characteristics of pan evaporation trends under the humid conditions of northeast India. *Agric. For. Meteorol.* **2009**, *149*, 763–770. [CrossRef]
63. Jhajharia, D.; Dinpashoh, Y.; Kahya, E.; Choudhary, R.R.; Singh, V.P. Trends in temperature over Godavari River basin in Southern Peninsular India. *Int. J. Climatol.* **2014**, *34*, 1369–1384. [CrossRef]
64. McVicar, T.R.; Roderick, M.L.; Donohue, R.J.; Li, L.T.; van Niel, T.G.; Thomas, A.; Grieser, J.; Jhajharia, D.; Himri, Y.; Mahowald, N.M.; *et al.* Global review and synthesis of trends in observed terrestrial near-surface wind speeds: Implications for evaporation. *J. Hydrol.* **2012**, *416–417*, 182–205. [CrossRef]
65. Schewe, J.; Heinke, J.; Gerten, D.; Haddeland, I.; Arnell, N.W.; Clark, D.B.; Dankers, R.; Eisner, S.; Fekete, B.M.; Colón-González, F.J.; *et al.* Multimodel assessment of water scarcity under climate change. *PNAS* **2014**, *111*, 3245–3250. [CrossRef] [PubMed]
66. Immerzeel, W.W.; Droogers, P.; de Jong, S.M.; Bierkens, M.F.P. Large-scale monitoring of snow cover and runoff simulation in Himalayan river basins using remote sensing. *Remote Sens. Environ.* **2009**, *113*, 40–49. [CrossRef]
67. Masood, M.; Yeh, P.J.-F.; Hanasaki, N.; Takeuchi, K. Model study of the impacts of future climate change on the hydrology of Ganges–Brahmaputra–Meghna (GBM) basin. *Hydrol. Earth Syst. Sci. Discuss.* **2014**, *11*, 5747–5791. [CrossRef]
68. Mukhopadhyay, B. Signature and hydrologic consequences of climate change within the upper-middle Brahmaputra Basin. *Hydrol. Process.* **2013**, *27*, 2126–2143. [CrossRef]
69. Meenu, R.; Rehana, S.; Mujumdar, P. Assessment of hydrologic impacts of climate change in Tunga–Bhadra river basin, India with HEC-HMS and SDSM. *Hydrol. Process.* **2013**, *27*, 1572–1589. [CrossRef]
70. Kauffeldt, A.; Halldin, S.; Rodhe, A.; Xu, C.-Y.; Westerberg, I.K. Disinformative data in large-scale hydrological modelling. *Hydrol. Earth Syst. Sci.* **2013**, *17*, 2845–2857. [CrossRef]
71. Hagemann, S.; Chen, C.; Haerter, J.O.; Heinke, J.; Gerten, D.; Piani, C. Impact of a statistical bias correction on the projected hydrological changes obtained from three GCMs and two hydrology models. *J. Hydrometeorol.* **2011**, *12*, 556–578. [CrossRef]
72. Pechlivanidis, I.G.; Arheimer, B.; Donnelly, C.; Hundecha, Y.; Huang, S.; Aich, V.; Samaniego, L.; Eisner, S.; Shi, P. Analysis of hydrological extremes at different hydro-climatic regimes under present and future conditions. *Clim. Chang.* **2016**. submitted.
73. Wagner, P.D.; Kumar, S.; Schneider, K. An assessment of land use change impacts on the water resources of the Mula and Mutha Rivers catchment upstream of Pune, India. *Hydrol. Earth Syst. Sci.* **2013**, *17*, 2233–2246. [CrossRef]

10

Hydrological Evaluation of Lake Chad Basin using Space Borne and Hydrological Model Observations

Willibroad Gabila Buma *, Sang-Il Lee and Jae Young Seo

Department of Civil and Environmental Engineering, Dongguk University, Seoul 04620, Korea; islee@dongguk.edu (S.-I.L.); dabbi2011@naver.com (J.Y.S.)
* Correspondence: willibroad@naver.com.

Academic Editors: Yingkui Li and Michael A. Urban

Abstract: Sustainable water resource management requires the assessment of hydrological changes in response to climate fluctuations and anthropogenic activities in any given area. A quantitative estimation of water balance entities is important to understand the variations within a basin. Water resources in remote areas with little infrastructure and technological knowhow suffer from poor documentation, rendering water management difficult and unreliable. This study analyzes the changes in the hydrological behavior of the Lake Chad basin with extreme climatic and environmental conditions that hinder the collection of field observations. Total water storage (TWS) from the Gravity Recovery and Climate Experiment (GRACE), lake level variations from satellite altimetry, and water fluxes and soil moisture from Global Land Data Assimilation System (GLDAS) were used to study the spatiotemporal variability of the hydrological parameters of the Lake Chad basin. The estimated TWS varies in a similar pattern as the lake water level. TWS in the basin area is governed by the lake's surface water. The subsurface water volume changes were derived by combining the altimetric lake volume with the TWS over the drainage basin. The results were compared with groundwater outputs from WaterGAP Global Hydrology Model (WGHM), with both showing a somewhat similar pattern. These results could provide an insight to the availability of water resources in the Lake Chad basin for current and future management purposes.

Keywords: GRACE; Lake Chad basin; GLDAS; WGHM; groundwater; altimetry; hydrological changes

---

1. Introduction

In some developing parts of the world, very limited and low quality ground water data often hinder proper water management studies [1]. Moreover, the estimation of large-scale water balance using these limited ground-based measurements is prone to inaccuracies [2]. Sometimes, obtaining these datasets from the appropriate authorities involves lengthy administrative procedures, rendering studies extremely difficult.

Some, if not all of these, are associated with the Lake Chad Basin (LCB). Its scale and lack of modern infrastructure are major challenges for data collection, analysis and management [3]. Under these circumstances, satellite gravimetric, altimetry and hydrological models have proven useful in the study of these water bodies.

The Gravity Recovery and Climate Experiment (GRACE) is a joint mission between Deutsche Forschungsanstalt fur Luft und Raumfahrt (DLR) and National Aeronautics and Space Administration (NASA) that was launched in 2002. It records the Earth's time variable gravity field with a temporal and spatial resolution usually within a few hundreds of kilometers. These products together with the products [data] from satellite altimetry, Global Land Data Assimilation System (GLDAS), and WaterGAP Global Hydrology Model (WGHM) were used for this study. Satellite and hydrologic model products have widely been used for ground-based hydrological measurements and studies, and they also serve as inputs to land surface and atmospheric models [4,5]. They are also used to verify these models.

Lake Chad Basin (LCB) extends between latitude 6° N and 24° N, and between longitude 8° E and 24° E (Table 1). It covers an area of about 2,400,000 km^2, which is equivalent to 8% of the total area of the African Continent. About 20% of this total area is the conventional basin, which is under the mandate of the Lake Chad Basin Commission (LCBC).

Table 1. Morphometric data for Lake Chad.

Parameter	Lake Chad Basin
Location	6° N and 20° N, 7° E and 25° E
Catchment area	2.4×10^6 km^2
Conventional Basin	427,500 km^2
Lake area	1350 km^2

Lake Chad itself occupies the central region of the LCB. It is a closed lake, predominantly fed by two perennial rivers (the Chari and the Logone) and an ephemeral one (the Komadugu Yobe) (Figure 1). It serves as a source of freshwater and fish, and also aids pastoral and agricultural land for a population of 30 million across the basin by offering a relatively easy and permanent access to water [6].

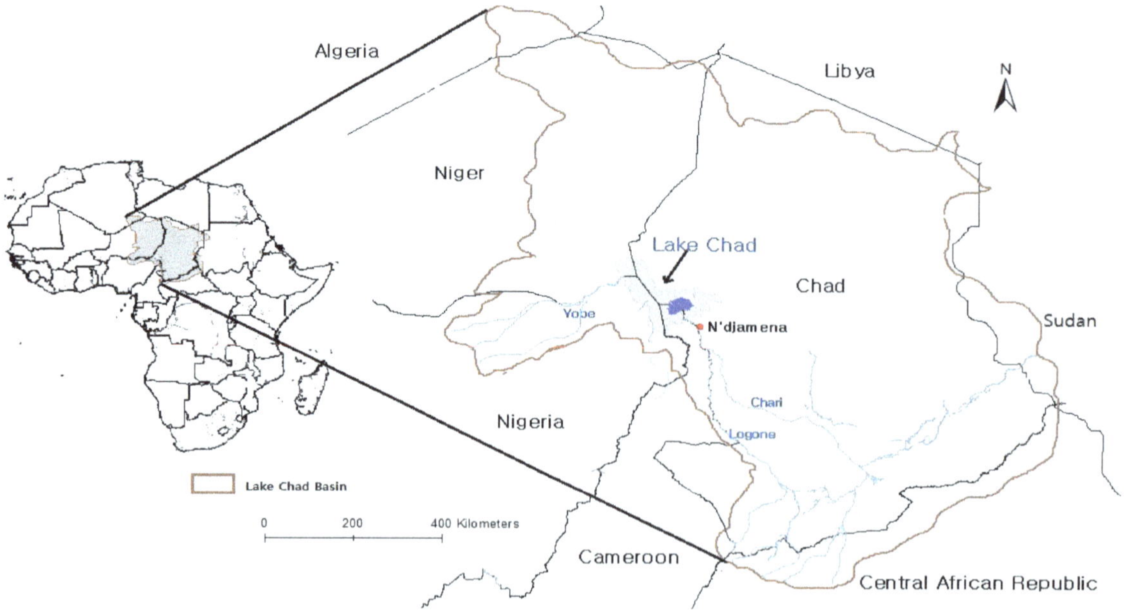

Figure 1. Locations of the Lake Chad Basin.

Increase in population, dam constructions, and irrigation development facilities during the last four decades have caused the surface area of Lake Chad to shrink from 24,000 km^2 to 1300 km^2 [7,8] (Figure 2). Studies have shown that the decrease was due to persistent drought and irrigation activities in the area [9–11].

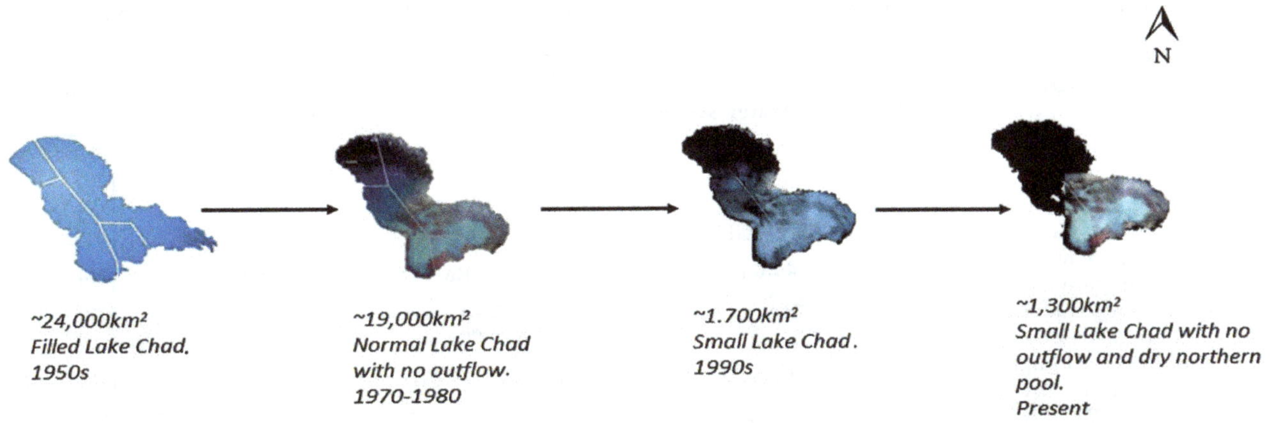

Figure 2. Schematics of the state of Lake Chad from Landsat 5 images; courtesy of NASA. Modified after [12].

In an attempt to manage and reduce the persistent droughts in this area, the water transfer project, whose main objective is halting the shrinkage of Lake Chad through an inflow of water coming from the Ubangi River, was introduced by the LCBC.

Water scarcity triggers food insecurity, poverty, migration and conflicts. As such, it is very important for a population or nations as a whole to secure stable and reliable water resource management techniques. A step towards this would be to understand the changes experienced by the water body in their vicinity. With limited and unreliable *in situ* data collection, understanding and documenting these changes can be very challenging and costly prompting researchers to rely on the use of satellite gravimetric, altimetry and hydrological models in monitoring water resources in such remote areas (Table 2).

Previous studies have applied remote sensing and satellite data to the Lake Chad region-some of which focused on the changes in stream flow patterns connected to the lake [9,13–15], Leblanc *et al.* [16] reported on the existence of a mega-lake Chad. They used satellite images from Landsat and Moderate Resolution Imaging Spectrometer (MODIS) for their studies. Thermal remote sensing techniques, such as the Meteosat thermal maximum composite data, was used to account for the variability of inundated areas within the lake under flooded vegetation [17]. Satellite imagery and GRACE data were used to study the regional hydrogeology and made an attempt at estimating the actual evapotranspiration over the LCB [5]. Altimetry data and ground-based information were used to predict the downstream lake and marsh heights using imperial regression techniques [18].

Few studies have used remote sensing data for investigating groundwater recharge around this area like [3,6] used satellite images (Meteosat thermal data) combined with hydrogeological data to identify the thermal change of groundwater in the depression zones, and then estimated values of recharge and discharge of the area. We try to define time series data of groundwater depending on the nature and fluctuation of this property in time and space.

In this study, we combined updated remotely sensed and hydrological model datasets. These datasets have been used to study the diverse aspects of basin hydrology within the continent [6,7,11,19–24]. Some of these point out the lack of readily available *in situ* data for these studies [5,22], some cases validated the *in situ* measurements in Lake Chad and other parts of Africa (Table 2).

Table 2. Literature review of remotely sensed data sets used in the studies of some watershed in Africa.

Study Area	Data Products			Reference
	Terrestrial Water Storage	Rainfall	Lake Height	
Lake Chad	GRACE [1]	GPCP [2]	–	[5]
Lake Chad	–	–	Sat. Alt. [3]	[25]
Lake Chad	–	NOAA [4], TRMM [5]	–	[12]
East African Great Lake	GRACE, WGHM [6]	GPCP	Sat. Alt.	[25]
Lake Victoria, Malawi and Tamganyika	GRACE	GLDAS, TRMM	Sat. Alt.	[26]
Okavango catchment	GRACE	TRMM	Sat. Alt.	[27]
Congo river basin	GRACE	GLDAS, TRMM	Sat. Alt.	[26]
Lake Victoria, Tamganyika and Malawi	GRACE, WGHM	GLDAS, TRMM	Sat. Alt.	[28]

[1] GRACE: Gravity Recovery and Climate Experiment; [2] GPCP: Global Precipitation Climatology Project; [3] Sat. Alt.: Satellite Altimetry; [4] NOAA: National Oceanic and Atmospheric Administration; [5] TRMM: Tropical Rainfall Measuring Mission; [6] WGHM: WaterGap Hydrological Model.

We utilized satellite gravimetric, altimetric and hydrological models products over the Lake Chad basin to characterize the spatiotemporal and multiscale variability in its hydrological cycle, to infer the effect of rainfall on water storage in this region, and, finally, to investigate subsurface water variations within this region and perform comparison with groundwater outputs from a global hydrological model.

2. Materials and Methods

2.1. Terrestrial Water Storage (TWS) from GRACE

GRACE is known for estimating high-precision time-varying gravity field and the changes of Earth's surface mass at a high degree of accuracy on a time scale ranging from months to a decade [29,30]. These variations are mainly due to redistribution of water mass in the surface fluid envelopes of the earth. It provides estimates of TWS, which encompasses surface water, soil moisture, groundwater and snow. However, experimental errors while using GRACE increase rapidly and concurrently as the degree of the spherical harmonic coefficients, causing inaccurate results at higher degree terms of the spherical harmonic coefficients [31,32]. Spatial averaging functions are normally used to reduce the high degree of noise in the GRACE gravity field. This provides researchers with accurate surface mass changes. An additional de-stripping averaging filter is used for suppressing the "N–S" stripping noise in the GRACE data. There is also a leakage effect, which is caused by the spatial averaging functions. This causes some signals of the GRACE mass anomalies to leak outside the region of interest. The accuracy of GRACE is high enough to detect surface mass variations corresponding to hydrological loads of 1 cm at monthly and longer time scales, with horizontal dimensions of hundreds of kilometers and larger [30].

Numerous studies on the reliability of its data sets has been carried out by comparing its TWS products to that of Land Surface Models or *in situ* land observations-India [31], the Korean Peninsula [33,34], the East African lakes [25], Mali in Africa [35]. It has also been widely used in the studies of lakes around the world [36–39].

The GRACE Level 3 (Release 05) is the latest and more accurate of GRACE products. It provides processed time variability gravity field products. These products are provided as sets of spherical harmonic coefficients averaged on a monthly scale.

For this study, we used the monthly land mass grid observations (Level 3) provided by the Center of Space Research (CSR), University of Texas, at Austin from January 2003 to December 2013. The data are available as monthly $1^\circ \times 1^\circ$ grids of TWS over our study area [40]. The data set was truncated at 60 degrees and smoothed with the Gaussian filter of 300 km. GRACE data enhancement techniques provided [30] were also included to improve the accuracy of these TWS estimates.

For the time frame of our study, we encountered six months of missing data during these dates: June 2011, May and October 2012, and March, August, and November 2013. For these missing datasets, the temporal linear interpolation was done between them since they were not contiguous.

2.2. *Lake Height from Altimetry*

Generally, satellite altimeters are nadir-pointing instruments that record the average surface "spot" height directly below the satellite as it transverses over the Earth's surface. It basically determines the distance from the satellite to a target surface by measuring the satellite-to-surface round-trip time of a radar pulse. The altimeter emits a radar wave and analyzes the return signal that bounces off the surface. The surface height is calculated as the difference between the satellite's positions on orbit with respect to an arbitrary reference surface, *i.e.*, the Earth's center is represented by a reference ellipsoid and the range between the satellite and the surface is obtained by calculating the time taken for the signal to return. From this, the measurements of the sea surface height and other characteristics of oceans, lakes, floodplains, and rivers can be obtained. A lot of information can be extracted from satellite altimetry.

Institutions like the Foreign Agricultural Services (FAS) of the United States Department of Agriculture, Hydroweb, and European Space Agency (ESA) have been making available the up-to-date and reliable user-friendly data sets.

Lakes, rivers and oceans have all been monitored over the years using these data sets [9,11,41–44]. Surface water level data sets are sometimes given in the form of graphs and tables for major water bodies based on combination of various radar altimetry sensors. These data sets are made available free of charge via web applications, such as USDA's Global Reservoir and Lake Monitor (http://www.pecad.fas.usda.gov), Hydroweb of Geodesy, Oceanography and Hydrology from Space (GOHS; http://www.legos.obs-mip.fr/), and River and Lake system provided by ESA (http://tethys.eaprs.cse.dmu.ac.uk). Repeat track methods used in the derivation of time series of the lake surface height variation uses the reference lake height profile. This is derived from averaging all height profiles across the lake within a given time span. This effectively smoothens out any varying effects of tide and wind set-up. These resulting time series of height variations are expected to have an accuracy of about 20 cm root mean square (RMS) for lakes with minimal tides and limited dynamic variability.

For our study area, the satellite altimetry data has widely been used in the studies of the lake in which *in situ* datasets were compared with these altimetry products. The results showed accurate water level variations for Lake Chad in the two data sets [11,40,41]. Altimetry missions with a 10-day repeat track, such as TOPEX/Poseidon (1992–2006), Jason-1 (2001–2013), and Jason-2 (2008–present), or those with a 35-day repeat track, such as ERS-2 (1995–2000) and ENVISAT (2002–2010), can be used to extract lake height variations. ENVISAT altimetry estimates were used for this study.

2.3. *Soil Moisture from GLDAS*

Operated by NASA and the National Oceanic and Atmospheric Administration (NOAA), GLDAS is a land surface simulation system that aims to ingest satellite and ground-based observational data products, using advanced land surface modeling and data assimilation techniques, in order to generate optimal fields of land surface state (e.g., soil moisture and surface temperature) and flux (e.g., evaporation and sensible heat flux) products [4]. It adopts four advanced land surface models (LSMs): The Community Land Model (CLM), Mosaic, Noah, and Variable Infiltration Capacity (VIC). GLDAS executes spatial resolutions globally at both 0.25° and 1.0°, with temporal resolutions of three hours and monthly products since 1979. GLDAS data are widely used for land-surface flux simulations. As such, the simulation accuracy using GLDAS dataset is largely contingent upon the accuracy of the GLDAS dataset. The data are available from the Goddard Earth Sciences Data and Information Services Center (GES DISC).

In this study, we used Noah 1.0° grid data which has four layers of vertical soil moisture. The monthly average soil moisture is computed as the sum of all the layers [45].

2.4. Groundwater Estimates from the WaterGap Hydrological Model

The WaterGAP Global Hydrological Model (WGHM) is a submodel of the global water use and availability model WaterGAP 2.2 It computes groundwater recharge, surface runoff and river discharge as well as storage variations of water in canopy, rivers, soil, lakes, wetlands, groundwater and snow at a spatial resolution of 0.5° [46].

WGHM is based on the best global data sets currently available, and it is able to simulate variations in water bodies. It computes the water storage in the snow pack, rooted soil zone, groundwater, on vegetation surfaces, and in surface water reservoirs. Here, the simulated estimates provided by [47] were used. In order to obtain a reliable estimate of water availability, they tuned the model against the observed discharge at 1235 gauging stations, which represent 50% of the global land area and 70% of the actively discharging area. In Africa, most basins north of the equator do not perform well [47]. Detailed information about the modelling concept and its corresponding assumptions can be found in [48].

Model outputs assessment performance was not carried out in the LCB due to limited data availability. However, the Chari and Komadougou Yobe river basins, which predominantly feed the lake, were included in the calibration scheme of this region. Too much water was modeled for both basins. The Chari-Logone river system, which supplies most of the water into the southern part of LC, has a Nash-Sutcliffe efficiency of around 0.6, which is quite good. On other hand, inland water bodies in Africa showed a good match with the WGHM model output-for instance, the East African great lakes as reported by [46].

For this study, outputs from the WaterGap 2.2a model forced with precipitation from the Global Precipitation Climatology Centre (GPCC) and data from the European Center for Medium-Range Weather Forecast (ECMWF) integrated forecast system were used. These outputs include; global-scaled gsssroundwater storage, total water storage, baseflow, and groundwater recharge (diffused and below surface water bodies). There is no data available after 2009 (Table 3). This can be found on the website (https://www.uni-frankfurt.de/49903932/7_GWdepletion) [49].

Table 3. Summary of data sets used for this study.

Variable	Dataset	Resolution		Period
		Spatial	Temporal	
Terrestrial Water Storage	GRACE	$1° \times 1°$	1 month	2003–2013
Lake Height	Sat. Alt.	$1° \times 1°$	30 days	2003–2013
Rainfall	GLDAS	$1° \times 1°$	1 month	2003–2013
Soil moisture	GLDAS	$1° \times 1°$	1 month	2003–2013
Groundwater	WGHM	$0.5° \times 0.5°$	1 month	2003–2009

2.5. Data Processing

2.5.1. Variability in TWS and Lake Height

Seasonal-Trend Decomposition Procedure based on Loess (STL) method is a filtering procedure that decomposes a time series into its additive components of variation (trend, seasonal and the remainder components) by the application of Loess smoothing models [50]. This was used to model GRACE monthly storage variations as well as the time series of altimetric lake height.

In brief, the steps performed during STL decomposition are as follows:

1. Cycle-subseries smoothing: series are built for each seasonal component, and smoothed separately.

2. Low-pass filtering of smoothed cycle-subseries: the subseries are put together again, and smoothed.
3. Detrending of the seasonal series.
4. Deseasonalizing the original series, using the seasonal component calculated in the previous steps; and Smoothing the deseasonalized series to get the trend component.

In R statistical software, the STL algorithm is available through the *stl* function. We use it with its default parameters. The degrees for the Loess fitting are $d = 1$ in steps (iii) and (iv), and $d = 0$ in step (ii).

The parameter values must be chosen by the data analyst. We assume each observation X_i in time series is the sum of these components:

$$X_i = T_i + S_i + I_i \tag{1}$$

where T_i = Trend, S_i = Seasonality and I_i = Interannual components.

Often, six parameters determine the degree of smoothing in trend and seasonal components. For detailed information on method and parameters, consult [50] paper on STL methods. For our study, these parameters are:

$n_{(p)}$: The number of observations in each seasonal cycle, = 12 months (yearly periodicity with monthly data);
$n_{(i)}$: The number of passes through the inner loop (usually set to equal one or two) = 1 month;
$n_{(o)}$: The number of robustness iterations of the outer loop (Values qual one or two) = po robustness while a zero value has no robustness iteration) = 5 months;
$n_{(l)}$: The span of the loess window for the low-pass filter (computed as the next odd number to $n_{(p)}$) = 13 months;
$n_{(s)}$: The smoothing parameter for the seasonal component, = 12 months (seasonal length is same as the periodic length);
$n_{(t)}$: The smoothing parameter for the trend component, = 22 months.

$$n_{(t)} \geqslant \left[\frac{1.5n_{(p)}}{1 - 1.5n_{(s)}^{-1}} \right] \times 2 \tag{2}$$

For this analysis, R statistical software was used [51]. It is a free software environment and a programming language for statistical computing and graphics. It is widely used among statistics and data miners for developing statistical software and data analysis. MS excel was also used for subsequent data representation and analysis.

2.5.2. Subsurface Water Volume Change

Subsurface water volume (Groundwater + Soil moisture) was investigated. GRACE data provides changes in total water storage, which includes Lake water storage (LS), Snow water equivalent storage (SWES), soil moisture storage (SMS), and groundwater storage (GWS) within the basin. With satellite and model-based estimates of LS and SMS, subsurface water volume can be estimated. SWES was ignored for our study area since this area is humid.

Estimates of the subsurface water changes was evaluated using the following disaggregation equation,

$$\Delta SSW = \Delta SM + \Delta GW = \Delta TWS - \Delta LS \tag{3}$$

Here, ΔSSW = Subsurface Water, ΔGW = Groundwater, ΔLS = Lake water, ΔTWS = Terrestrial Water Storage, ΔSM = Soil Moisture.

In an attempt to express ΔSSW and ΔLS in terms of volume, both were multiplied by the LCB area and Lake Area, respectively.

3. Results and Discussions

3.1. TWS and Altimetry Lake Height

Based on our study period, the STL trend of the time series of monthly GRACE TWS shows a decrease in average TWS of the Lake Chad basin (Figure 3) for the periods 2003–2005 and 2009–2010 with the latter being the lowest water estimates at −0.54 cm/year. There is an increase in TWS concentrations from 2006 to 2008, and 2010 through 2013, with the latter being the highest storage estimates of 0.69 cm/year.

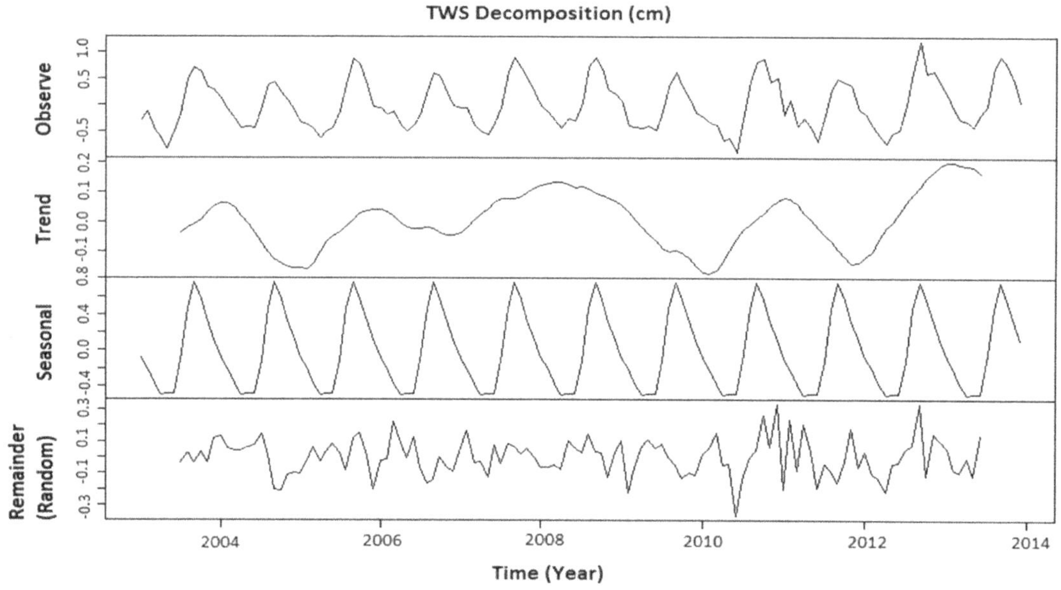

Figure 3. STL decomposition of the time series of monthly GRACE TWS.

The STL trend of the time series of altimetric lake height (Figure 4) shows a decrease in lake level from 2003 through 2005 and a steady increase until after 2008 with an average height of about 0.3 m/year. From this point, it begins to slope down towards 2010. From 2010 to 2012, the Lake experiences its lowest height averaging to about 0.23 m/year. Different rates are shown in Table 4.

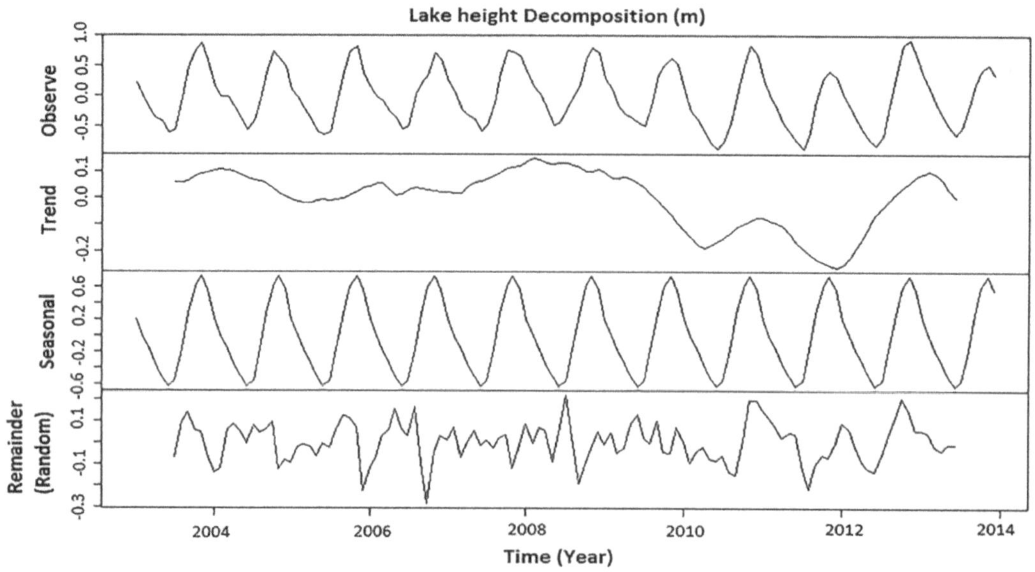

Figure 4. STL decomposition of the time series of monthly lake altimetric height.

Table 4. STL fitted trend of the time series of TWS and lake height.

Period	TWS (cm)	Lake Height (m)
2003–2005	0.25	0.32
2006–2009	0.84	0.62
2010–2013	0.38	−0.78

The STL decomposition plot of the monthly TWS estimates and lake altimetric height shows a similar pattern with their seasonal components suggesting an annual increase from the months of July–September as well as a decrease from October–June. This implies the Lake's height follows the seasonal pattern of the rainfall cycle around this area. There is a correlation (>80%) between them which points out the similarity in their pattern (Figure 5a).

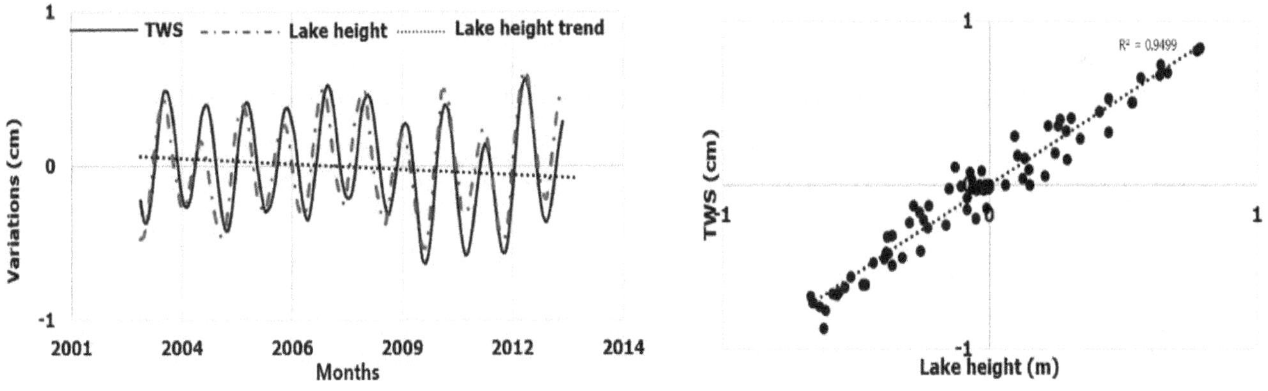

(a) (b)

Figure 5. (**a**) Represents TWS and lake height after smoothing with a six-month window; (**b**) Represents the autocorrelation between TWS and lake height.

Their seasonal component suggests an average annual increase in September and a main annual drop exists in November. This is due to the rainfall regime that exists over the Lake region. This relationship will be discussed in Section 3.2. The largeness and uniform size of the seasonal cycle of Lake Chad means that, over the years, Lake Chad has a fast water renewal process.

3.2. TWS and Rainfall

Rainfall estimates from GLDAS were compared to GRACE TWS over the LCB. During the wet season (July–September), there is an increase in seasonal pattern of rainfall over the study region with 2012 having the maximum annual averages. Figure 6 shows a comparison between the time series of the monthly estimated GLDAS rainfall and the change in GRACE TWS and, as expected, both curves show a good agreement during most of the study period in terms of pattern.

Based on Figure 6, we can clearly see the existence of a phase shift between GLDAS rainfall and GRACE TWS. This phase shift is about a month and a half. Their lagged correlation was also high (>0.9). From trend analysis, rainfall that precedes TWS increases throughout the study period. Its seasonal cycle goes ahead to confirm this phase shift that exists between rainfall and TWS in this region.

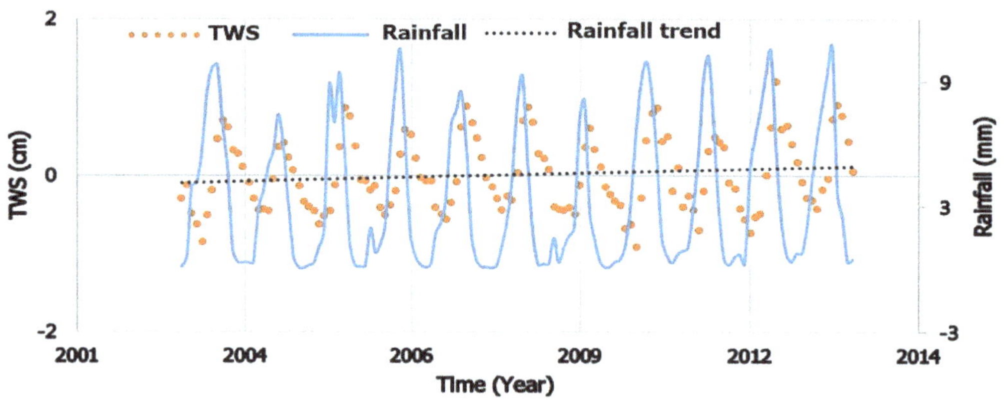

Figure 6. Comparisons between the yearly estimates of rainfall and TWS.

Figure 7 also confirms the bimodal rainfall regime that exists within this region with most of its heavy rainfall occurring between July–September and shorter rains from October–December.

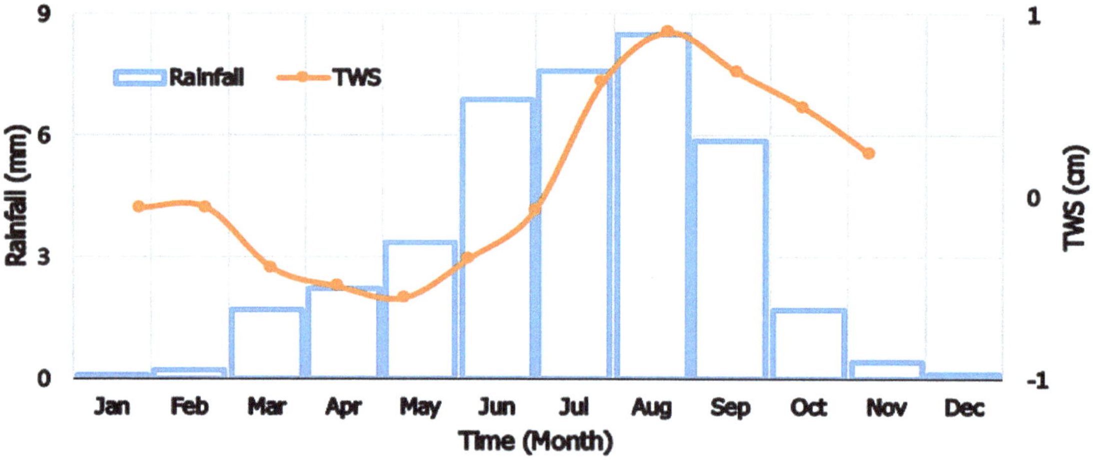

Figure 7. Comparisons between the monthly average rainfall and TWS.

3.3. TWS and Soil Moisture

Terrestrial water storage (TWS) of the Lake Chad Basin and the lake's surface water volume change from altimetry are critically analyzed. Due to its improved precision, GRACE can detect gravitational changes within an area of 200,000 km^2 or larger [44]. Hence, it is suitable for our study area which is 427,500 km^2.

Figure 8 shows a good agreement between altimetric lake water volume with GRACE basin water volume with a correlation of about 73%. There is also a good agreement in their seasonal cycles (Figure 9). We can infer that surface water of Lake Chad governs the volume of water in the basin area.

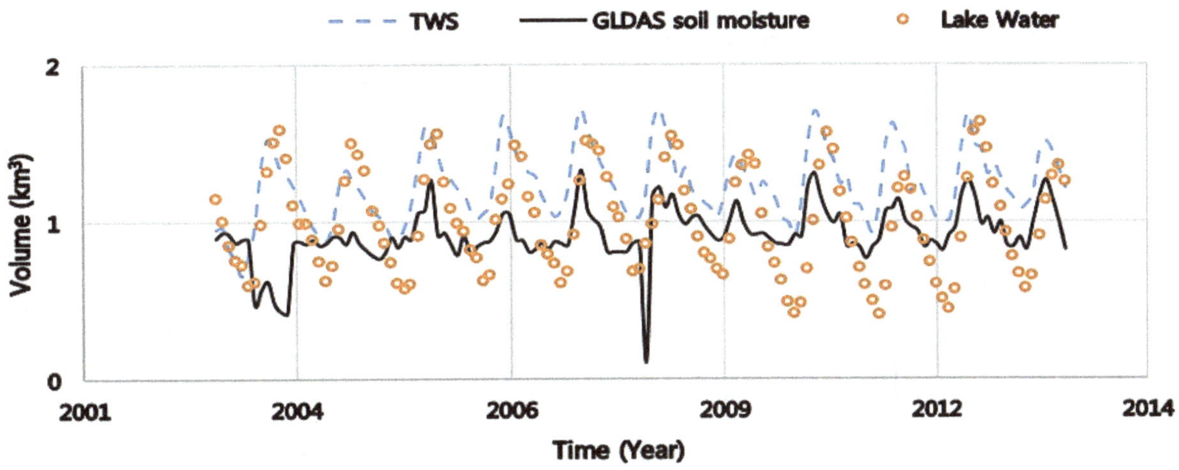

Figure 8. Effect of altimetric lake water and soil moisture on TWS.

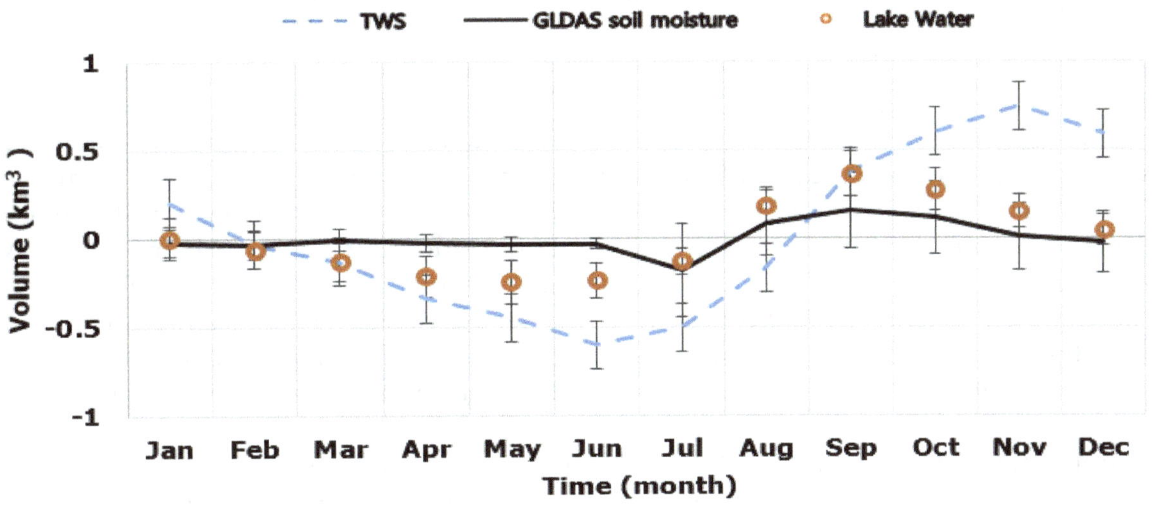

Figure 9. Monthly averaged seasonal cycle of the Lake Chad. Error bars represent the standard deviation for each month.

GLDAS outputs of soil moisture estimates can also be seen in Figure 8. It has a weak agreement with basin water volume with a correlation of about 35%. Hence, soil moisture content plays little or no role in the variability in water volume stored in the Lake Chad Basin. It is solely governed by surface water of the Lake Chad.

Additionally, investigated were the changes in subsurface water volume, which comprises of groundwater and soil moisture (Section 2.5.2). In early 2003 and 2007, there is a dissimilarity between volume estimates of WGHM groundwater and subsurface water estimates. The results obtained were compared with WGHM outputs (Figure 10). It shows two peak periods at approximately mid-2005 and 2006. Both curves have a somewhat similar pattern with a correlation coefficient of about 47% between them.

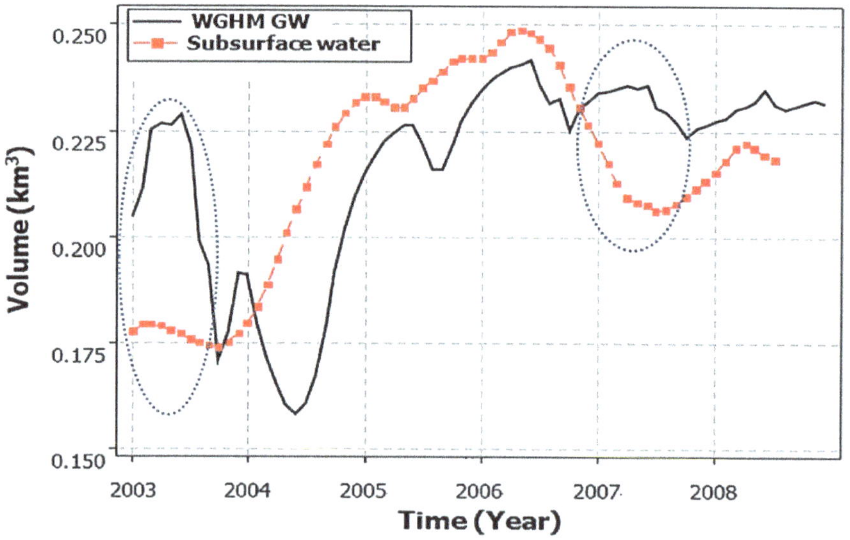

Figure 10. Changes in subsurface water (groundwater and soil moisture) and WGHM GW outputs. The annual signal is removed and data are smoothed with a six-month window.

4. Conclusions

In this study, we see how the use of GRACE TWS data along with other data sources could be of great value in the hydrological studies of remote areas. We investigated hydrological variability associated within the LCB from January 2003 to December 2013. GRACE TWS and altimetric lake volume showed a similar pattern for the majority of the study period. After comparing TWS with soil moisture content, we found that the discharge in this area is governed by surface water of the lake. This is the first attempt in which these remotely sensed datasets were used to study the varying patterns of the lake's hydrology. Deriving this characteristic spatiotemporal analysis of surface mass anomalies across the LCB, future improvements can be made in the management of water resources in this area.

For much of the study period, GRACE TWS variations within the basin show a similar pattern of variation as the averaged lake height variation from altimetry. A trend analysis showed increasing precipitation with maximum annual average increase in August but decreasing water level in the lake from altimetry with the minimum annual average occurring in 2012 with a value of 0.2 m. Our study also showed that altimetry-based volume with TWS from GRACE provides information on soil moisture and groundwater. This could help in detecting subsurface water storage changes in relation to climate variability or anthropogenic activities especially in situations where *in situ* measurements are not available. This could also lead to new research prospects, where researchers could try to find out the main cause behind the lake's decreasing trend.

This characterization can help in the proposed Water Transfer Project from the Ubangi River to Lake Chad in a number of ways. For instance, the primary objective of this project is to halt the shrinkage of Lake Chad through an inflow of water coming from the Congo basin. The association between LC level and precipitation will enable managers to plan for the total water volume that can be released and retained based on future forecasts.

Acknowledgments: This research was supported by a grant (15AWMP-B079625-02) from the Water Management Research Program funded by the Ministry of Land, Infrastructure and Transport of Korean government and the Basic Science Research Program through the National Research Foundation of Korea (NRF) funded by the Ministry of Education (NRF-2015R1D1A1A09060690).

Author Contributions: Willibroad G. Buma developed the methodology and conducted the work under the supervision and review of Sang-Il Lee. Jae Young provided technical assistance and helped in developing the manuscript.

References

1. Swenson, S.; Wahr, J. Monitoring the water balance of Lake Victoria, East Africa, from space. *J. Hydrol.* **2009**, *370*, 163–176. [CrossRef]
2. Allen, R.; Joseph, J.; Kevin, L.; Ahjond, S. Adaptive management for a turbulent future. *Environ. Manag.* **2011**, *92*, 1339–1345. [CrossRef] [PubMed]
3. Leblanc, M.; Leduc, C.; Razack, M.; Lemoalle, J.; Dagorne, D.; Mofor, L. Applications of remote sensing and GIS for groundwater modelling of large semiarid areas: Example of the Lake Chad Basin, Africa. *IAHS Publ.* **2003**, *278*, 186–192.
4. Rodell, M.; Houser, P.R.; Jambor, U.; Gottschalck, J.; Mitchell, K.; Meng, C.-J.; Arsenault, K.; Cosgrove, B.; Radakovich, J.; Bosilovich, M.; *et al.* The global land data assimilation system. *Am. Meteorol. Soc.* **2004**, *85*, 381–394. [CrossRef]
5. Boronina, A.; Ramillien, G. Application of AVGRR imagery and GRACE measurements for calculation of actual evapotranspiration over the Quaternary aquifer (Lake Chad basin) and validation of groundwater models. *Hydrol. J.* **2008**, *348*, 98–109. [CrossRef]
6. Leblanc, M.; Favreau, G.; Tweed, S.; Leduc, C.; Razack, M.; Mofor, L. Remote sensing for ground water modelling in large semiarid areas: Lake Chad basin, Africa. *Hydrol. J.* **2007**, *15*, 97–100.
7. Gao, H.; Bohn, T.J.; Podest, E.; McDonald, K.C.; Lettenmaier, D.P. On the causes of the shrinking of Lake Chad. *Environ. Res. Lett.* **2011**, *6*, 034021. [CrossRef]
8. Singh, A.; Diop, S.; M'mayi, P.L. Africa's Lakes: Atlas of Our Changing Environment (Nairobi: UNEP). Available online: www.unep.org/dewa/AfricaAfricaAtlas/ (accessed on 10 May 2016).
9. Campbell, R.W. Earthshots: Satellite Images of Environmental. Available online: http://earthshots.usgs.gov/ (accessed on 10 May 2016).
10. Coe, M.T.; Foley, J.A. Human and natural impacts on the water resources of the Lake. *Geophys. Res. Atmos.* **2001**, *106*, 3349–3356. [CrossRef]
11. Birkett, C.M. Synergistic remote sensing of Lake Chad: Variability of basin inundation. *Remote Sens. Environ.* **2000**, *72*, 218–236. [CrossRef]
12. Churchill, O.; Belay, D.; Sium, G. Characteristics of Lake Chad Level Variability and Links to ENSO, precipitation and river discharge. *Sci. World J.* **2014**, *13*, 145893.
13. Li, K.Y.; Coe, M.T.; Ramankutty, N.; De Jong, R. Modeling the Hydrological Impact of Land-Use Change in West Africa. *J. Hydrol.* **2007**, *337*, 258–268. [CrossRef]
14. Li, B.; Rodell, M.; Zaitchik, B.F.; Reichle, R.H.; Koster, R.D.; Tonie, M.A. Assimilation of Grace Terrestrial Water Storage into a Land Surface Model: Evaluation and Potential Value for Drought Monitoring in Western and Central Europe. *J. Hydrol.* **2012**, *446–447*, 103–115. [CrossRef]
15. Le Coz, M.; Delclaux, F.; Genthon, P.; Favreau, G. Assessment of Digital Elevation Model (Dem) Aggregation Methods for Hydrological Modeling: Lake Chad Basin, Africa. *Comp. Geosci.* **2009**, *35*, 1661–1670. [CrossRef]
16. Leblanc, M.; Favreau, G.; Maley, J.; Nazoumou, Y.; Leduc, C.; Stagnitti, F.; van Oevelen, P.; Delclaux, F.; Lemoalle, J. Reconstruction of Megalake Chad using Shuttle Radar Topographic Mission data. *Palaeogeogr. Palaeoclimatol. Palaeoecol.* **2006**, *239*, 16–27. [CrossRef]
17. Leblanc, M.; Razack, M.; Dagorne, D.; Mofor, L.; Jones, C. Application of Meteosat thermal data to map soil infiltrability in the central part of the Lake Chad basin, Africa. *Geophys. Res. Lett.* **2003**, *30*. [CrossRef]
18. Coe, M.T.; Birkett, C.M. Calculation of river discharge and prediction of lake height from satellite radar altimetry: Example for the Lake Chad basin. *Water Resour. Res.* **2004**, *40*. [CrossRef]
19. Isiorho, S.A.; Matisoff, G. Groundwater seepage and its implication on the water resources planning and management in the Chad Basin. *Water Resour. Res.* **1989**, *1*, 210–215.
20. Isiorho, S.A.; Matisoff, G. Groundwater recharge from Lake Chad. *Limnol. Oceanogr.* **1990**, *35*, 931–938. [CrossRef]
21. Leduc, C.; Sabljak, S.; Taupin, J.D.; Marlin, C.; Favreau, G. Estimation de la recharge de la nappe quaternaire dans le Nord-Ouest du bassin du lac Tchad (Niger oriental) à partir de mesures isotopiques. *Earth Planet. Sci. Lett.* **2000**, *330*, 355–361. [CrossRef]
22. Goni, I.B. Tracing stable isotope values from meteoric water to groundwater in the southwestern part of the Chad basin. *Hydrogeol. J.* **2006**, *14*, 742–752. [CrossRef]

23. Ngounou, N.B.; Jacques, M.; Jean, S.R. Groundwater Recharge from Rainfall in the Southern Border of Lake Chad in Cameroon. *World App. Sci. J.* **2007**, *2*, 125–131.
24. Lemoalle, J.; Leblanc, M.; Bader, J.C.; Tweed, S.; Mofor, L. Thermal remote sensing of water under flooded vegetation: New observations of inundation patterns for the Small Lake Chad. *Hydrol. Res.* **2011**, *404*, 87–98.
25. Becker, M.; LLovel, W.; Cazenave, A.; Guntner, A.; Crétaux, J.F. Recent hydrological behaviour of the East African great lakes region inferred from GRACE, satellite altimetry and rainfall observations. *Comptes Rendus Geosci.* **2010**, *342*, 223–233. [CrossRef]
26. Jean-Paul, B.; Jacques, H.; Caroline, D.L. Retrieval of Large-Scale Hydrological Signals in Africa from GRACE Time-Variable Gravity Fields. *Pure Appl. Geophys.* **2012**, *169*, 1373–1390.
27. Milzow, C.; Krogh, P.E.; Bauer-Gottwein, P. Combining satellite radar altimetry, SAR surface soil moisture and GRACE total storage changes for hydrological model calibration in a large poorly gauged catchment. *Hydrol. Earth Syst. Sci.* **2011**, *15*, 1729–1743. [CrossRef]
28. Hassan, A.A.; Jin, S. Lake level change and total water discharge in East Africa Rift Valley from satellite-based observations. *Global Planet Change* **2014**, *117*, 79–90. [CrossRef]
29. Tapley, B.D.; Bettadpur, S. The gravity recovery and climate experiment: Mission overview and early results. *Geophys. Res. Lett.* **2004**, *31*. [CrossRef]
30. Wahr, J.; Swenson, S.; Zlotnicki, V.; Velicogna, I. Time-variable gravity from GRACE: First results. *Geophys. Res. Lett.* **2004**, *31*. [CrossRef]
31. Swenson, S.; Chambers, D.; Wahr, J. Estimating geocenter variations from a combination of GRACE and ocean model output. *J. Geophys. Res. Solid Earth* **2008**, *113*. [CrossRef]
32. Rodell, M.; Velicogna, I.; Famiglietti, J.S. Satellite-based estimates of groundwater depletion in India. *Nature* **2009**, *460*, 999–1002. [CrossRef] [PubMed]
33. Lee, S.I.; Kim, J.S.; Lee, S.K. Estimation of average terrestrial water storage changes in the Korean Peninsula using GRACE satellite gravity data. *J. Korea Water Resour. Assoc.* **2012**, *8*, 805–814. [CrossRef]
34. Lee, S.I.; Seo, J.Y.; Lee, S.K. Validation of terrestrial water storage change estimates using hydrologic simulation. *J. Water Resour. Ocean Sci.* **2014**, *4*, 5–9. [CrossRef]
35. Henry, C.M.; Allen, D.M.; Huang, J. Groundwater storage variability and annual recharge using well-hydrograph and GRACE satellite data. *Hydrogeol. J.* **2011**, *19*, 741–755. [CrossRef]
36. Joseph, L.A.; Sharifi, M.A.; Ogonda, G.; Wickert, J.; Grafarend, E.W.; Omulo, M.A. The Falling Lake Victoria Water Level: GRACE, TRIMM and CHAMP Satellite Analysisof the Lake Basin. *Water Resour. Manag.* **2008**, *22*, 775–796.
37. Song, C.; Huang, B.; Ke, L. Modeling and analysis of lake water storage changes on the Tibetan Plateau using multi-mission satellite data. *Remote Sens. Environ.* **2013**, *135*, 25–35. [CrossRef]
38. Longuevergne, L.; Wilson, C.R.; Scanlon, B.R.; Crétaux, J.-F. GRACE water storage estimates for the Middle East and other regions with significant reservoir and lake storage. *Hydrol. Earth Syst. Sci.* **2013**, *17*, 4817–4830. [CrossRef]
39. Tourian, M.J.; Elmi, O.; Chen, Q.; Devaraju, B.; Roohi, Sh.; Sneeuw, N. A spaceborne multisensor approach to monitor the desiccation of Lake Urmia in Iran. *Remote Sens. Environ.* **2015**, *156*, 349–360. [CrossRef]
40. NASA MEaSUREs Program. Available online: http://grace.jpl.nasa.gov (accessed on 10 May 2016).
41. Crétaux, J.-F.; Birkett, C.M. Lake studies from satellite radar altimetry. *Comptes Rendus Geosci.* **2006**, *338*, 1098–1112. [CrossRef]
42. Kouraev, A.V.; Semovski, S.V.; Shimaraev, M.N.; Mognard, N.M.; Légresy, B.; Remy, F. Observations of Lake Baikal ice from satellite altimetry and radiometry. *Remote Sens. Environ.* **2007**, *108*, 240–253. [CrossRef]
43. Kropacek, J.; Braun, A.; Kang, S.C.; Feng, C.; Ye, Q.H.; Hochschild, V. Analysis of lake level changes in Nam Co in central Tibet utilizing synergistic satellite altimetry and optical imagery. *Int. J. Appl. Earth Obs. Geoinf.* **2012**, *17*, 3–11. [CrossRef]
44. Duan, Z.; Bastiaanssen, W.G.M. Estimating water volume variations in lakes and reservoirs from four operational satellite altimetry products and satellite imagery data. *Remote Sens. Environ.* **2013**, *134*, 403–416. [CrossRef]
45. Goddard Earth Sciences Data and Information Services Center. Available online: http://grace.jpl.nasa.gov (accessed on 10 May 2016).

46. Döll, P.; Kaspar, F.; Bernhard, L. A global hydrological model for deriving water availability indicators: Model tuning and validation. *J. Hydrol.* **2003**, *270*, 105–134. [CrossRef]
47. Döll, P.; Muller, H.S.; Schuh, C.; Portmann, F.T. Global-scale assessment of groundwater depletion and related groundwater abstractions: Combining hydrological modeling with information from well observations and GRACE satellites. *Water Resour. Res.* **2014**, *50*, 5698–5720. [CrossRef]
48. Güntner, A.; Stuck, J.; Döll, P.; Schulze, K.; Merz, B. A global analysis of temporal and spatial variations in continental water storage. *Water Resour. Res.* **2007**, *43*. [CrossRef]
49. Data source for the groundwater depletion. Available online: https://www.uni-frankfurt.de/49903932/7_GWdepletion (accessed on 10 May 2016).
50. Cleveland, R.B.; Clevelan, W.S.; McRae, J.E.; Thacker, E.L. STL: A seasonal-trend decomposition procedure based on loess. *J. Off. Stat.* **1990**, *6*, 3–73.
51. A language and environment for statistical computing. Available online: http://www.Rproject.org/ (accessed on 10 May 2016).

11

Assessment of Climate Change Impacts on Water Quality in a Tidal Estuarine System using a Three-Dimensional Model

Wen-Cheng Liu [1,2,*] and Wen-Ting Chan [1]

1 Department of Civil Disaster Prevention Engineering, National United University, Miaoli 36063, Taiwan; wen_ting1988@hotmail.com
2 Taiwan Typhoon and Flood Research Institute, National Applied Research Laboratories, Taipei 10093, Taiwan
* Correspondence: wcliu@nuu.edu.tw.

Academic Editor: Yingkui Li

Abstract: Climate change is one of the key factors affecting the future quality and quantity of water in rivers and tidal estuaries. A coupled three-dimensional hydrodynamic and water quality model has been developed and applied to the Danshuei River estuarine system in northern Taiwan to predict the influences of climate change on water quality. The water quality model considers state variables including nitrogen, phosphorus, organic carbon, and phytoplankton as well as dissolved oxygen, and is driven by a three-dimensional hydrodynamic model. The hydrodynamic water quality model was validated with observational salinity distribution and water quality state variables. According to the analyses of statistical error, predictions of salinity, dissolved oxygen, and nutrients from the model simulation quantitatively agreed with the observed data. The validated model was then applied to predict water quality conditions as a result of projected climate change effects. The simulated results indicated that the dissolved oxygen concentration was projected to significantly decrease whereas nutrients will increase because of climate change. Moreover, the dissolved oxygen concentration was lower than 2 mg/L in the main stream of the Danshuei River estuary and failed to meet the water quality standard. An appropriate strategy for effective water quality management for tidal estuaries is needed given the projected persistent climate trends.

Keywords: climate change; model assessment; numerical modeling; SELFE; water quality; tidal estuarine system

1. Introduction

Estuaries are among the world's vital aquatic resources. They provide food resources and a habitat for ecologically and economically important fish and shellfish species, recreational regions, educational and scientific experiences, and other important ecosystem services [1–5]. For example, the Guandu Natural Park in Taipei city, which is located at the confluence of the Danshuei River and the Keelung River, serves as an educational purpose and scientific experience [6]. Ecosystem services are fundamental life-support processes upon which all organisms depend [7]. Two ecosystem services that estuaries provide are water filtration and habit protection. However, adverse impacts on the estuarine ecosystem by environmental perturbations (e.g., anthropogenic nutrient loading, land use change, hydrological modification) have been widely reported [8–10]. The adverse impacts include impaired water quality, habitat loss, and diminished resources [11]. These perturbations result in declining water quality and deleterious changes in ecosystem structure and tropic dynamics [12,13]. The deleterious water quality subsequently produces the problems of odor, aesthetics, human pathogens, and increased public health risk. For example, McKibben *et al.* [14] reported that harmful algal blooms are proliferations of microscopic algae that harm the environment by producing toxins that accumulate in shellfish or fish, or through the accumulation of biomass that in turn affects co-occurring organisms and alters food webs in negative ways. Impacts include human illness and mortality following direct consumption or indirect exposure to toxic shellfish or toxins in the environment.

Climate change occurs naturally, but human population growth and associated land-cover deforestation and burning of fossil fuel have substantially accelerated the increase in greenhouse gases (CO_2, CH_4, N_2O, *etc.*). The elevated concentration of CO_2 and other greenhouse gases from anthropogenic activities have caused warming of the global climate by modifying radiative forcings, and continued changes will result in climate shifts [15–18]. Feng *et al.* [19] used model-projected future surface temperature and precipitation to examine the change/shifts of climate types over the global land area. They concluded that compared to the present-day condition, the boreal winter temperature over the global land area is projected to increase by 3–12 °C by 2071–2100 under a high emission scenario. Strong warming (>8 °C) appears along the Arctic coastal regions, moderate warming (5–7 °C) appears in the mid-latitude of the Northern Hemisphere, while the warming in the tropical and the Southern Hemisphere is relatively smaller (<5 °C). The projected warming in the boreal summer is much weaker (3–6 °C). Xin *et al.* [20] studied climate change projections over East Asia under various representative concentration pathway (RCP) scenarios using simulations conducted with the Beijing Climate Center Climate System Model for the Coupled Model Intercomparison Project phase 5. Under all RCPs, including RCP2.6, RCP4.5, RCP6.0, and RCP8.5, the East Asian climate is found to be warmer and wetter in the 21st century than the present climatology (1986–2005). For 2080–2099, the East Asian mean surface air temperature is higher than for present climatology by 0.98 °C (4.4%) under RCP2.6, 1.89 °C (7.7%) under RCP4.5, 2.47 °C (7.1%) under RCP6.0, and 4.06 °C (9.1%) under RCP8.5.

The impacts of climate change on human health have been widely reported [21–23]. Numerous studies project greater morbidity and mortality from direct exposure, as well as greater health risks due to decreased air quality, water-borne disease, and other infectious diseases [23,24]. Impacts of climate change on river and estuarine systems provide a subject of active research [25–27] because of the importance of water resources for human activities. Potential impacts of climate change on hydrology cover changes in runoff discharge, river flow, and groundwater storage [28]. Impacts on water quality include many factors (physical, including temperature, turbidity; chemical, including pH and concentration; biological, including biodiversity and species abundance across the entire food web from microbial pools and macrophytes up to fishes). With respect to water quality, most climate change impacts can be attributed to changes in either discharge—which controls dilution, flow velocity, and residence time—or water temperature. The impact of climate change on river and estuarine water quality is also heavily dependent on the future evolution of human activities (pollutions, withdrawals, *etc.*), so the direct influence of climate change may end up being relatively small [29].

The impact of climate change on estuaries has been reviewed by Robins *et al.* [30], who reported that potential changes to physical processes include flooding and coastal squeeze, caused by increased sea level rise, changing surge and wave climates, and changing river flow events. Sea level rise will cause a shift towards net sediment accretion, but with reduced transport in UK estuaries. Turbulent mixing that is critical for water quality and coastal ecology is controlled by river flow variability. Therefore, alterations to river flows will change the estuarine fronts, stratification, and mixing. The combination of sea level rise and longer dry periods in summer will cause negative impacts on eutrophication, harmful algal blooms, and hypoxia.

Numerical water quality models are useful in assisting the understanding of biological processes and the assessment of the influences of climate change on water quality conditions in aquatic systems [31–38]. For example, Tu [39] used a GIS-based watershed simulation model, AVGWLF, to simulate the future changes in streamflow and nitrogen load under different climate change and land use change scenarios at a watershed in eastern Massachusetts, USA. AVGWLF simulates daily streamflows and monthly nitrogen loads. As a result, the historical observed daily streamflow and nitrogen loads have to be used for model calibration and validation. The AVGWLF model tracks monthly streamflow and nitrogen load well in both calibration and validation. The coefficient of determination (R2) and Nash-Sutcliffe coefficient (NS) values in calibration for streamflow in most of the watersheds are higher than 0.7, and for nitrogen load are higher than 0.6. The R2 and NS values in validation of streamflow and nitrogen loads are even higher than the corresponding values in calibration. The validated model was used to project the impact of different climate scenarios (A1B, B1, and A2) on streamflow ad nitrogen load. The results revealed that the monthly streamflows in late fall and winter increase, whereas those in the summer months decrease, mainly as a result of climate change. Simulated nitrogen loads in late fall and winter months increase greatly, whereas those in spring and summer months have mixed responses affected by both climate and land use changes [39]. Rehana and Mujumdar [40] adopted a water quality model, QUAL2K, to simulate the water quality responses of six climate change scenarios covering different streamflow, air temperature, and water temperature at different stations. The simulated results suggested that all climate change scenarios would cause impairment in water quality. It was found that there was a significant decrease in dissolved oxygen levels owing to the impact of climate change on temperature and flows. For example, Luo *et al.* [41] applied the Soil and Water Assessment Tool (SWAT) to evaluate and enhance the watershed modeling approach in characterizing climate change impacts on water supply and ecosystem stressors. The SWAT was applied to headwater drainage basins in the northern Costal Ranges and Sierra Nevada mountain range in California. SWAT parameters for hydrological simulation were initialized within the ArcSWAT interface. Input data for watershed morphology have a 12-km spatial resolution. Input parameters mainly include the SCS runoff curve number (CN), snowmelt-related parameters, channel hydraulic conductivity, and parameters for groundwater recharge. The model was calibrated with daily streamflow at different selected stations. The calibrated model was then applied to project the effects of climate change. They concluded that the hydrological cycle and water quality of headwater drainage basins in California, especially their seasonality, were very sensitive to projected climate change.

These kinds of numerical models used to resolve one-dimensional and two-dimensional issues cannot well represent the spatial variations in three dimensions. For examples, Wan *et al.* [42] documented the development, calibration, and verification of a three-dimensional water quality model for the St. Lucie Estuary, a small and shallow estuary located on the east coast of south Florida. Modeling results revealed that high algae concentrations in estuaries are likely caused by excessive nutrient and algae supply in freshwater inflows. Cerco and Noel [43] applied the CE-QUAL-ICM (Corps of Engineers Integrated Compartment Water Quality Model) eutrophication model to simulate a 21-year (1985-2005) water quality model of Chesapeake Bay. The most significant finding was the influence of physical processes, notably stratification and associated effects (e.g., anoxic volume), on computed water quality. Li *et al.* [44] developed a three-dimensional hydrodynamic model coupled with a water quality model to determine the environmental capacity of nitrogen and phosphorus in Jiaozhou Bay, China. The model was calibrated based on data collected in 2003. The proposed water quality model effectively reproduced the spatiotemporal variability in nutrient concentration. However, few studies have emphasized the impacts of climate change on estuarine water quality using three-dimensional hydrodynamics and water quality coupling models.

This study aims to apply a coupled three-dimensional hydrodynamic and water quality (SELFE-WQ) model to characterize the water quality conditions in the estuarine system and assess the impacts of climate change scenarios on water quality in the Danshuei River estuarine system in northern Taiwan. The model was validated with observational salinity and water quality state variables. The validated water quality model was then applied to project the water quality conditions in estuarine system responses to climate change scenarios under the low flow condition.

2. Materials and Methods

The Danshuei River, with its tributaries, is the largest river system in northern Taiwan; its watershed encompasses 2726 km^2, with a combined length of 158.7 km (Figure 1). The regional climate is subtropical with the temperature varying between 10 and 35 °C, and the annual precipitation in the region ranges between 1500 mm and 2500 mm, with the majority falling in late spring (May) to early fall (October). The long-term average annual river flow rate is 6.6×10^9 m^3/y. The contributions of freshwater from the three major tributaries are, on average, 27% from the Keelung River, 31% from the Tahan Stream, and 37% from the Hsintien Stream. In addition to the mainstream of the Danshuei River, the lower reaches of the three major tributaries are also affected by tide. The principal tidal constituents of the estuary lean toward semi-diurnal tides, with a mean tidal range of 2.1 m and a spring tidal range of 3.5 m. Seawater intrusion reaches into all three tributaries except during periods of very high river inflows. In general, saltwater intrusion reaches 25–30 km from the Danshuei River mouth. The hydrodynamic characteristics in the system are mainly controlled by tide, river inflow, and the density gradient induced by the mixing of saline and freshwater [45,46]. The average flushing time of the Danshuei River is 2–4 days [47].

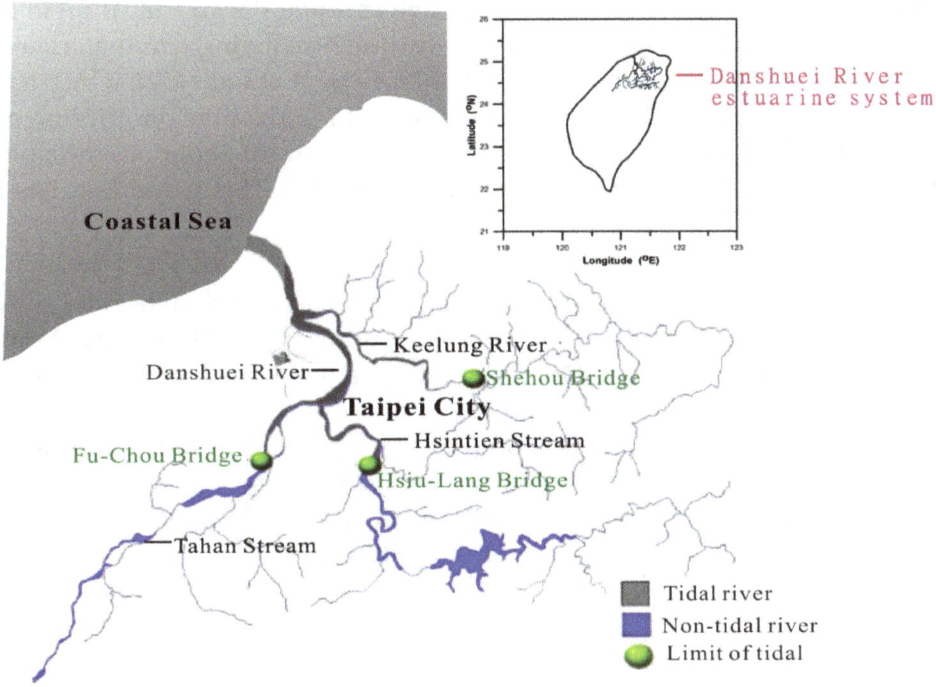

Figure 1. Danshuei River estuarine system and watershed.

The Danshuei River flows through the metropolitan area of Taipei, which has a population of approximately 6 million. A huge amount of treated and untreated domestic sewage was discharged into the river system and resulted in low dissolved oxygen and high nutrients. Viable biological activities are observed only in the lowest reach of the estuary, where the pollutant concentrations are reduced as a result of dilution by seawater [48,49].

3. Materials and Methods

3.1. Hydrodynamic Model

The numerical modeling of ocean circulation at scales ranging from estuaries to ocean basins is maturing as a field. Most modern oceanic and estuarine circulation codes solve for some form of the three-dimensional Navier–Stokes equations and can be complemented with conservation equations for a given water volume and salt concentration. In this paper, a three-dimensional, semi-implicit Eulerian-Lagrangian finite element model (SELFE, Zhang and Baptisa [50]) was implemented to simulate the Danshuei River estuarine system and its adjacent coastal sea. SELFE solves the Reynolds stress-averaged Navier–Stokes equations, which use conservation laws for mass, momentum, and salt with hydrostatic and Boussinesq approximations, to determine the free-surface elevation, three-dimensional water velocity, and salinity.

Unlike most 3D models using finite-difference/finite-volume schemes, SELFE is based on a finite-element scheme. No model splitting was used in SELFE, thus eliminating the errors associated with the splitting between internal and external modes [51]. Semi-implicit schemes were applied to all the equations to enhance the stability and maximize the efficiency of the system. An Eulerian-Lagrangian method was used to treat advection in the momentum equation, thus permitting the use of large time steps without compromising on stability. The horizontal space was discretized in the form of an unstructured grid of triangular elements, whereas the hybrid vertical coordinates—partly terrain-following S coordinates and partly Z coordinates—were used in the vertical direction. The wetting and drying algorithm was incorporated into the model. The minimum depth criterion for wetting and drying simulation was set to be 0.05 m.

Because turbulent mixing plays a critical role in determining the stratification in the tidal estuary, several reports have documented the model results of turbulence mixing parameterization. SELFE uses the generic length scale (GLS) turbulence closure of Umlauf and Burchard [52], which has the advantage of encompassing most of the 2.5-equation closure model (K–Ψ). A detailed description of the turbulence closure model, the vertical boundary conditions for the momentum equation, the numerical solution methods, and the numerical stability parameters can be found in Zhang and Baptista [50].

3.2. Water Quality Model

The water quality model used in this study was based on a three-dimensional conventional water quality analysis simulation program called WASP5, originally developed by Ambrose *et al.* [53]. It constitutes a complicated system of four interacting parts: dissolved oxygen, nitrogen cycle, phosphorus cycle, and phytoplankton dynamics. Eight water quality components are included: dissolved oxygen (DO), phytoplankton as carbon (PHYT), carbonaceous biochemical oxygen demand (CBOD), ammonium nitrogen (NH_4), nitrate and nitrite nitrogen (NO_3), organic nitrogen (ON), ortho-phosphorus or inorganic phosphorus (OP), and organic phosphorus (OP). The conceptual framework for the water quality model is presented in Figure 2.

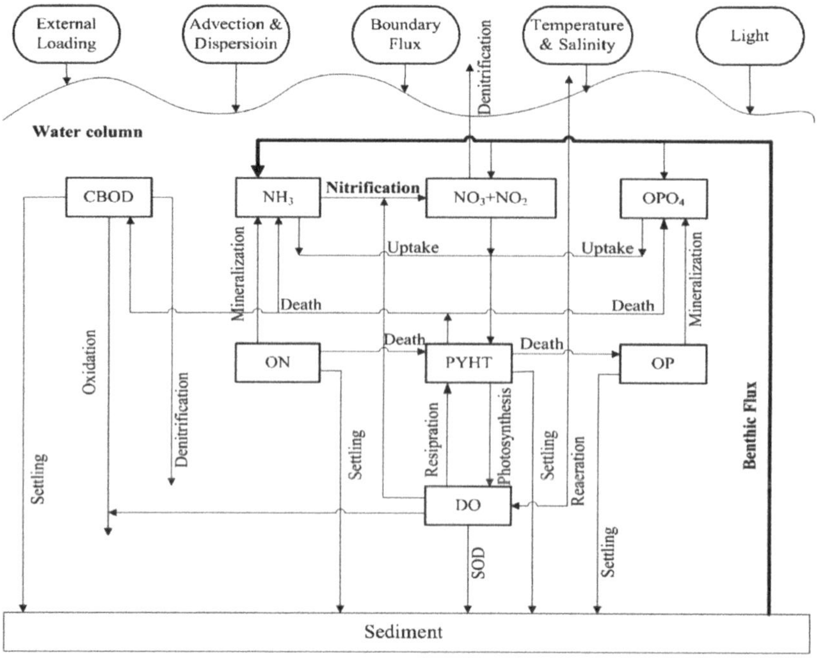

Figure 2. Schematic of water quality model.

A mathematical formulation of the conservation of mass can be written as follows:

$$\frac{\partial C}{\partial t}+\frac{\partial (uC)}{\partial x}+\frac{\partial (vC)}{\partial y}+\frac{\partial (wC)}{\partial z}=\frac{\partial}{\partial x}(A_h\frac{\partial C}{\partial x})+\frac{\partial}{\partial y}(A_h\frac{\partial C}{\partial y})+\frac{\partial}{\partial z}(K_v\frac{\partial C}{\partial z})+S_e+S_i \quad (1)$$

where C is the concentration of water quality components; u, v, and w are the water velocity components corresponding to a Cartesian coordinate system (x, y, z); A_h and K_v are the coefficients of horizontal viscosity and vertical eddy diffusion, respectively; S_e is the time rate of external additional (withdrawal) across the boundaries; and S_i is the time rate of internal increase/decrease by biogeochemical reaction processes.

Equation (1) gives the distribution of each state variable using the physical parameters determined from the hydrodynamic model. The last two terms, S_e and S_i represent, respectively, the external and internal sources (or sinks), the latter being primarily due to biogeochemical processes.

The present model of DO includes the following processes: source from photosynthesis, reaeration through surface and external loading, and sinks due to decay of CBOD, nitrification, algae respiration, and SOD. The mathematical representation is:

$$S_i=-K_cCBOD-a_{no}\frac{K_{n23}N_2}{K_{h23}+N_2}\frac{DO}{DO+K_{nit}}+a_ca_{co}(PQ\cdot G-\frac{R}{RQ})Chl \quad (2)$$

$$S_e=(1-\lambda_1)K_r(DO_s-DO)-\frac{SOD}{\Delta z}\frac{DO}{DO+K_{DO}}+\frac{WDO}{V} \quad (3)$$

where a_c = ratio of carbon to chlorophyll in phytoplankton (mg C/μg Chl); a_{co} = ratio of oxygen demand to organic carbon recycled = 2.67; a_{no} = ratio of oxygen consumed per unit of ammonia nitrogen nitrified = 4.57; $CBOD$ = concentration of carbonaceous of biochemical oxygen demand (mg/L); Chl = concentration of chlorophyll a (μg/L); DO = concentration of dissolved oxygen (mg/L); DO_s = saturation concentration of DO (mg/L); G = growth rate of phytoplankton (1/day); K_c = first-order decay rate of $CBOD$ (1/day); K_{DO} = half-saturation concentration for benthic flux of $CBOD$ (mg/L); K_{h23} = half-saturation concentration for nitrification (mg/L); K_{n23} = nitrification rate of ammonia nitrogen to nitrite-nitrate nitrogen (mg/L/day); K_{nit} = half-saturation concentration for oxygen limitation of nitrification (mg/L); K_r = reaeration rate (1/day); N_2 = concentration of ammonia nitrogen; PQ = photosynthesis quotient (mole O_2/mole C); R = respiration rate of phytoplankton (1/day); RQ = respiration quotient (mole CO_2/mole O_2); V = layer volume (cm^3); WDO = external loading of DO (mg/day) including point and nonpoint sources; Δz = layer thickness (cm); and $\lambda_1 = 0$ for $k = 1$ (at top layer), $\lambda_1 = 1$ for $2 \leqslant k \leqslant N$, and N is the number of layers.

The sediment oxygen demand (SOD) in Equation (3) is the rate of oxygen consumption exerted by the bottom sediment and the overlay water due to the respiration of the benthic biological communities and the biochemical degradation of organic matter. The SOD is a major component of the dissolved oxygen (DO) budget and a key parameter to be determined through the model validation in the water quality model.

According to previous study implemented by Chen *et al.* [54], the component of phytoplankton species in the Danshuei River estuary includes diatoms, green algae, and others; therefore, these three major species are taken into account in the model simulation.

3.3. Model Schematization and Implementation

In the present study, the horizontal resolutions, 200 m × 200 m and 40 m × 40 m, of the bathymetric and topographical data in the Taiwan Strait and Danshuei-River estuarine system were obtained from the Ocean Data Bank and Water Resources Agency, Taiwan. The deepest point within the study area is 110 m (below the mean sea level) near the northeast corner of the computational domain (Figure 3). The model mesh for the Danshuei-River estuarine system and its adjacent coastal sea consists of 5119 elements (Figure 3). To meet the accuracy requirements, fine-grid resolution was used locally, and coarse resolution was implemented away from the region of interest. In this computational domain, the mesh size varied from 6000 m in the Taiwan Strait down to 40 m in the upper reach of Danshuei River estuary. The mesh size (40 m) used in the upper reach of Danshuei River estuary would be an appropriate resolution because the bathymetric and topographic data in 40 m × 40 m were only obtained.

In the vertical direction, ten z-levels and ten evenly spaced S-levels were specified at each horizontal grid, *i.e.*, the thickness of the cell depended on the bottom elevation of each grid. The vertical resolutions in the coastal sea and Danshuei River estuary range from 10–20 m and 0.015–1.2 m, respectively. A 120-s time step was used in our simulations without any signs of numerical instability.

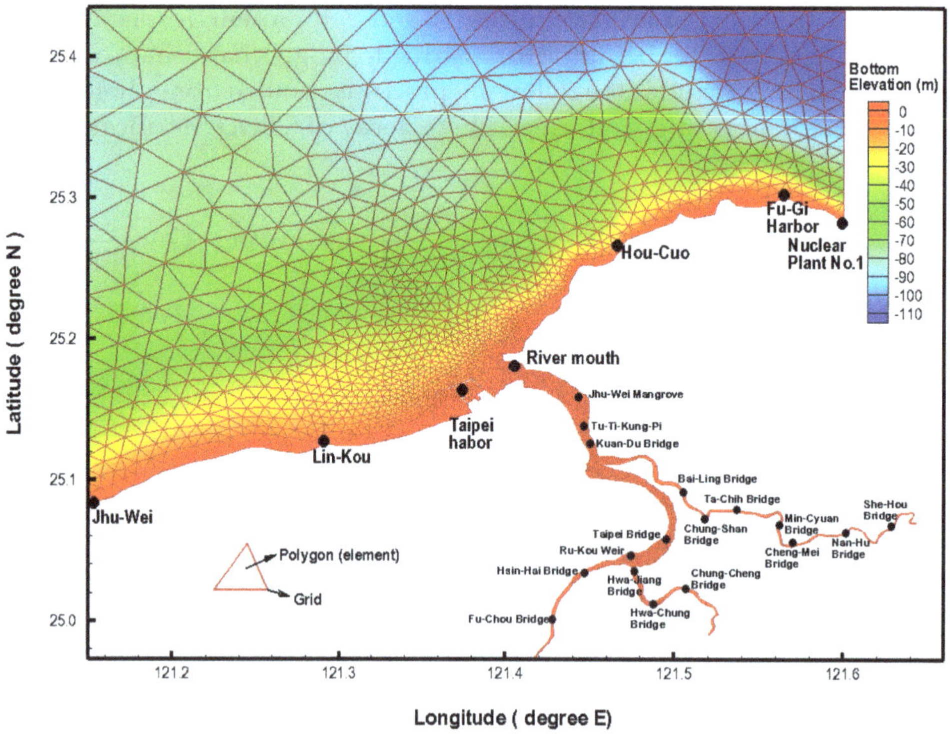

Figure 3. The topography of the Danshuei River estuarine system and its adjacent coastal and unstructured grid for the computational domain.

4. Model Validation

4.1. Salinity Distribution

Salinity distributions reflect the combined results of all processes, including density circulation and mixing processes. These processes in turn control the density circulation and modify the mixing processes [45]. In the present study, the salinity distribution along the Danshuei River-Tahan Stream collected by the Water Resources Agency, Taiwan, was used for model validation. Liu *et al.* [55] reported that a five-constituent tide (*i.e.*, M_2, S_2, N_2, K_1, and O_1) is sufficient to represent the tidal components in the Taiwan Strait. The five-constituent tide was adopted in the model simulation as a forcing function at the coastal sea boundaries. The model was run for a two-year simulation. The salinity of the open boundaries in the coastal sea was set to 35 *ppt*. The upstream boundary conditions at the three tributaries (Tahan Stream, Hsintien Stream, and Keelung River) were specified with daily freshwater discharges; therefore, the salinity at the upstream boundaries was set to be 0 *ppt*.

The simulated salinity distribution compared favorably to the salinity measurements along the Danshuei River–Tahan Stream during the flood and ebb tides on 26 November 2010, shown in Figure 4. The measured salinity during the flood and ebb tides means that the salinity was measured at instantaneous flood and ebb tides. Note that the field data of salinity were measured 0.5 m below the water surface and then every 1.0 m below the water surface 0.5 m, and the simulated salinity was presented with the top layer and bottom layer. The measured salinity shown in Figure 4 presents the mean salinity in vertical direction plus/minus one standard deviation. The absolute mean error and root mean square error of the difference between the measured salinities and the computed salinity on 26 November 2010 are 2.71 *ppt* and 3.72 *ppt*, respectively, during the flood tide. The absolute mean error and root mean square error are 0.49 *ppt* and 0.67 *ppt*, respectively, during the ebb tide. It can be seen that the modeling performance for the ebb tide is better than that for the flood tide. This may be the reason that the higher horizontal eddy diffusion is calculated according to 2.5-equation closure model, resulting in salinity diffusion to the upstream region during the flood tide.

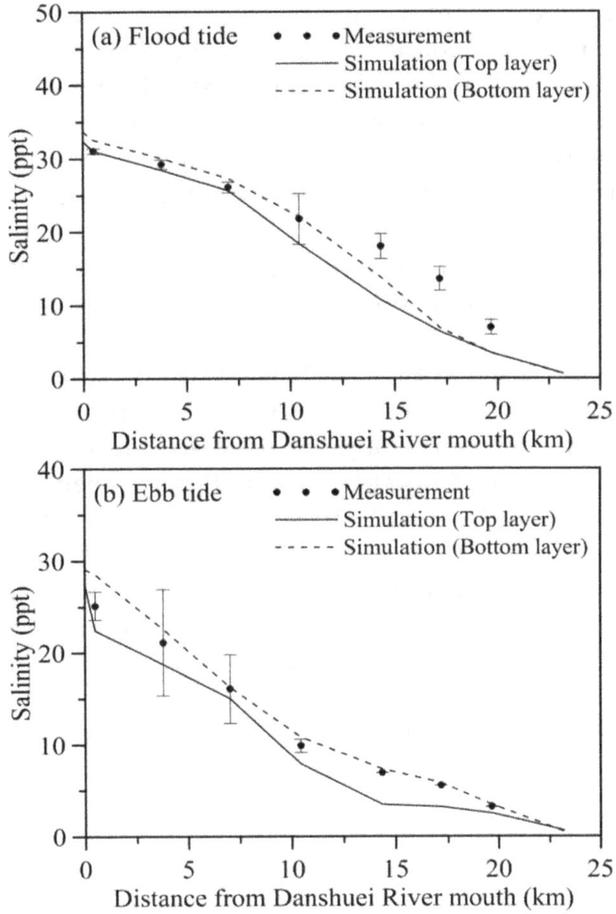

Figure 4. The comparison between the measured and simulated salinities along the Danshuei River–Tahan Stream on 26 November 2010 during (**a**) flood tide and (**b**) ebb tide.

4.2. Water Quality Distribution

Chen *et al.* [56] implemented a comprehensive field sampling and lab analysis program for the Danshuei River to collect the data in 2009 and 2010. They found that adjacent to the metropolitan Taipei City, the spatial trend of the deteriorated water quality is mostly attributed to the wastewaters directly discharged into the river channels. Efforts were made successively to estimate the point source loadings by Montgomery Watson Harza (MWH) [57] adopted in the following water quality simulations. The freshwater discharges in 2010 and 2011 were adopted at the upstream boundaries at the Tahan Stream, the Hsintien Stream, and the Keelung River. The five-constituent tide used to generate the time-series tidal level was employed at the ocean boundaries. Concentrations of water quality state variables ammonium nitrogen, total nitrogen, total phosphorus, carbonaceous biochemical oxygen demand, dissolved oxygen, and chlorophyll *a* at the river boundaries and at the ocean boundaries were established based on the monthly measurement by the Taiwan Environmental Protection Administration (TEPA). The model was conducted with two-year simulation.

Eight measured datasets were collected on 1 March, 1 June, 3 September, 2 December in 2010, 3 March, 1 June, 5 September, and 1 December in 2011 and were used for model validation. The model parameters were initially estimated from the literature [58]. These were adjusted and tuned until a reasonable reproduction of field data at observation stations was obtained. The coefficients adopted for water quality simulations are listed in Table 1. The longitudinal water quality distributions predicted by the water quality model on 3 March 2011 for the Danshuei River–Tahan Stream, the Hsintien Stream, and the Keelung River are shown in Figures 5–7 respectively. Water quality distributions of the dissolved oxygen, carbonaceous biochemical oxygen demand, ammonium nitrogen, and total phosphorus concentrations at the top and bottom layers along the river channels are presented in the figures, together with the observations from monitoring stations. Both the model-predicted and observed dissolved oxygen concentrations along the Danshuei River–Tahan Stream show a decrease from the Danshuei River mouth to Hsin-Hai Bridge and an increase at the Fu-Chou Bridge (Figure 5a). In the lower estuary, the dissolved oxygen concentrations increase toward the river mouth as a result of seawater dilution. It also shows that quite low dissolved oxygen concentrations occur at the Chong-Yang Bridge, Chung-Siao Bridge, and Hsin-Hai Bridge. Carbonaceous biochemical oxygen demand, ammonium nitrogen, and total phosphorus all show the same spatial trends along the river channel from the Tahan Stream to the Danshuei River (Figure 5b–d). The concentrations increase from the Danshuei River mouth to Hsin-Hai Bridge, reach a maximum at the Hsin-Hai Bridge, and then gradually decrease toward the Fu-Chou Bridge. The maximum concentrations of all three occur at the Hsin-Hai Bridge, resulting in low dissolved oxygen. The figure shows that the model generally captured the spatial trends of the observed longitudinal distributions.

The ratio of nitrogen to carbon in different water bodies has been documented in reports [59–61]. The ratio ranges from 0.02–0.25 mg N/mg C. However, we set this ratio to 0.01 mg N/mg C in the model simulation, which is lower than the suggested value. This is the reason that the concentration of ammonium nitrogen (NH_4) in the Danshuei River estuarine system is quite high compared to other estuaries [10,42,62]. If we adopted the higher ratio of nitrogen to carbon in the model, the simulation results of CBOD would be too high and DO would be too low to compare with the measured data.

Table 1. Coefficients used in the water quality model.

Coefficients	**Value**	**Unit**
Deoxygenation rate at 20 °C	0.16	day^{-1}
Nitrification rate at 20 °C	0.13	day^{-1}
Phytoplankton respiration rate at 20 °C	0.6	day^{-1}
Denitrification rate at 20 °C	0.09	day^{-1}
Organic nitrogen mineralization at 20 °C	0.075	day^{-1}
Organic phosphorus mineralization at 20 °C	0.22	day^{-1}
Optimum phytoplankton growth rate at 20 °C	2.5	day^{-1}
Optimal temperature for growth of phytoplankton	16	°C
The morality rate of phytoplankton at 20 °C	0.003	day^{-1}
Half-saturation constant for oxygen limitation of carbonaceous deoxygenation	0.5	mg O_2 L^{-1}
Half-saturation constant for oxygen limitation of nitrification	0.5	mg O_2 L^{-1}
Half-saturation constant for uptake of inorganic nitrogen	25	μg N L^{-1}
Half-saturation constant for uptake of inorganic phosphorus	1	μg P L^{-1}
Half-saturation constant for oxygen limitation of denitrification	0.1	mg O_2 L^{-1}
Half-saturation constant of phytoplankton limitation of phosphorus recycle	1	mg C L^{-1}
Sediment oxygen demand at 20 °C	3.5	g/m^2 day
Optimal solar radiation rate	250	langleys/day
Total daily solar radiation	300	langleys/day
Ratio of nitrogen to carbon in phytoplankton	0.25	mg N/mg C
Ratio of phosphorus to carbon in phytoplankton	0.025	mg P/mg C

Table 1. *Cont.*

Coefficients	Value	Unit
Ratio of phytoplankton to carbon	0.04	mg Phyt/mg C
Organic carbon (as CBOD) decomposition rate at 20 °C	0.21	day^{-1}
Anaerobic algae decomposition rate at 20 °C	0.01	day^{-1}
Denitrification rate at 20 °C	0.01	day^{-1}
Organic nitrogen decomposition rate at 20 °C	0.01	day^{-1}
Organic phosphorus decomposition rate at 20 °C	0.01	day^{-1}
Benthic NH_4 flux	0.04	mg N day^{-1}
Benthic NO_3 flux	0.003	mg N day^{-1}
Benthic PO_4 flux	0.005	mg P day^{-1}
Ratio of nitrogen to carbon	0.01	mg N/mg C
Ratio of phosphorus to carbon	0.01	mg P/mg C

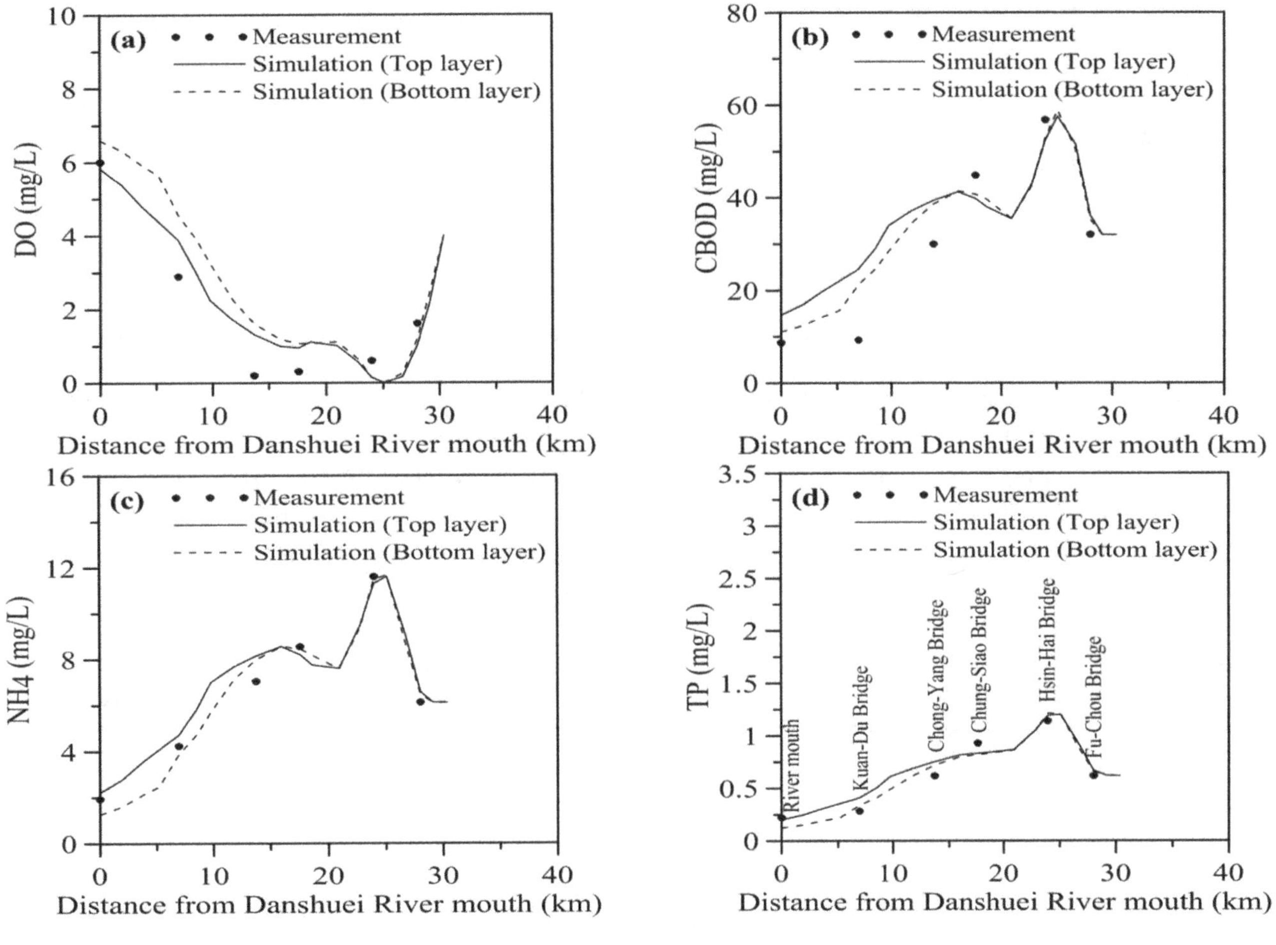

Figure 5. The comparison between the measured and simulated water quality distributions along the Danshuei River to the Tahan Stream on 3 March 2011 (**a**) DO; (**b**) CBOD; (**c**) NH_4; and (**d**) TP.

The longitudinal water quality distributions along the Hsintien Stream are illustrated in Figure 6. Both the observation data and model predictions show that the water quality conditions degrade as the river reach approaches the Hsintien Stream mouth, where it joins the main stream of the Danshuei River; the dissolved oxygen decreases, and the organic carbon, ammonium nitrogen and total phosphorus increase monotonically. The model faithfully represents the observed carbonaceous biochemical oxygen demand, ammonium nitrogen, and total phosphorus along the Hsintien Stream.

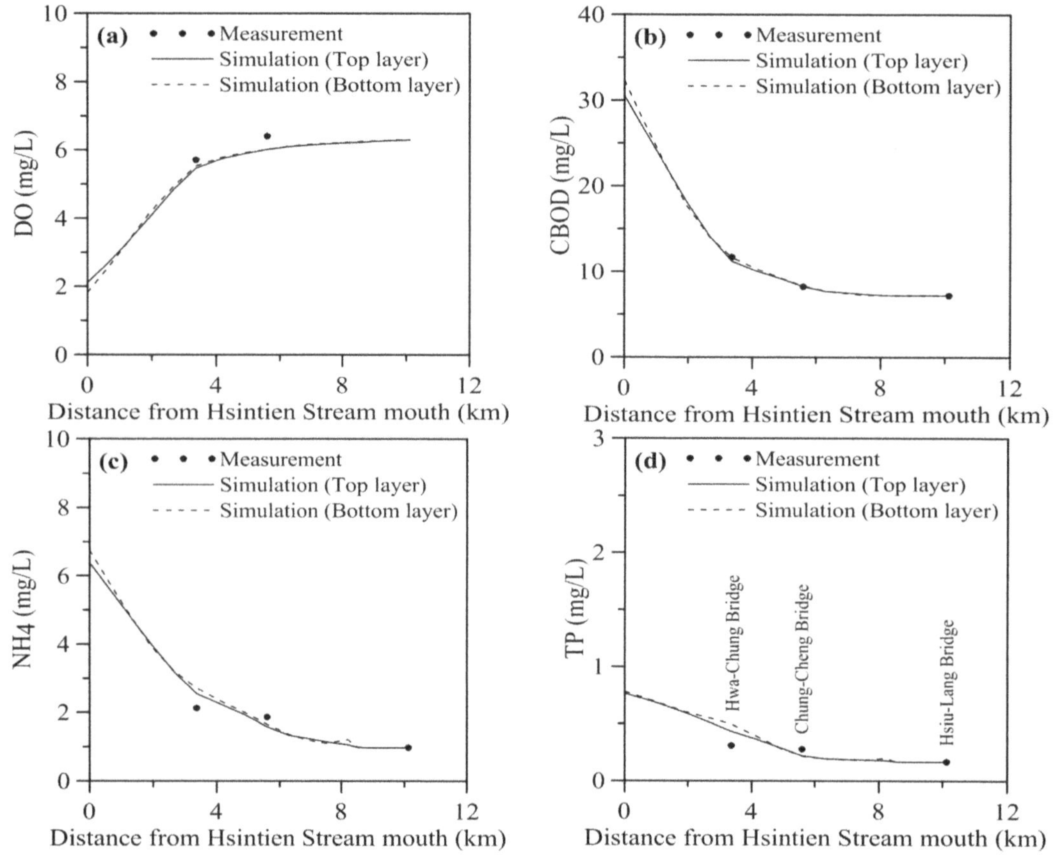

Figure 6. The comparison between the measured and simulated water quality distributions along the Hsintien Stream on 3 March 2011 (**a**) DO; (**b**) CBOD; (**c**) NH_4; and (**d**) TP.

The longitudinal water quality distribution along the Keelung River (Figure 7) shows that significant pollution loadings were discharged into the river section around the Bai-Ling Bridge and the Chung-Shan Bridge, where the lowest dissolved oxygen in the Keelung River was observed. The model can realistically mimic the observed dissolved oxygen. The model was also revealed to match the observed carbonaceous biochemical oxygen demand, ammonium nitrogen, and total phosphorus very well along the Keelung River. Due to the page limitation, the statistical errors, including the absolute mean error and root mean square error on 2 December 2010, 3 March, 1 June, 5 September 2011 are shown only in Tables 2–5.

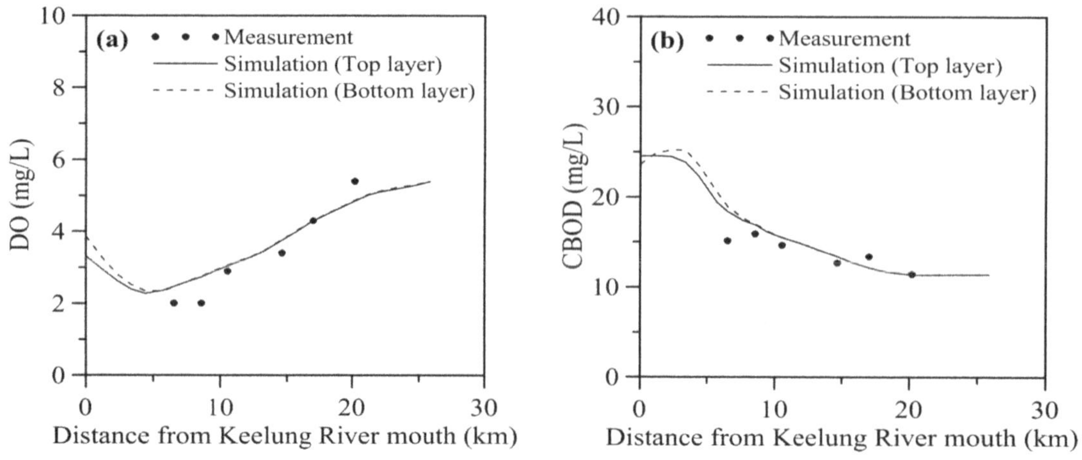

Figure 7. *Cont.*

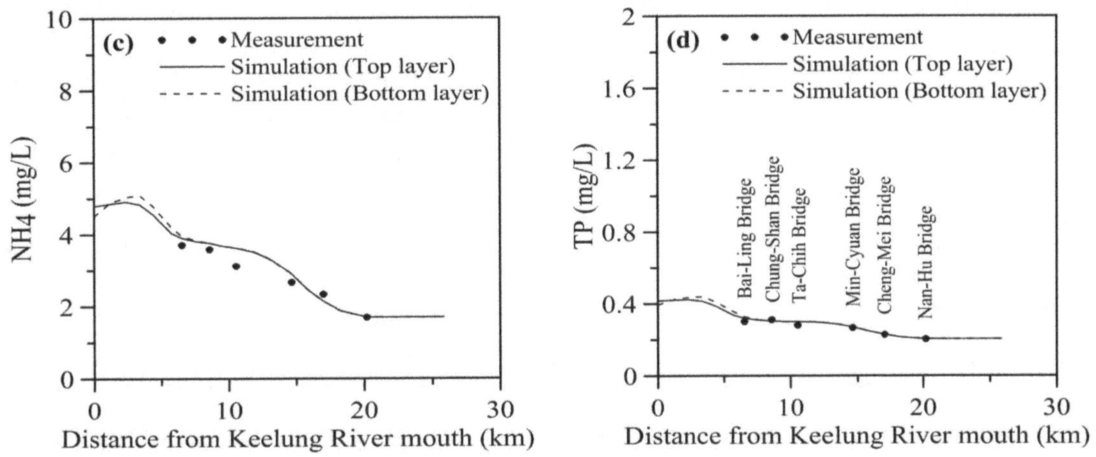

Figure 7. The comparison between the measured and simulated water quality distributions along the Keelung River on 3 March 2011 (**a**) DO; (**b**) CBOD; (**c**) NH_4; and (**d**) TP.

Table 2. Statistical error between simulated and measured water quality state variables on 2 December 2010.

Water Quality Variable	Danshuei River–Tahan Stream		Hsintien Stream		Keelung River	
	AME (mg/L)	RMSE (mg/L)	AME (mg/L)	RMSE (mg/L)	AME (mg/L)	RMSE (mg/L)
Dissolved oxygen	0.63	0.88	2.17	3.47	0.91	0.98
Carbonaceous biochemical oxygen demand	2.21	2.79	1.79	2.29	2.02	2.51
Ammonium nitrogen	0.36	0.54	0.52	0.74	0.15	0.19
Total phosphorus	0.07	0.10	0.08	0.12	0.008	0.01

Note that AME $= \frac{1}{N}\sum_{i=1}^{N}|(C_p)_i - (C_o)_i|$, RMSE $= \sqrt{\frac{1}{N}\sum_{i=1}^{N}[(C_p)_i - (C_o)_i]^2}$, where C_p is the predicted water quality concentration; and C_o is the observed water quality concentration.

Table 3. Statistical error between simulated and measured water quality state variables on 3 March 2011.

Water Quality Variable	Danshuei River–Tahan Stream		Hsintien Stream		Keelung River	
	AME (mg/L)	RMSE (mg/L)	AME (mg/L)	RMSE (mg/L)	AME (mg/L)	RMSE (mg/L)
Dissolved oxygen	0.74	0.85	0.86	1.18	0.38	0.45
Carbonaceous biochemical oxygen demand	6.56	7.5	0.11	0.16	1.25	1.67
Ammonium nitrogen	0.35	0.46	0.25	0.32	0.23	0.27
Total phosphorus	0.08	0.08	0.07	0.09	0.01	0.01

Table 4. Statistical error between simulated and measured water quality state variables on 1 June 2011.

Water Quality Variable	Danshuei River–Tahan Stream		Hsintien Stream		Keelung River	
	AME (mg/L)	RMSE (mg/L)	AME (mg/L)	RMSE (mg/L)	AME (mg/L)	RMSE (mg/L)
Dissolved oxygen	0.83	0.97	0.29	0.44	0.68	0.76
Carbonaceous biochemical oxygen demand	2.48	2.84	2.23	2.85	1.60	1.86
Ammonium nitrogen	0.52	0.80	0.44	0.67	0.35	0.42
Total phosphorus	0.99	0.10	0.08	0.14	0.06	0.07

Table 5. Statistical error between simulated and measured water quality state variables on 5 September 2011.

Water Quality Variable	Danshuei River–Tahan Stream		Hsintien Stream		Keelung River	
	AME (mg/L)	RMSE (mg/L)	AME (mg/L)	RMSE (mg/L)	AME (mg/L)	RMSE (mg/L)
Dissolved oxygen	1.41	1.65	0.67	0.83	1.04	1.19
Carbonaceous biochemical oxygen demand	1.10	1.32	006	0.07	0.80	0.91
Ammonium nitrogen	0.47	0.78	0.03	0.04	0.56	0.71
Total phosphorus	0.04	0.06	0.01	0.01	0.02	0.02

5. Model Project Responses to Climate Change Impact

The future climate scenarios frequently used in Taiwan have been based on the Intergovernmental Panel on Climate Change (IPCC) Special Report on Emissions Scenarios (SRES) A1B and A2 scenarios. The Water Resources Agency [58] projected the streamflow in the Danshuei River basin due to the climate change scenarios in year 2039 (*i.e.*, short term). The projected results in streamflow during the dry seasons based on different scenarios are summarized in Table 6. These results indicate that the decreasing rates of streamflows in the Tahan Stream, the Hsintien Stream, and the Keelung River are 45.54%, 4.15%, and 45.65%, respectively, for the A2 scenarios, whereas they are 19.05%, 3.44%, and 24.32%, respectively, for the A1B scenario.

Table 6. The streamflows for the present condition and under climate change scenarios during Q75 low flow.

River	Q_{75} Low Flow under Present Condition (m^3/s)	Decreasing Rate under A2 Scenario (%)	Q_{75} Low Flow under A2 Scenario (m^3/s)	Decreasing Rate under A1B Scenario (%)	Q_{75} Low Flow under A1B Scenario (m^3/s)
Tahan Stream	3.36	45.54	1.83	19.05	2.72
Hsintien Stream	14.23	4.15	13.64	3.44	13.74
Keelung River	3.33	45.65	1.81	24.32	2.52

To perform the model prediction of the water quality in the estuarine system, the five-constituent tide at the ocean boundaries was used to force the model simulation. Concentrations of the water quality state variables ammonium nitrogen, total nitrogen, total phosphorus, carbonaceous biochemical oxygen demand, dissolved oxygen, and chlorophyll *a* at the river boundaries and at the ocean boundaries were established based on mean values calculated from the measured water quality data collected from 2003 to 2013 as observed by TEPA. The river discharges at the tidal limits of the three major tributaries—the Tahan Stream, the Hsintien Stream, and the Keelung River—were conducted using the Q75 low flow condition, where Q75 flow is the flow that is equaled or exceeded

75% of the time. The Q75 river flows at the upstream reaches of the Tahan Stream, the Hsintien Stream, and the Keelung River are 3.36, 14.23, and 3.33 m^3/s, respectively, for the present condition. For the A2 and A1B scenarios, the Q75 river flows at the upstream reaches of the Tahan Stream, the Hsintien Stream, and the Keelung River are presented in Table 6.

The predicted water quality distribution for the present condition and the A2 climate change scenario under Q75 low flow along the Danshuei River to the Tahan Stream, the Hsintien Stream, and the Keelung River, respectively, is shown in Figures 8–10. A comparison of the present condition with the A2 climate change scenario reveals that the dissolved oxygen concentration decreased by a maximum of 1.75 mg/L and that the carbonaceous biochemical oxygen demand, ammonium nitrogen, and total phosphorus increased by a maximum of 6.4, 1.1, and 0.04 mg/L, respectively, in the Danshuei River–Tahan Stream (Figure 8). The dissolved oxygen concentration decreased by a maximum of 0.15 mg/L, and the carbonaceous biochemical oxygen demand, ammonium nitrogen, and total phosphorus increased by a maximum of 0.33, 0.14, and 0.01 mg/L, respectively, in the Hsintien Stream (Figure 9). The dissolved oxygen concentration decreased by a maximum of 1.50 mg/L, and the carbonaceous biochemical oxygen demand, ammonium nitrogen, and total phosphorus increased by a maximum of 0.85, 0.48, and 0.03 mg/L, respectively, in the Keelung River (Figure 10). We found that the dissolved oxygen concentration was lower than 2 mg/L in the Danshuei River-Tahan Stream and did not meet the minimum requirement of TEPA. The maximum rate of dissolved oxygen, carbonaceous biochemical oxygen demand, ammonium nitrogen, and total phosphorus under climate change scenarios A2 and A1B is summarized in Table 7. The maximum rate refers to the maximum values determined by the formula represented by $\frac{C_p - C_c}{C_p} \times 100\%$, where C_p is the water quality concentration at the present time and C_c is the water quality concentration under climate change.

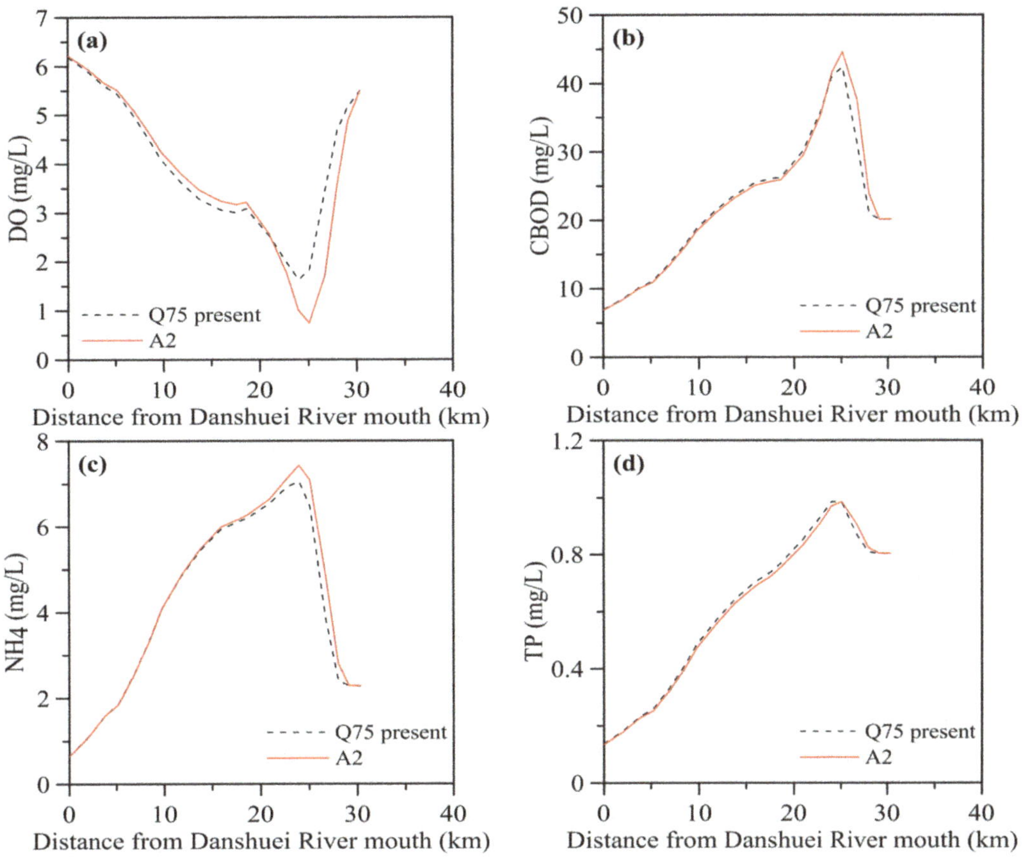

Figure 8. Predicting water quality distributions for present and climate change (A2 scenario) conditions under Q75 low flow along the Danshuei River to the Tahan Stream (**a**) DO; (**b**) CBOD; (**c**) NH4; and (**d**) TP.

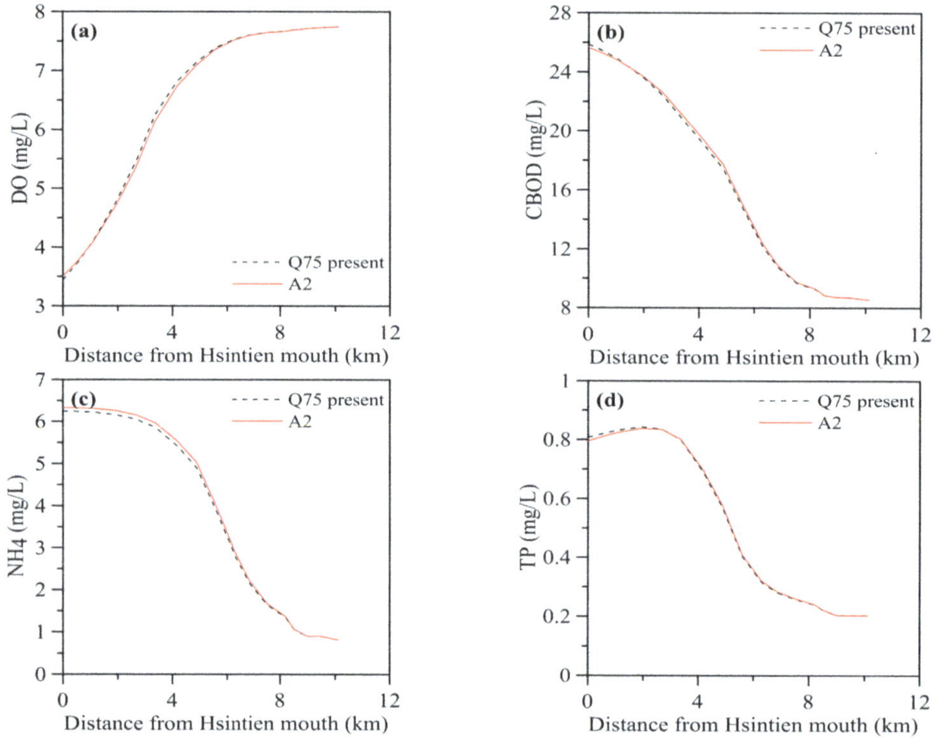

Figure 9. Predicting water quality distributions for present and climate change (A2 scenario) conditions under Q75 low flow along the Hsintien Stream (**a**) DO; (**b**) CBOD; (**c**) NH4; and (**d**) TP.

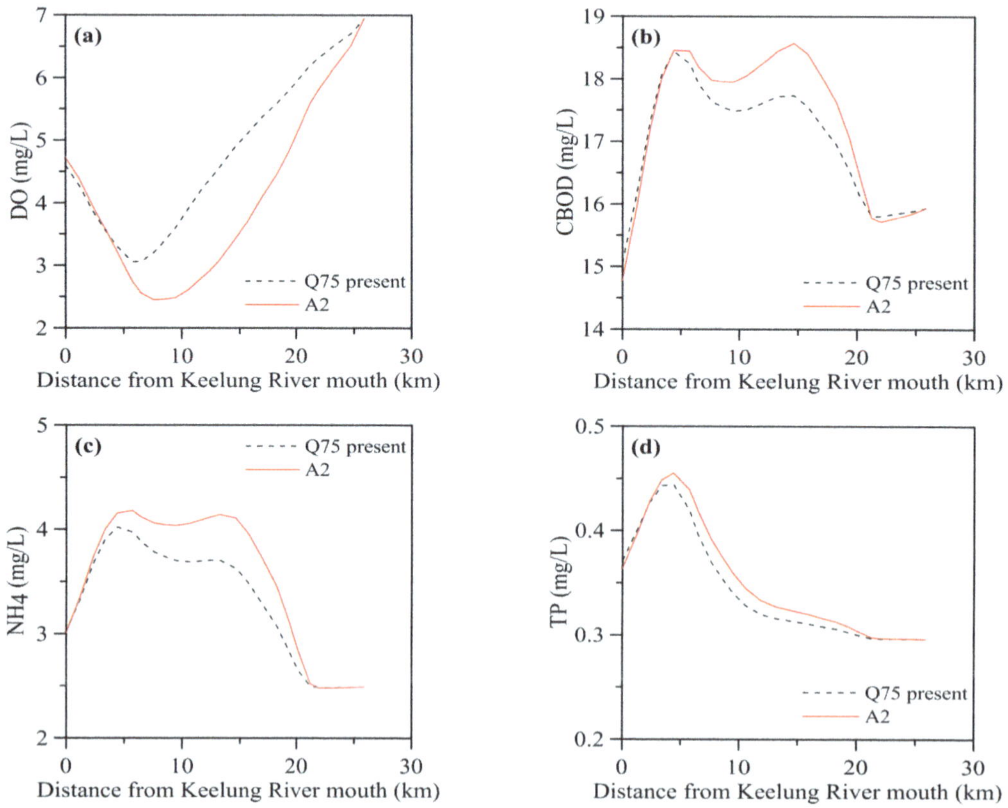

Figure 10. Predicting water quality distribution for present and climate change (A2 scenario) conditions under Q75 low flow along the Keelung River (**a**) DO; (**b**) CBOD; (**c**) NH4; and (**d**) TP.

Table 7. Maximum rate of water quality state variables under climate change scenarios A2 and A1B.

River	Maximum Rate under Climate Change A2 Scenario				Maximum Rate under Climate Change A1B Scenario			
	DO (%)	CBOD (%)	NH_4 (%)	TP (%)	DO (%)	CBOD (%)	NH_4 (%)	TP (%)
Danshuei River–Tahan Stream	−59.4	20.46	26.9	4.4	−25.8	7.5	9.8	1.8
Hsintien Stream	−2.0	1.9	3.8	1.7	−1.9	1.6	3.0	1.6
Keelung River	−33.7	4.9	13.8	6.2	−14.5	2.3	6.2	3.3

Note: minus and plus represent a decrease and increase, respectively.

The vertical distributions of monthly (in August) average salinity, dissolved oxygen, carbonaceous biochemical oxygen demand, ammonium nitrogen, and total phosphorus in the Danshuei River-Tahan Stream under the present condition and A2 scenario, respectively, are shown in Figures 11 and 12. It can be seen that the limit of salt water intrusion for the A2 scenario (Figure 12a) moves further upriver compared with the present condition (Figure 11a). The limit of salt water intrusion for the present condition, A2 scenario, and A1B scenario in the Danshuei River-Tahan Stream, the Hsintien Stream, and the Keelung River is illustrated in Table 8. The differences in the limit of salt water intrusion between the A2 scenario and present condition are 1.52 km, 0.25 km, and 1.33 km, respectively, in the Danshuei River-Tahan Stream, the Hsintien Stream, and the Keelung River. According to Figures 11a and 12a, we can observe the vertical stratification in salinity exhibited in the lower Danshuei River estuary. The dissolved oxygen concentration for the A2 scenario (Figure 12b) in an estuarine system decreases compared to the present condition (Figure 11b), while the concentrations of carbonaceous biochemical oxygen demand, ammonium nitrogen, and total phosphorus (Figure 12c–e) increase (Figure 11c–e). No significant vertical stratification in water quality was found in the Danshuei River estuary.

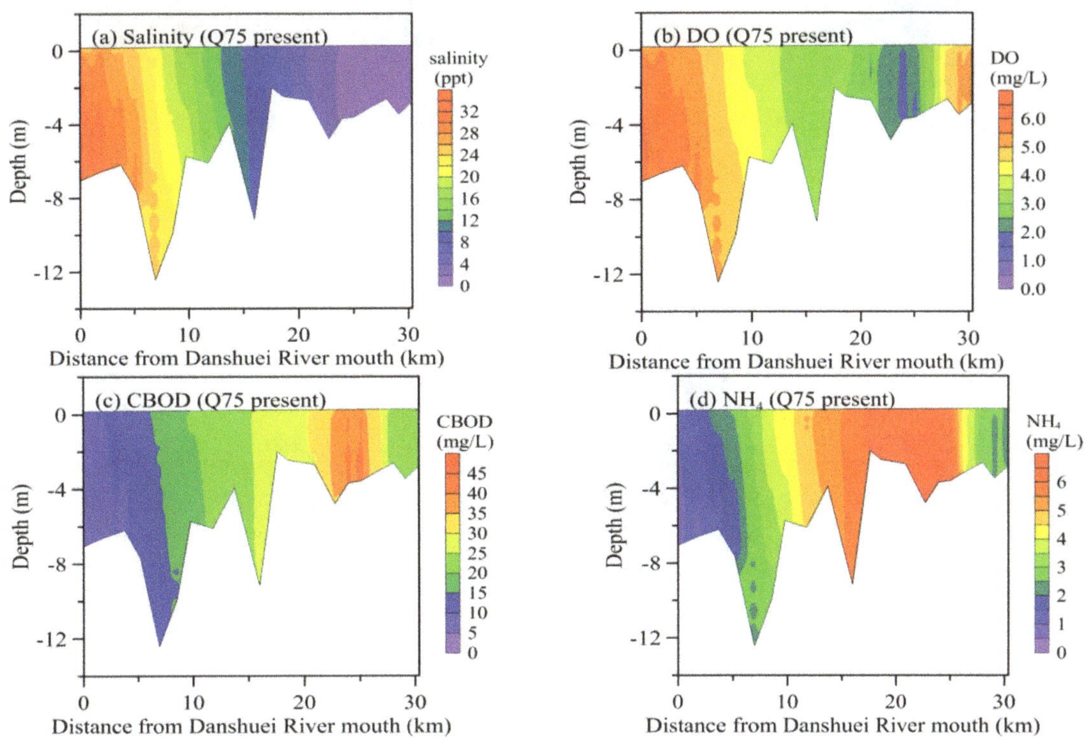

Figure 11. *Cont.*

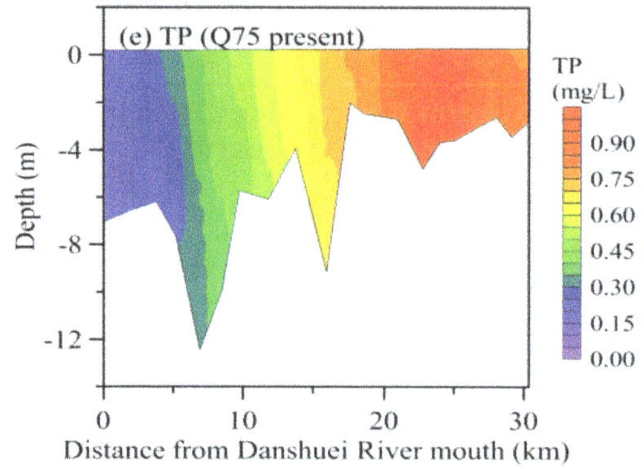

Figure 11. The vertical distribution of the monthly average water quality concentration in the Danshuei River to Tahan Stream under Q75 flow for the present condition (**a**) Salinity; (**b**) DO; (**c**) CBOD; (**d**) NH4; and (**e**) TP.

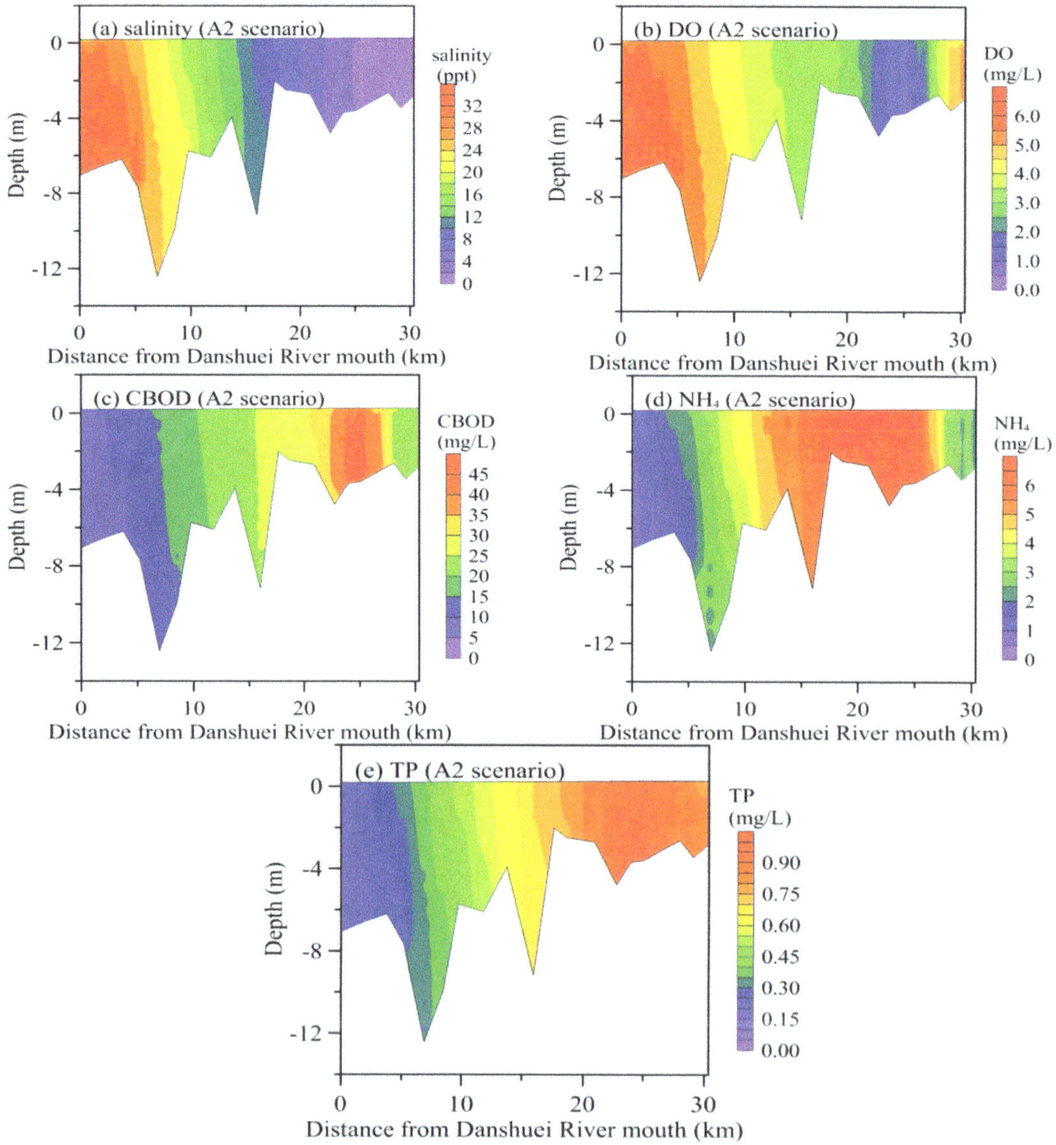

Figure 12. The vertical distribution of the monthly average water quality concentration in the Danshuei River to Tahan Stream under Q75 flow for the A2 Scenario (**a**) Salinity; (**b**) DO; (**c**) CBOD; (**d**) NH4; and (**e**) TP.

Table 8. The limit of salt water intrusion in the Danshuei River estuarine system under different scenarios.

River	Present Condition (km)	A2 Scenario (km)	A1B Scenario (km)
Danshuei River–Tahan Stream	24.67	26.19	25.24
Hsintien Stream	3.06	3.31	3.20
Keelung River	11.29	12.62	11.80

6. Discussion

Water quality studies utilizing a coupled hydrodynamic and water quality model tend to contain some limitations and assumptions. These limitations and assumptions exist in both the data and model. The major data used in this study were climate change scenarios. The climate change scenarios (*i.e.*, A1B and A2) used in this study were obtained from the Water Resources Agency, Taiwan, which projected the streamflow during dry seasons in the Danshuei River basin. We know that climate change is a non-stationary and dynamic problem; however, the streamflow projected from the climate change model and used in this study is a steady-state condition. This could lead to bias in future streamflow estimates that result in more uncertainty in the modeling results of water quality.

As mentioned in the Section "Water Quality Model", there are many parameters in the water quality model. The model was validated with two-year measured data. The model parameters after validation are kept for modeling future conditions without adjustment under future scenarios. However, future climate change might change the parameters. All of these parameters might reduce the accuracy of the modeling results. Nevertheless, after considering the aforementioned limitations and assumptions, the modeling results of water quality are relevant and reliable under the current climate change scenarios. The approaches are useful for assessing the impact of climate change on estuarine water quality.

Some literature has stated that climate change causes the degradation of water quality. For example, Tung *et al.* [59] evaluated the effects of climate change on sustainable water quality management and proposed a systematic assessment procedure including a weather generation model, the streamflow component of GWLF, QUAL2E, and an optimization model. Their studies indicated that streamflows may likely increase in humid seasons and decrease in arid seasons. The reduction of streamflow in arid seasons might further degrade water quality and assimilation capacity. Our study also demonstrated that the dissolved oxygen would decrease as a result of climate change, which reduces the streamflow during dry seasons. Wetz and Yoskowitz [27] reported that drought coupled with burgeoning population growth in coastal watersheds places a serve strain on freshwater supplies and greatly reduces freshwater inflows to estuaries, especially when coincident with seasonal peaks in human freshwater demand. Freshwater contains nutrients and organic matter that upon delivery to the coastal zone, fuels the rich productivity of coastal ecosystems and shapes critical fish habitats through its effects on salinity gradients and stratification. Low freshwater inflow events have the potential to significantly alter the water quality and ecosystem structure. In this study, we found that the dissolved oxygen would decrease and nutrients would increase for the low flow condition as a result of climate change. The decreased dissolved oxygen would result in malodor, fish mortality, and microbial proliferation, which causes the issue of public health.

In future research, future climate scenarios will be performed with a global climate model (GCM model) combined with a rainfall-runoff model to project time-series streamflow, which can be incorporated into the hydrodynamic and water quality model. The impact of sea-level rise on the estuarine water quality can be investigated. A long-term early warning system triggering proper adaptations to reduce climate change effects can also be studied.

7. Conclusions

A coupled three-dimensional hydrodynamic water quality model was applied to predict the water quality conditions in the Danshuei River estuarine system due to the projected effects of climate change. The model was validated against salinity distribution and water quality state variables including dissolved oxygen, carbonaceous biochemical oxygen demand, ammonium nitrogen, and total phosphorus. The simulated results using the three-dimensional hydrodynamic water quality model revealed that the computed salinity and water quality state variables well reproduced the observed data. The overall performance of the model is in qualitative agreement with the available field data.

The validated model was then used to assess the effects of climate change on water quality in the Danshuei River estuarine system during the low flow condition. Two climate change scenarios, A2 and A1B, were considered for model simulation. The simulated results indicated that the dissolved oxygen concentration has significantly decreased and the concentrations of carbonaceous biochemical oxygen demand, ammonium nitrogen, and total phosphorus have obviously increased because of climate change. Moreover, the dissolved oxygen concentration would be lower than 2 mg/L in the main stream of the Danshuei River estuary and would fail to meet the minimum requirement of TEPA. The deleterious water quality would produce other issues related to human pathogens and public health.

The simulated results may vary depending on the estuarine system, climate scenario, water quality model, and parameters considered. Considering the limitations of this study, the results are valid only under current climate change scenarios in the study area. However, the results and methodologies in this study still have implications for future water quality management in the estuarine system for the study area and other regions facing similar stresses from climate change.

Acknowledgments: This study was supported in part by the Ministry of Science and Technology (MOST) Taiwan, under grant No. 102-2625-M-239-002. This financial support was greatly appreciated. The authors express their appreciation to the Taiwan Water Resources Agency and Environmental Protection Administration for providing the observed data used in our model validation. The authors sincerely thank three anonymous reviewers for their valuable comments to substantially improve this paper.

Author Contributions: Wen-Cheng Liu supervised the progress of the MOST project and served as a general editor. Wen-Ting Chan performed the data collection, model establishment, and model simulations and discussed the results with Wen-Cheng Liu. All authors read and approved the final manuscript.

References

1. Hobbie, J.E. *Estuarine Science—A Synthetic Approach to Research and Practice*; Island Press: Washington, DC, USA, 2000.
2. Pendleton, L.H. *The Economic and Market Value of Coasts and Estuaries: What's at Stake*; Restore America's Estuaries: Arlington, VA, USA, 2008; p. 182.
3. Yoskowitz, D.W.; Santos, C.; Allee, B.; Carollo, C.; Henderson, J.; Jordan, S.; Ritchie, J. Ecosystem Services in the Gulf of Mexico. In Proceedings of the Gulf of Mexico Ecosystem Services Workshop, Bay St. Louis, MS, USA, 16–18 June 2010; Harte Research Institute for Gulf of Mexico Studies: Corpus Christi, TX, USA, 2010; p. 16.
4. Skerratt, J.; Wild-Allen, K.; Rizwi, F.; Whitehead, J.; Coughanowr, C. Use of a high resolution 3D fully coupled hydrodynamic, sediment and biogeochemical model to understand estuarine nutrient dynamics under various water quality scenarios. *Ocean Coast. Manag.* **2013**, *83*, 52–66. [CrossRef]
5. Yoskowitz, D.W.; Werner, S.R.; Carollo, C.; Santos, C.; Washburn, T.; Isaksen, G.H. Gulf of Mexico offshore ecosystem services: Relative evaluation by stakeholders. *Mar. Policy* **2015**. in press. [CrossRef]
6. Lee, R.W. Research of Construct Wetland Environment Model for Guandu Natural Park. Master's Thesis, National Taiwan Ocean University, Keelung, Taiwan, 2012.
7. Daily, G.C. *Introduction: What are Ecosystem Services? Nature's Services: Societal Dependence on Natural Ecosystems*; Island Press: Washington, DC, USA, 1997.
8. Chau, K.W. An unsteady three-dimensional eutrophication model in Tolo harbour, Hong Kong. *Mar. Pollut. Bull.* **2005**, *51*, 1078–1084. [CrossRef] [PubMed]
9. McGlathery, K.J.; Sundback, K.; Anderson, I.C. Eutrophication in shallow coastal bay and lagoons: The role of plants in the coastal filter. *Mar. Ecol. Prog. Ser.* **2007**, *348*, 1–18. [CrossRef]
10. Wild-Allen, K.; Skerratt, J.; Whitehead, J.; Rizwi, F.; Parslow, J. Mechanisms driving estuarine water quality: A 3D biogeochemical model for informed management. *Estuar. Coast. Shelf Sci.* **2013**, *135*, 33–45. [CrossRef]

11. Kennish, M.J. Environmental threats and environmental future of estuaries. *Environ. Conserv.* **2002**, *29*, 78–107. [CrossRef]
12. Paerl, H.W.; Valdes, L.M.; Peierls, B.L.; Adolf, J.E.; Harding, L.W. Anthropogenic and climatic influences on the eutrophication of large estuarine ecosystem. *Limnol. Oceanogr.* **2006**, *51*, 448–462. [CrossRef]
13. Rabalais, N.N.; Turner, R.E.; Diaz, R.J.; Justic, D. Global change and eutrophication of coastal waters. *ICES J. Mar. Sci.* **2009**, *66*, 1528–1537. [CrossRef]
14. McKibben, S.M.; Watkins-Brandt, K.S.; Wood, A.M.; Hunter, M.; Forster, Z.; Hopkins, A.; Du, X.; Eberhart, B.T.; Peterson, W.T.; White, A.E. Monitoring Oregon coastal harmful algae: Observations and implications for a harmful algal bloom-monitoring project. *Harmful Algae* **2015**, *50*, 32–44. [CrossRef]
15. Stone, M.C.; Hotchkiss, R.H.; Hubbard, C.M.; Fontaine, T.A.; Mearns, L.O.; Arnold, J.G. Impacts of climate change on Missouri River Basin water yield. *J. Am. Water Resour. Assoc.* **2001**, *37*, 1119–1129. [CrossRef]
16. Chang, H.; Knight, C.G.; Staneva, M.P.; Kostov, D. Water resource impacts of climate change in southwestern Bulgaria. *Geojournal* **2002**, *57*, 159–168. [CrossRef]
17. Moss, R.H.; Edmonds, J.A.; Hibbard, K.A.; Manning, M.R.; Ross, S.K.; Van Vuuren, D.P.; Carter, T.R.; Emori, S.; Kainuma, M.; Kram, T.; *et al.* The next generation of scenarios for climate change research and assessment. *Nature* **2010**, *463*, 747–756. [CrossRef] [PubMed]
18. Trenberth, K.E. Framing the way to relate climate extremes to chimate change. *Clim. Change* **2012**, *115*, 283–290. [CrossRef]
19. Feng, S.; Hu, Q.; Hung, W.; Ho, C.H.; Li, R.; Tang, Z. Projected climate regime shift under future global warming from multi-model, multi-scenario CMIP5 simulations. *Glob. Planet Change* **2014**, *112*, 41–52. [CrossRef]
20. Xin, X.; Zhang, L.; Zhang, J.; Wu, T.; Fang, Y. Climate change projection over East Asia with BCC_CSM1.1 climate model under RCP scenarios. *J. Meteorl. Soc. Jpn.* **2013**, *91*, 413–429. [CrossRef]
21. Braks, M.; van Ginkel, R.; Wint, W.; Sedda, L.; Sprong, H. Climate change and public health policy: Translating the science. *Int. J. Environ. Res. Public Health* **2014**, *11*, 13–29. [CrossRef] [PubMed]
22. Kendrovski, V.; Spasenovska, M.; Menne, B. The public health impacts of climate change in the former Yugoslav Republic of Mecedonia. *Int. J. Environ. Res. Public Health* **2014**, *11*, 5975–5988. [CrossRef] [PubMed]
23. Petkova, E.P.; Ebi, K.L.; Culp, D.; Redlener, I. Climate change and health on the U.S. Gulf Coast: Public health adaption is needed to address future risks. *Int. J. Environ. Res. Public Health* **2015**, *12*, 9342–9356. [CrossRef] [PubMed]
24. Luber, G.; Prudent, N. Climate change and human health. *Trans. Am. Clin. Climatol. Assoc.* **2009**, *120*, 113–117. [PubMed]
25. Whitehead, P.; Wade, A.; Butterfield, D. Potential impacts of climate change on water quality and ecology in six UK rivers. *Hydrol. Res.* **2009**, *40*, 113–122. [CrossRef]
26. Wu, Y.; Liu, S.; Gallant, A.L. Predicting impacts of increased CO_2 and climate change on the water cycle and water quality in the semiarid James River Basin of the Midwestern USA. *Sci. Total Environ.* **2012**, *430*, 150–160. [CrossRef] [PubMed]
27. Wetz, M.S.; Yoskowitz, D.W. An "extreme" future for estuaries? Effects of extreme climate events on estuarine water quality and ecology. *Mar. Pollut. Bull.* **2013**, *69*, 7–18. [CrossRef] [PubMed]
28. Middelkoop, H.; Daamen, K.; Gellens, D.; Grabs, W.; Kwadijk, J.C.J.; Lang, H.; Parmet, B.W.; Schadler, B.; Schulla, J.; Wilke, K. Impact of climate change on hydrological regimes and water resources management in the Rhine Basin. *Clim. Change* **2001**, *49*, 105–128. [CrossRef]
29. Ducharne, A.; Baubion, C.; Beaudoin, N.; Benoit, M.; Billen, G.; Brisson, N.; Garnier, J.; Kieken, H.; Lebonvallet, S.; Ledoux, E.; *et al.* Long term prospective of the Seine River system: Confronting climate and direct anthropogenic changes. *Sci. Total Environ.* **2007**, *375*, 292–311. [CrossRef] [PubMed]
30. Robins, P.E.; Skov, M.W.; Lewis, M.J.; Gimenez, L.; Davies, A.G.; Malham, S.K.; Neill, S.P.; McDonald, J.E.; Whitton, T.A.; Jackson, S.E.; *et al.* Impact of climate change on UK estuaries: A review of past trends and potential projections. *Estuar. Coast. Shelf Sci.* **2016**, *169*, 119–135. [CrossRef]
31. Garcia, A.; Juanes, J.A.; Alvarez, C.; Revilla, J.A.; Medina, R. Assessment of the response of a shallow macrotidal estuary to changes in hydrological and wastewater inputs through numerical modeling. *Ecol. Model.* **2010**, *221*, 1194–1208. [CrossRef]
32. Zhou, N.; Westrich, B.; Jiang, S.; Wang, Y. A coupling simulation based on hydrodynamics and water quality model of the Pearl River Delta, China. *J. Hydrol.* **2010**, *396*, 267–276. [CrossRef]

33. Piliwal, R.; Patra, R.R. Applicability of MIKE 21 to assess temporal and spatial variation in water quality of an estuary under the impact of effluent from an industrial estate. *Water Sci. Technol.* **2011**, *63*, 1932–1943. [CrossRef]
34. Long, T.Y.; Wu, L.; Meng, G.H.; Guo, W.H. Numerical simulation for impacts of hydrodynamic conditions on algae growth in Chongqing Section of Jialing River, China. *Ecol. Model.* **2011**, *222*, 112–119. [CrossRef]
35. Gao, G.; Falconer, R.A.; Lin, B. Modelling effects of a tidal barrage on water quality indicator distribution in the Severn Estuary. *Front. Environ. Sci. Eng.* **2013**, *7*, 211–218. [CrossRef]
36. Hanfeng, Y.E.; Guo, S.; Li, F.; Li, G. Water quality evaluation in tidal river reaches of Liaohe River estuary, China using a revised QUAL2K model. *Chin. Geogr. Sci.* **2013**, *23*, 301–311.
37. Sun, J.; Lin, B.; Jiang, G.; Li, K.; Tao, J. Modelling study on environmental indicators in an estuary. *Proc. Inst. Civil Eng. Water Manag.* **2014**, *167*, 141–151. [CrossRef]
38. Seo, D.; Song, Y. Application of three-dimensional hydrodynamics and water quality model of the Youngsan River, Korea. *Desalin. Water Treat.* **2015**, *54*, 3712–3720. [CrossRef]
39. Tu, J. Combined impact of climate and land use changes in streamflow and water quality in eastern Massachusetts, USA. *J. Hydrol.* **2009**, *379*, 268–293. [CrossRef]
40. Rehana, S.; Mujumdar, P.P. River water quality response under hypothetical climate change scenarios in Tunga-Bhadra river, India. *Hydrol. Process.* **2011**, *25*, 3372–3386. [CrossRef]
41. Luo, Y.; Ficklin, D.L.; Liu, X.; Zhang, M. Assessment of climate change impacts on hydrology and water quality with a watershed modeling approach. *Sci. Total Environ.* **2013**, *450*, 72–82. [CrossRef] [PubMed]
42. Wan, Y.; Ji, Z.G.; Shen, J.; Hu, G.; Sun, D. Three dimensional water quality modeling of a shallow subtropical estuary. *Mar. Environ. Res.* **2012**, *82*, 76–86. [CrossRef] [PubMed]
43. Cerco, C.F.; Noel, M.R. Twenty-year simulation of Chesapeake Bay water quality using the CE-QUAL-ICM eutrophication model. *J. Am. Water Resour. Assoc.* **2013**, *49*, 1119–1133. [CrossRef]
44. Li, K.; Zhang, L.; Li, Y.; Zhang, L.; Wang, X. A three-dimensional water quality model to determine the environmental capacity of nitrogen and phosphorus in Jiaozhou Bay, China. *Mar. Pollut. Bull.* **2015**, *91*, 306–316. [CrossRef] [PubMed]
45. Hsu, M.H.; Kuo, A.Y.; Kuo, J.T.; Liu, W.C. Procedure to calibrate and verify numerical models of estuarine hydrodynamics. *J. Hydraul. Eng.* **1999**, *125*, 166–182. [CrossRef]
46. Liu, W.C.; Chen, W.B.; Hsu, M.H. Influence of discharge reductions on salt water intrusion and residual circulation in Danshuei River Estuary. *J. Mar. Sci. Technol.* **2011**, *19*, 596–606.
47. Wang, C.F.; Hsu, M.H.; Kuo, A.Y. Residence time of Danshuei River estuary, Taiwan. *Estuar. Coast. Shelf Sci.* **2004**, *60*, 381–393. [CrossRef]
48. Liu, W.C.; Liu, S.Y.; Hsu, M.H.; Kuo, A.Y. Water quality modeling to determine minimum instream flow for fish survival in tidal rivers. *J. Environ. Manag.* **2005**, *76*, 293–308. [CrossRef] [PubMed]
49. Wang, C.F.; Hsu, M.H.; Liu, W.C.; Hwang, J.S.; Wu, J.T.; Kuo, A.Y. Simulation of water quality and plankton dynamics in the Danshuei River Estuary, Taiwan. *J. Environ. Sci. Health A* **2007**, *42*, 933–953. [CrossRef] [PubMed]
50. Zhang, Y.L.; Baptista, A.M. SELFE: A semi-implicit Eulerian-Lagrangian finite-element model for cross-scale ocean circulation. *Ocean Model.* **2008**, *21*, 71–96. [CrossRef]
51. Shchepetkin, A.F.; McWilliams, J.C. A method for computing horizontal pressure-gradient force in an oceanic model with a nonaligned vertical coordinate. *J. Geophys. Res.* **2003**, *108*. [CrossRef]
52. Umlauf, L.; Buchard, H. A generic length-scale equation for geophysical turbulence models. *J. Mar. Res.* **2003**, *61*, 235–265. [CrossRef]
53. Ambrose, R.B., Jr.; Wool, T.A.; Martin, J.L. *The Water Quality Analysis Simulation Program, WASP5, Part A: Model Documentation*; U.S. Environmental Protection Agency: Athens, GA, USA, 1993; p. 202.
54. Chen, W.B.; Liu, W.C.; Hsu, M.H. Water quality modeling in a tidal estuarine system using a three-dimensional model. *Environ. Eng. Sci.* **2011**, *28*, 443–459. [CrossRef]
55. Liu, W.C.; Chen, W.B.; Kuo, J.T. Modeling residence time response to freshwater discharge in a mesotidal estuary, Taiwan. *J. Mar. Syst.* **2008**, *74*, 295–314. [CrossRef]
56. Chen, C.H.; Lung, W.S.; Yang, C.H.; Lin, C.F. Spatially variable deoxygenation in the Danshui River: Improvement in model calibration. *Water Environ. Res.* **2013**, *85*, 2243–2253. [CrossRef] [PubMed]

57. Montgomery Watson Harza (MWH). *Evaluation of Hydrodynamics and Water Quality Monitoring and Management in the Danshuei River Basin of New Taipei City*; Report to New Taipei City Government; Montgomery Watson Harza (MWH): Broomfield, CO, USA, 2011.
58. Bowie, G.L.; Mills, W.B.; Porcella, D.B.; Campbell, J.R.; Pagenkopf, J.R.; Rupp, G.L.; Johnson, K.M.; Chan, P.W.H.; Gherini, S.A. *Rates, Constants, and Kinetics Formulations in Surface Water Quality Modeling*, 2nd ed.; EPA/600/3–85/040; Environmental Research Laboratory, Office of Research and Development, U.S. Environmental Protection Agency: Athens, GA, USA, 1985; p. 455.
59. Gray, K.R.; Biddlestone, A.J. Composing-process parameters. *Chem. Eng.* **1973**, *2*, 71–76.
60. Meyers, P.A.; Ryoshi, I. Lacustrine organic goechemistry—An overview of indicators of organic matter sources and diagenesis in lake sediments. *Org. Geochem.* **1993**, *20*, 867–900. [CrossRef]
61. Emerson, S.; Hedges, J. Sediment diagenesis and benthic flux. In *Treatise on Geochemistry*, 2nd ed.; Henry, E., Heinrich, D.H., Karl, K.T., Eds.; Elsevier: Amsterdam, The Netherlands, 2013; pp. 293–319.
62. Kim, T.; Sheng, Y.P.; Park, K. Modeling water quality and hypoxia dynamics in Upper Charlotte Harbor, Florida, U.S.A. during 2000. *Estuar. Coast. Shelf Sci.* **2010**, *90*, 250–263. [CrossRef]
63. Water Resources Agency. *Strengthening Sustainable Water Resources Utilization and Adaptive Capability to Climate Change*; Final Report; Water Resources Agency: Nantun, Taiwan; Taichung, Taiwan, 2008. (In Chinese)
64. Tung, C.P.; Lee, T.C.; Liao, W.T.; Chen, Y.J. Climate change impact assessment for sustainable water quality management. *Terr. Atmos. Ocean. Sci.* **2012**, *23*, 565–576. [CrossRef]

12

The Impact of Climate Change on the Duration and Division of Flood Season in the Fenhe River Basin, China

Hejia Wang [1,2], Weihua Xiao [2,*], Jianhua Wang [2,*], Yicheng Wang [2], Ya Huang [2,3], Baodeng Hou [2] and Chuiyu Lu [2]

1 Department of Hydraulic Engineering, Tsinghua University, Haidian District, Beijing 100084, China; sxjcwhj@126.com
2 State Key Laboratory of Simulation and Regulation of Water Cycle in River Basin, China Institute of Water Resources and Hydropower Research, 1-A Fuxing Road, Haidian District, Beijing 100038, China; wangych@iwhr.com (Y.W.); hygccw@163.com (Y.H.); houbaodeng@163.com (B.H.); cylu@iwhr.com (C.L.)
3 College of Civil Engineering and Architecture, Guangxi University, Nanning 530004, China
* Correspondence: jhwangiwhr@126.com (J.W.); xwsen998@163.com (W.X.).

Academic Editor: Yingkui Li

Abstract: This study analyzes the duration and division of the flood season in the Fenhe River Basin over the period of 1957–2014 based on daily precipitation data collected from 14 meteorological stations. The Mann–Kendall detection, the multiscale moving *t*-test, and the Fisher optimal partition methods are used to evaluate the impact of climate change on flood season duration and division. The results show that the duration of the flood season has extended in 1975–2014 compared to that in 1957–1974. Specifically, the onset date of the flood season has advanced 15 days, whereas the retreat date of the flood season remains almost the same. The flood season of the Fenhe River Basin can be divided into three stages, and the variations in the onset and retreat dates of each stage are also examined. Corresponding measures are also proposed to better utilize the flood resources to adapt to the flood season variations.

Keywords: flood season division; flood limit level; climate change; precipitation; Fenhe River Basin

---

1. Introduction

Global climate change is inevitable [1], causing frequently occurring extreme weather and climate events and unevenly distributed rainfall. Fenhe River, the second largest tributary of the Yellow River in Northern China, is experiencing such changes. According to historical records, it had a 5-year-long wet season from 1963 to 1967, and had a 15-year-long dry season from 1979 to 1993. It has been suggested that the climate, especially precipitation, in the Fenhe River Basin has been affected by human activities and natural phenomena such as El Niño-Southern Oscillation and Pacific Decadal Oscillation [2]. Therefore, it is of critical importance to study the variation in the flood season and evaluate the impact of climate change.

Several studies have investigated the duration of flood seasons. Odekunle [3] and Sâmia *et al.* [4] determined the onset and retreat time of the flood seasons in Nigeria and South American monsoon areas, respectively, and both of their methods were effective. Hachigonta *et al.* [5] found that the onset date of the flood season in Zambia has a significant spatial variation. Other studies [6–10] have analyzed temporal variations in flood seasons. All of these studies indicate that the onset and retreat time of the flood season have fluctuations that are affected by natural variations over long periods.

Although numerous researchers have investigated the onset and retreat times of the flood season, few studies explore the variation of the flood season in light of climate change. In this paper, we detect the abrupt changing point of the climate based on the annual precipitation in 1957–2014 in the Fenhe River Basin, estimate the onset and retreat time of flood seasons divided by this abrupt changing point, and, finally, divide the flood season into three stages to analyze the temporal variation of each stage. Based on the results, we propose measures to alleviate the discord between flood control and flood resources utilization.

2. Study Area

The Fenhe River is the largest river in the province of Shanxi, China and also the second largest tributary of the Yellow River. The Fenhe River Basin covers an area of 39471 km^2 (110°30′–113°32′ E; 35°20′–39°00′ N), accounting for 25.3% of the area of Shanxi. The Fenhe River Basin lies between Mt. Lvliang and Mt. Taihang and irregularly distributed as a wide stripe in West-Central Shanxi. It is located in the mid-latitude continental monsoon climate zone with arid and semiarid climates. During 1957–2014, the mean annual precipitation was 472.4 mm. The drainage map of the Fenhe River Basin is shown in Figure 1.

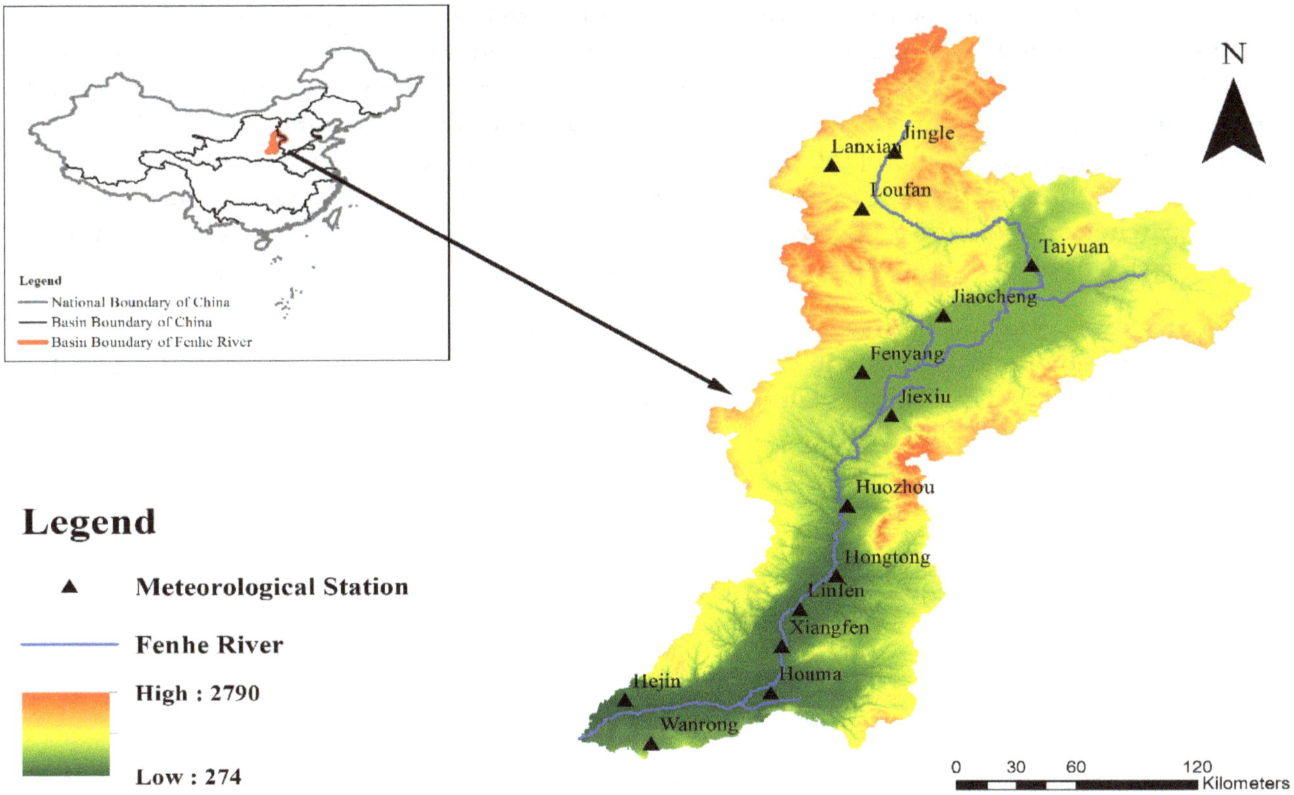

Figure 1. Plane graph of the Fenhe River Basin.

The data used for this study include daily precipitation data from 1957 to 2014 at 14 meteorological stations within the basin, and are obtained from the China Meteorological Administration [11]. Some stations have incomplete and false data records for some years. We use the inverse distance weighted method to fill and replace these missing or false data. It should be stated that the precipitation data during 1957–1979 are separated by rain and snow, but they are no longer separated and have not been since 1980. In this study, the snow data before 1980 are converted to the corresponding rainfall.

3. Materials and Methods

3.1. Climate Abrupt Change Detection

Mann–Kendall detection is a nonparametric statistical test method which is widely used in time series data trend tests [12]. Samples are not needed to obey a certain distribution in this way, which is also unacted on the interference of a few outliers. Therefore, it is suitable for hydrology, meteorology, and other non-normal distribution data. In this paper, the annual precipitation during 1957–2014 are chosen as the time series data to detect the abrupt point. The specific steps of Mann–Kendall detection are then further explored [13].

3.2. The Onset and Retreat Dates of Flood Season

Currently, the onset and retreat time of flood season are mostly defined by experience, which is inevitably subjective. The accurate detect of the two dates will provide scientific support to relevant sectors to forecast the flood season and to regulate the reservoirs for the purpose of flood control. The moving *t*-test is suitable for the multiscale detection of the change of flood season; the beginning of flood season is a symbol that the precipitation varies from less to more, at which time the abrupt point may exist. The situation is just the opposite when it comes to the end of flood season. Fraedrich *et al.* took the historical flood level of the River Nile, for example, and found three distinct epochs from AD 622 to AD 1470 [14]. Wang *et al.* detected the beginning and ending dates of rainy season in Dalian from 1951 to 1998 by applying the moving *t*-test method, which showed that it is more objective and exact than the empirical method [15]. Jiang *et al.* redistricted the coherently dry/wet episode of the Nile River and found the results were coincided with the historical disaster and famine of Egypt [16]. Other researchers also obtained ideal results by using this method [17–19]. Therefore, the multiscale moving *t*-test is applied in this paper to define the onset and retreat dates of flood season.

The multiscale moving *t*-test is used to detect the differences between two subsamples before and after the abrupt point with equivalent sample sizes, namely, $n = n_1 = n_2$. The *t*-statistic of precipitation in the Fenhe River Basin can be calculated as follows:

$$t(n,i) = (\overline{x}_{i2} - \overline{x}_{i1})\, n^{1/2} \left(s_{i2}^2 + s_{i1}^2\right)^{-1/2} \tag{1}$$

where

$$\overline{x}_{i1} = \sum_{j=i-n}^{i-1} x_j/n;\ s_{i1}^2 = \sum_{j=i-n}^{i-1} (x_j - \overline{x}_{i1})^2/(n-1) \tag{2}$$

$$\overline{x}_{i2} = \sum_{j=i}^{j+n-1} x_j/n;\ s_{i2}^2 = \sum_{j=i}^{i+n-1} (x_j - \overline{x}_{i2})^2/(n-1) \tag{3}$$

The test of the multiscale abruption is realized by altering the n value. $N = 2, 3, \ldots, <N/2$, and $i = n+1, \mathrm{n+2}, \ldots, N\text{-}n$, where N is the number of days in a year (365 or 366), and n is the timescale. The confidence level of *t*-test at 0.01 is approximately equivalent to the confidence level of Yamamoto and the Mann–Kendall test at 0.05. In order to make the graphics visualized and the analysis convenient, the further calculation is as follows:

$$t_r(n,i) = t(n,i)/t_{0.01}(n) \tag{4}$$

$tr(n, i)$ can be treated as the threshold to judge whether there is an abrupt point. When $tr(n, i)$ is greater than 1.0, it means that the abruption of variation trend is rising; when $tr(n, i)$ is less than -1.0, it means the abruption of variation trend is descending. The center of extreme values of the $tr(n, i)$ absolute value means that the abruption is the most prominent when the time is i and the timescale is n.

3.3. Flood Season Division

Flood season division means dividing the flood season into several stages in the light of the distinct differences and the regularity of flood characteristics during the different periods. Accurate division is the prerequisite for controlling the flood limit level, which can bring huge benefits to agricultural irrigation, power generation, *etc.*, and alleviate the contradiction between flood control and flood resources utilization. Extensive researches on flood season division are conducted, and a mass of methods has been applied to flood season division. Liu *et al.* successfully divided the flood season in the Geheyan Reservoir by means of the changing point method [20]. Jin *et al.* used the fuzzy set analysis method with the city of Nanping and divided the flood season into three stages [21]. Many other methods are also ideal for solving this problem [22–27].

The essence of flood season division is a multidimensional orderly clustering analysis of time series. The Fisher optimal partition method is just a method of clustering of ordered samples and was used to divide seismicity period successfully by He *et al.* [28]. It has some basic characteristics: firstly, it can take multi-index into consideration; secondly, it can meet the time sequence of the flood season; finally, it can confirm the optimum number of flood season divisions. Therefore, the Fisher optimal partition method is applied in this paper to divide the flood season of the Fenhe River Basin.

We can define $\{x_1, x_2, \ldots, x_n\}$ as ordered samples, where every sample is m-dimensional vector. The symbol $B(n, k)$ indicates that n ordered samples are divided into k parts. The division can be expressed as follows:

$$P_1 = \{i_1, i_1+1, \ldots, i_2-1\}\,; P_2 = \{i_2, i_2+1, \ldots, i_3-1\}\,; \ldots P_k = \{i_k, i_k+1, \ldots, n\}\,, \tag{5}$$

where division points are $1 = i_1 < i_2 < \ldots < i_k < n = i_{k+1} - 1$.

Supposing a part P contains samples $\{x_i, x_{i+1}, \ldots, x_j\}$, the mean value can be denoted as follows:

$$\overline{x_p} = \frac{1}{j-i+1}\sum_{t=1}^{j} x_t \tag{6}$$

Therefore, the sum of the squares of deviations of the part can be denoted as follows:

$$D\,(i,j) = \sum_{t=i}^{j} \left(x_t - \overline{x_p}\right)^T \left(x_t - \overline{x_p}\right) \tag{7}$$

The purpose of the Fisher optimal partition method is to find division points, which make the sum of the squares of deviations of every part the minimum. Consequently, the optimal division can be depicted as the following formula:

$$B\,(n,k) = \min \sum_{t=1}^{k} D\,(i_t, i_{t+1} - 1\,) \tag{8}$$

The theorem says that the optimal division $B(n, k)$ must be equal to the optimal division $B(n-1, k-1)$ plus the remaining part. According to the theorem above, the optimal division $B(n, 2)$ can be denoted as follows:

$$B\,(n,2) = \min_{2 \leqslant i \leqslant n} \{D\,(1, i-1) + D\,(i,n)\} \tag{9}$$

Similarly, the optimal division $B(n, k)$ can be denoted as follows:

$$B\,(n,k) = \min_{2 \leqslant i \leqslant n} \{B\,(i-1, k-1) + D\,(i,n)\} \tag{10}$$

Then, we can obtain the optimal solution by finding the division points in descending order. The division point i_k must meet the requirement of the following equation:

$$B(n,k) = \{B(i_k - 1, k - 1) + D(i_k, n)\} \quad (11)$$

Then, the part k can be obtained: $P_k = \{i_k, i_k + 1, \ldots, n\}$. Similarly, the remaining parts $P_{k-1}, P_{k-2}, \ldots, P_2, P_1$ can be obtained successively.

In this process, we need to draw the $B(n, k) - k$ curve according to the results, and then calculate the absolute value of slope of every division point $f(k)$. Next, the $f(k) - k$ curve can be drawn—if $f(k)$ is greater, the division into k parts is better than $k-1$ parts. When $f(k)$ is close to zero, there is no need for division. Generally, k is the optimal division number of parts when $f(k)$ meets the maximum, and the corner of $B(n, k) - k$ curve is also seen as the optimal division point.

4. Results

4.1. The Detection of Climate Abrupt Point

Figure 2 displays the results of Mann–Kendall detection of annual precipitation during 1957–2014. Here, UF(K) and UB(K) are statistics sequence which obey standard normal distribution. A intersection exists between UF(K) line and UB(K) line, and it just lies between the two critical lines whose confidence level are 0.05. The corresponding time of the intersection is 1974. Therefore, we can divide the period 1957–2014 into two periods, namely, 1957–1974 and 1975–2014. The annual precipitation is on the rise during the former period, while the trend declines during the latter period.

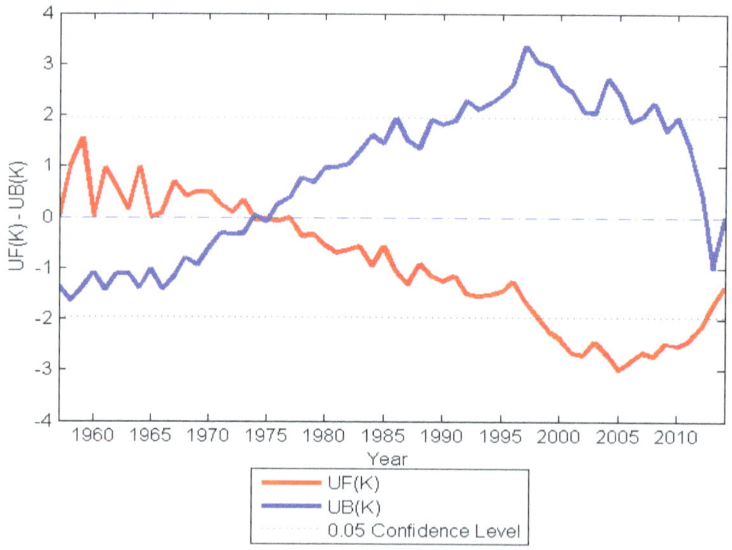

Figure 2. Mann-Kendall detection of the precipitation in the Fenhe River Basin for the period 1957–2014.

4.2. The Onset and Retreat Dates of Flood Season

Taking the year 1979 as an example, the result of the multiscale moving t-test of the Fenhe River Basin precipitation is shown in Figure 3. The color scale of the contour map shows that the larger the value, the deeper the color. The maximum value of $tr(n, i)$ is 2.2, which is greater than 1.0. The corresponding day is 165 and the corresponding timescale is 65 days. It means that the onset date of flood season in 1979 is Day 165, on which the abruption is the most conspicuous in the timescale of 65 days. Similarly, the retreat date of flood season in 1979 is Day 255, on which the abruption is the most conspicuous in the timescale of 63 days. Figure 4 shows the variation trend during 1957–2014, the red line represents the onset date of flood season in the Fenhe River Basin, and the blue line indicates the retreat date. Consequently, the annual average flood season of the two period 1957–1974 and

1975–2014 can be obtained. The onset and retreat dates of flood season during 1957–1974 are 170 and 253, respectively, and the flood season lasts 84 days long. In a similar way, the onset and retreat dates of flood season during 1975–2014 are 156 and 252, respectively, and the flood season is 13 days longer than that during 1957–1974. Out of convenience with respect to the flood season division, it was necessary to transform the timescale from day to pentad. According to the traditional Chinese calendar, a pentad is 5 days; one year contains 72 pentads. For instance, Day 170 can be converted to 4 June, which means the fourth pentad in June. Therefore, the flood season of 1957–1974 and 1975–2014 are 4 June–2 September and 1 June–2 September, respectively.

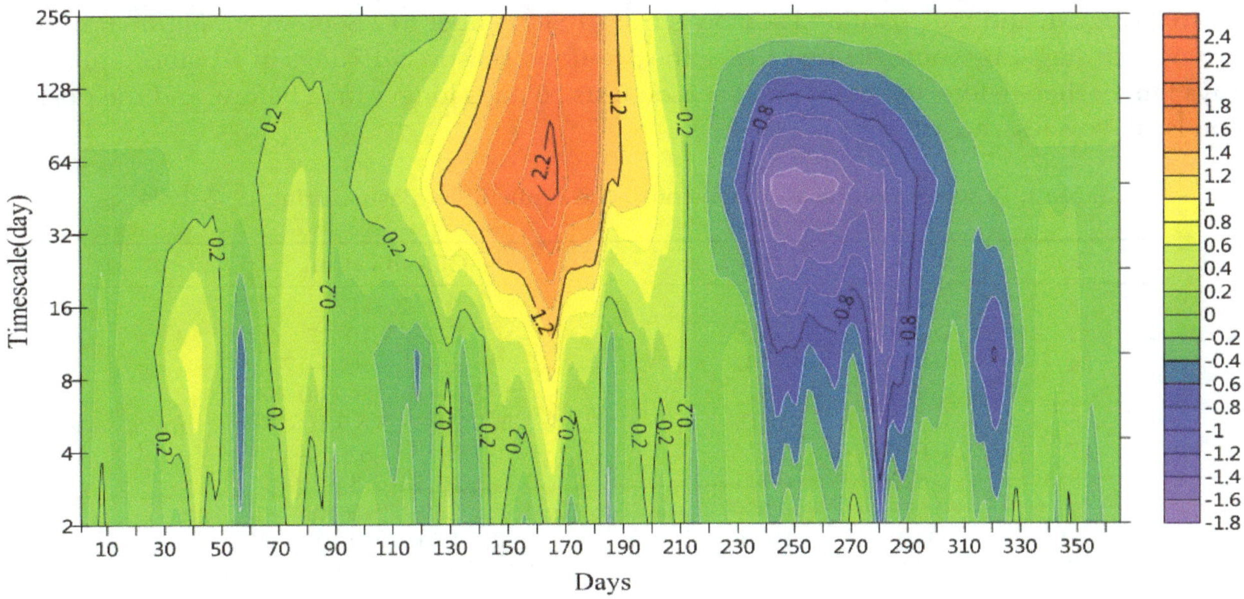

Figure 3. The multiscale moving *t*-test on the Fenhe River Basin daily mean precipitation in 1979.

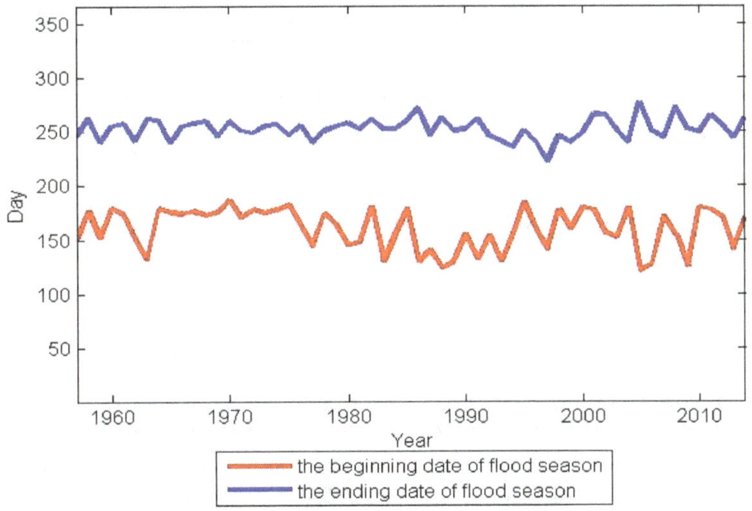

Figure 4. The onset and retreat dates of flood seasons for the period 1957–2014.

4.3. Flood Season Division

Five factors that can represent the seasonal rules of precipitation and flood are chosen as basic elements for flood season division: mean pentad precipitation, mean coefficient of variation of pentad precipitation, mean maximum 1-day-precipitation within a pentad, mean maximum 3-day-precipitation within a pentad, and number of hard rain days within a pentad. Tables 1 and 2 display the result of division of 1957–1974 and 1975–2014, respectively. Figure 5 shows the $B(n, k) - k$ curve and $f(k) - k$ curve of 1957–1974 and 1975–2014, respectively. The $f(k)$ is the maximum when k is equal to 3 in picture (b), and the curve $B(n, k) - k$ has a turning at the same time. Therefore, the optimal division number k is 3 during 1957–1974, which means the flood season can be divided into three stages during 1957–1974, namely, pre-flood season, main flood season, and post-flood season. The corresponding classification in Table 2 is 1–3, 4–9, 10–17, which means the pre-flood season is from 4 June to 6 June, the main flood season is from 1 July to 6 July, and the post-flood season is from 1 August to 2 September. Analogously, the flood season can be divided into three stages during 1975–2014. To be specific, the pre-flood season is from 1 July to 6 July, the main flood season is from 1 July to 6 August, and the post-flood season is from 1 September to 2 September. Figure 6 shows the length of each stage. By contrast, we can see the pre-flood season is 15 days in advance; thus, it lasts 15 days longer than before, the main flood season lasts 30 days longer than before, and the post-flood season is 30 days less, accordingly.

Table 1. Result of flood season division of the Fenhe River Basin during 1957–1974.

k	B(n,k)	f(k)	Classification
2	0.742		1–2,3–17
3	0.223	0.519	1–3,4–9,10–17
4	0.091	0.132	1–2,3–4,5–9,10–17
5	0.072	0.019	1–2,3–4,5–7,8–9,10–17
6	0.052	0.02	1–2,3–4,5–7,8,9,10–17
7	0.036	0.016	1–2,3,4,5–7,8,9,10–17
8	0.025	0.011	1–2,3,4,5,6–7,8,9,10–17
9	0.016	0.009	1–2,3,4,5,6–7,8,9,10,11–17
10	0.01	0.006	1–2,3,4,5,6,7,8,9,10,11–17
11	0.007	0.003	1–2,3,4,5,6,7,8,9,10,11–14,15–17
12	0.004	0.003	1–2,3,4,5,6,7,8,9,10,11–13,14,15–17
13	0.002	0.002	1–2,3,4,5,6,7,8,9,10,11–13,14,15–16,17
14	0	0.002	1–2,3,4,5,6,7,8,9,10,11–12,13,14,15–16,17
15	0	0	1,2,3,4,5,6,7,8,9,10,11-12,13,14,15–16,17
16	0	0	1,2,3,4,5,6,7,8,9,10,11,12,13,14,15–16,17

Table 2. Result of flood season division of the Fenhe River Basin during 1975–2014.

k	B(n,k)	f(k)	Classification
2	1.021		1–6,7–20
3	0.437	0.584	1–6,7–18,19–20
4	0.263	0.174	1–6,7–11,12,13–20
5	0.192	0.071	1–6,7–11,12,13–18,19–20
6	0.146	0.046	1–6,7–8,9–11,12,13-18,19–20
7	0.103	0.043	1–4,5–6,7–8,9–11,12,13–18,19–20
8	0.062	0.041	1–4,5–6,7–8,9–11,12,13–17,18,19–20
9	0.025	0.037	1–4,5–6,7–8,9–11,12,13–16,17,18,19–20
10	0.021	0.004	1–4,5–6,7–8,9–11,12,13,14–16,17,18,19–20
11	0.013	0.008	1–4,5–6,7–8,9–11,12,13,14,15–16,17,18,19–20
12	0.01	0.003	1–4,5–6,7–8,9–10,11,12,13,14,15-16,17,18,19–20
13	0.007	0.003	1–2,3–4,5–6,7–8,9–10,11,12,13,14,15–16,17,18,19–20
14	0.005	0.002	1–2,3,4,5–6,7–8,9–10,11,12,13,14,15–16,17,18,19–20
15	0.004	0.001	1–2,3,4,5–6,7–8,9–10,11,12,13,14,15,16,17,18,19–20
16	0.003	0.001	1–2,3,4,5–6,7,8,9–10,11,12,13,14,15,16,17,18,19–20
17	0.002	0.001	1–2,3,4,5–6,7,8,9–10,11,12,13,14,15,16,17,18,19,20
18	0.001	0.001	1–2,3,4,5–6,7,8,9,10,11,12,13,14,15,16,17,18,19,20
19	0	0.001	1,2,3,4,5–6,7,8,9,10,11,12,13,14,15,16,17,18,19,20

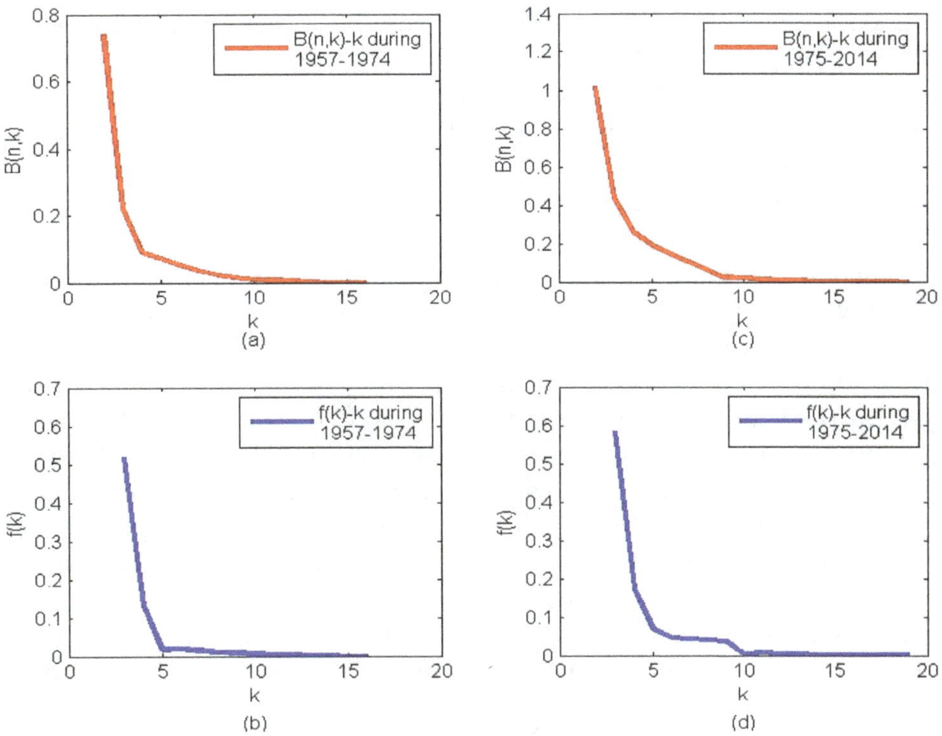

Figure 5. The $B(n, k) - k$ and $f(k) - k$ curves during 1957–1974 (**a**,**b**) and 1975–2014 (**c**,**d**).

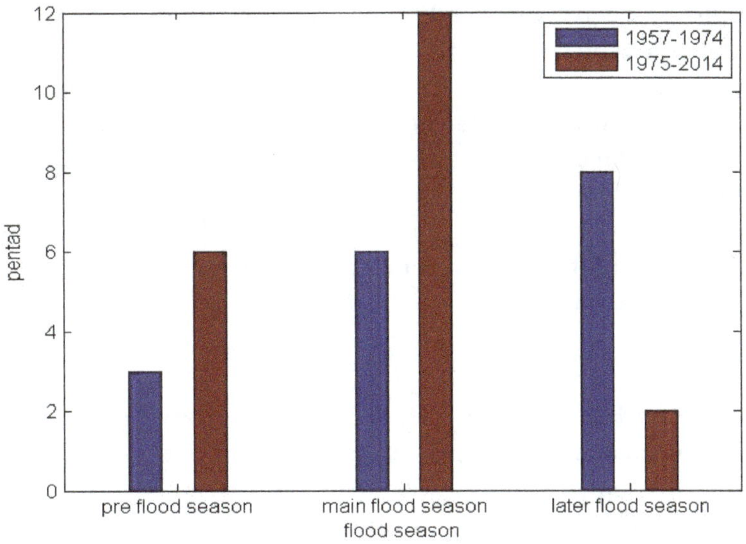

Figure 6. The bar graph of the length of each sub flood season.

5. Discussion

This research suggests that the flood season has advanced and extended from climate change since 1974, and it can be divided into three stages. Liu *et al.* also detected the abrupt point in the Fenhe River Basin, and they found that the precipitation changed abruptly in 1978 [29], which is approximately consistent with the results in this paper. The inconsistency between their article and this paper may derive from the inconsistency of the data—the period studied and the meteorological stations are different, while it is clear that the climate indeed changed significantly in the Fenhe River Basin.

With respect to the onset and retreat time of flood season, according to statistics, the precipitation during flood season account for more than 60% of the annual precipitation. Thus, the multiscale moving *t*-test method is objective and consistent with the characteristics of the precipitation. Pan *et al.* reconstructed the beginning times of flood season of the middle reaches of Yellow River according to the historical water level data from 1921 to 1950, and the results showed that they range from 6 July to 10 July. Our results indicate that the average onset dates of flood season during the period 1957–1974 and 1975–2014 in the Fenhe River Basin are 19 June and 5 June, respectively, which is inconsistent with Pan's results. Although the Fenhe River Basin lies in the middle reaches of the Yellow River, the terrain here is different from the mainstream of the Yellow River; the Fenhe River Basin is sandwiched between Mt. Lvliang and Mt. Taihang. The method and water level data used in Pan's paper are also different from the moving *t*-test method and the precipitation data used in this paper, which may have caused these differing results.

When it comes to the flood season division, Yang analyzed the diachronic distribution of storm floods and the regularity of flood season division in the Fenhe River Reservoir by recording the number of rainstorm days and the frequency of flood. The results indicated that the flood season can be divided into three stages [30], which is similar to the results of this paper. However, the Fisher optimal partition method focuses on the study from the angle of mathematics; it takes index into consideration more, and the time interval in this paper is 5 days, which is more accurate than 10 days in Yang's paper, so the result of the Fisher optimal partition method is more accurate.

Some countermeasures should be guaranteed to effectively utilize the flood water resources. We often set different flood limit levels during each stage [31]. In general, the prime task during the pre-flood season and the main flood season is flood prevention, so the flood limit levels in the pre- and main flood season should be a little lower. Further, the main purpose during the post-flood season is storing water, so the flood limit level during this period should be a little higher. As the pre-flood season has advanced 15 days and the main flood season has extended by 30 days, the flood limit level during this period should be decreased in advance and sustained longer so that the flood can flow safely through the basin. As for the shortening and delay of the post-flood season, the flood limit level should be elevated later so that flood can be transferred to water resources during this period.

Although some valuable findings have been obtained, further studies are needed to explore the flood season in greater detail. We used daily precipitation only to define the onset and retreat time of the flood season in the Fenhe River Basin. In fact, water level is also relative to the flood season, and the run-off also can impact the flood season. Future research should gather more data and consider more variables. If possible, we can take one reservoir as an example to evaluate the utilizable quantity of flood resources under climate change.

6. Conclusions

By using the Mann–Kendall detection, the multiscale moving *t*-test and the Fisher optimal partition methods, we present the onset and retreat dates of each stage in the Fenhe River Basin, which are characterized as the temporal variations from climate change. In the meantime, some useful conclusions of this study can be summarized as follows:

(1) The climate in the Fenhe River Basin has changed since 1974, and the annual precipitation illustrates a downward trend.

(2) The flood season in the Fenhe River Basin during 1975–2014 is 15 days longer than that during 1957–1974. Specifically, the pre-flood season is from 16 June to 30 June during 1957–1974, while it is from 1 June to 30 June during 1975–2014; the main flood season is from 1 July to 31 July during 1957–1974, while it is from 1 July to 31 August during 1975–2014; the post-flood season is from 1 August to 10 September during 1957–1974, whereas it is from 1 September to 10 September during 1975–2014.

(3) The flood limit level should be lowered at an earlier time to resist floods during the pre- and main flood season, and it should be raised at a later time so that more water for utilization can be stored after the flood season.

Author Contributions: Jianhua Wang contributed to design the methods; Yicheng Wang and Hejia Wang contributed ideas concerning the structure and content of the article; Baodeng Hou and Ya Huang contributed to data collection and processing; Chuiyu Lu drew the figures and maps; Weihua Xiao and Hejia Wang analyzed the results; Hejia Wang wrote the final manuscript.

References

1. International Panel of Climate Change (IPCC) Working Group I. *IPCC Fifth Assessment Report (AR5)*; IPCC: Stockholm, Sweden, 23–26 September 2013.
2. Ma, Z.G.; Shao, L.J. Relationship between dry/wet variation and the Pacific Decade Oscillation (PDO) in Northern China during the last 100 years. *Chin. J. Atmos Sci.* **2006**, *30*, 464–474.
3. Odekunle, T.O. Determining rainy season onset and retreat over Nigeria from precipitation amount and number of rainy days. *Theor. Appl. Climatol.* **2006**, *83*, 193–201. [CrossRef]
4. Sâmia, R.G.; Alan, J.P.C.; Mary, T.K. Revised method to detect the onset and demise dates of the rainy season in the South American Monsoon areas. *Theor. Appl. Climatol.* **2015**. [CrossRef]
5. Hachigonta, S.; Reason, C.J.C.; Tadross, M. An analysis of onset date and rainy season duration over Zambia. *Theor. Appl. Climatol.* **2008**, *91*, 229–243. [CrossRef]
6. Ding, J.L.; Xu, Z.S.; Fei, J.F.; Qiang, X.M. Analysis of the definition of the onset and ending dates of the annually first rainy season in South China and its interannual variation characteristics. *J. Trop. Meteorol.* **2009**, *25*, 59–65.
7. Gu, R.Y.; Zhou, W.C.; Bai, M.L.; Di, R.Q.; Yang, J. Influence of climate change on ice slush period at Inner Mongolia section of Yellow River. *J. Dese. Res.* **2012**, *32*, 1751–1756.
8. Pan, W.; Fei, J.; Man, Z.M.; Zheng, J.Y.; Zhuang, H.Z. The fluctuation of the beginning time of flood season in North China during AD1766-1911. *Quatern. Int.* **2015**, *380*, 377–381. [CrossRef]
9. Owusu, K.; Waylen, P.R. The changing rainy season climatology of mid-Ghana. *Theor. Appl. Climatol.* **2013**, *112*, 419–430. [CrossRef]
10. Ding, L.L.; Ge, Q.S.; Zheng, J.Y.; Hao, Z.X. Variation of starting date of pre-summer rainy season in South China from 1736 to 2010. *Acta. Geogr. Sin.* **2014**, *24*, 845–857.
11. China Meteorological Administration. Available online: http://data.cma.gov.cn/ (accessed on 15 March 2016).
12. Pettitt, A.N. A non-parametric approach to the change-point problem. *Appl. Stat.* **1979**, *28*, 126–135. [CrossRef]
13. Wei, F.Y. *Modern Climatic Statistical Diagnosis and Prediction Technology*; China Meteorological Press: Beijing, China, 1999; pp. 69–72.
14. Fraedrich, K.; Jiang, J.M.; Gerstengarbe, F.W.; Werner, P.C. Multiscale detection of abrupt climate changes: application to River Nile flood levels. *Int. J. Climatol.* **1997**, *17*, 1301–1315. [CrossRef]
15. Wang, L.L.; Zou, Y.R.; Sui, H.Q. An objective determination of the beginning and ending date of rainy season in Dalian. *Meteorol. Mon.* **2000**, *26*, 12–16.
16. Jiang, J.M.; Fraedrich, K.; Zou, Y.R. A scanning t test of multiscale abrupt changes and its coherence analysis. *Chin. J. Geophys.* **2001**, *44*, 31–39.
17. Jiang, J.M.; Mendelssohn, R.; Schwing, F.; Fraedrich, K. Coherency detection of multiscale abrupt changes in historic Nile flood levels. *Geophys. Res. Lett.* **2002**, *29*, 112. [CrossRef]
18. Schwing, F.; Jiang, J.M.; Mendelssohn, R. Coherency of multi-scale abrupt changes between the NAO, NPI, and PDO. *Geophys. Res. Lett.* **2003**, *30*, 325–348. [CrossRef]
19. Gu, X.Q.; Jiang, J.M. A new application of scanning *t*-test: Partition of dry/wet episodes in the western USA. *Quat. Sci.* **2006**, *26*, 742–751.
20. Liu, P.; Guo, S.L.; Li, W.; Xiong, H.K.; Zhang, W.X.; Guo, H.J.; Xu, D.L.; Wang, Z.X. Application of changing-point method for flood season stage in Geheyan Reservoir. *J. Yangtze River. Sci. Res. Inst.* **2007**, *24*, 8–11.
21. Jin, B.M.; Fang, G.H. Application of fuzzy set analysis method on flood stage study of Nanping. *Water Power* **2010**, *36*, 20–22.
22. Gao, B.; Liu, K.L.; Wang, Y.T.; Hu, S.Y. Application of system clustering method to dividing flood season of reservoir. *Water Resour. Hydropower Eng.* **2005**, *35*, 1–5.
23. Dong, Q.J.; Wang, X.J.; Wang, J.P.; Fu, C. Application of fractal theory in the stage analysis of flood season in Three Gorges Reservoir. *Resour. Environ. Yangtze Basin* **2007**, *16*, 400–404.

24. Xie, F.; Wang, W.S. Set pair analysis and its application to the division of flood period. *S-N Water Divers. Water Sci. Technol.* **2011**, *9*, 60–63. [CrossRef]
25. Wang, Z.Z.; Cui, T.T.; Wang, Y.T.; Yu, Z.B. Flood season division with an improved fuzzy c-mean clustering method in the Taihu Lake Basin in China. *Procedia Eng.* **2012**, *28*, 66–74. [CrossRef]
26. Wang, Z.Z.; Wang, Y.T.; Wu, H.Y.; Cui, T.T.; Xu, H.; Zhang, Y. Novel flood season division method based on fuzzy time series-effective cluster and its application to Taihu lake basin. *J. Hydroelectric Eng.* **2012**, *31*, 29–34.
27. Chen, L.; Singh, V.P.; Guo, S.L.; Zhou, J.Z.; Zhang, J.H.; Liu, P. An objective method for partitioning the entire flood season into multiple sub-seasons. *J. Hydrol.* **2015**, *528*, 621–630. [CrossRef]
28. He, H.W.; Zhang, A.L. The application of Fisher method to dividing seismicity period in Yunnan province. *J. Seismol. Res.* **1994**, *17*, 231–239.
29. Liu, Y.F.; Sun, H.; Yuan, Z.H.; Li, Y.R. Time series analysis of precipitation in flood season in fenhe river basin. *B. Soil Water Conserv.* **2011**, *31*, 121–125.
30. Yang, H.X. Research on the storm-flood characteristic of basin controlled by fenhe reservoir and its dividing by periods. *Shanxi Hydrotech.* **2008**, *3*, 40–41.
31. Gao, B.; Wang, Y.T.; Hu, S.Y. Adjustment and application of the limited level of reservoirs during the flood season. *Adv. Water Sci.* **2005**, *16*, 326–333.

13

Modeling Climate and Management Change Impacts on Water Quality and In-Stream Processes in the Elbe River Basin

Cornelia Hesse * and Valentina Krysanova

Potsdam-Institute for Climate Impact Research, Post Box 601203, Potsdam 14412, Germany; valentina.krysanova@pik-potsdam.de
* Correspondence: cohesse@pik-potsdam.de.

Academic Editors: Yingkui Li and Michael A. Urban

Abstract: Eco-hydrological water quality modeling for integrated water resources management of river basins should include all necessary landscape and in-stream nutrient processes as well as possible changes in boundary conditions and driving forces for nutrient behavior in watersheds. The study aims to assess possible impacts of the changing climate (ENSEMBLES climate scenarios) and/or land use conditions on resulting river water quantity and quality in the large-scale Elbe river basin by applying a semi-distributed watershed model of intermediate complexity (SWIM) with implemented in-stream nutrient (N+P) turnover and algal growth processes. The calibration and validation results revealed the ability of SWIM to satisfactorily simulate nutrient behavior at the watershed scale. Analysis of 19 climate scenarios for the whole Elbe river basin showed a projected increase in temperature (+3 °C) and precipitation (+57 mm) on average until the end of the century, causing diverse changes in river discharge (+20%), nutrient loads (NO_3-N: −5%; NH_4-N: −24%; PO_4-P: +5%), phytoplankton biomass (−4%) and dissolved oxygen concentration (−5%) in the watershed. In addition, some changes in land use and nutrient management were tested in order to reduce nutrient emissions to the river network.

Keywords: Elbe river basin; water quality modeling; in-stream processes; nutrients; SWIM; climate change impact assessment; ENSEMBLES; management change impacts

---

1. Introduction

Changes of the world's and Europe's climate and increased anthropogenic pressure on natural resources have already been detected in the past, and this development is likely to continue in the future [1–6].

Looking at the climate aspect, a global rise in mean temperature, change in precipitation pattern as well as an increase in intensity and frequency of extreme events can be recognized [1,2,7], impacting the water cycle [8–10], vegetation and biodiversity [11–14] as well as human health [15,16] and economy [17,18]. The potential warming in Europe could reach values from +1 to +6 °C by the end of this century [19], depending on the location. The annual mean precipitation is expected to increase in Northern Europe and decrease in most parts of Southern Europe and Mediterranean regions up to ±20% [20]. The catchment of the Elbe river, mainly located in Germany and the Czech Republic in Central Europe, is already experiencing changes in climate conditions as well as changes in extreme temperature and precipitation values, and this trend is expected to continue. During the last century (1882–2013) the average temperature in Germany increased by 1.2 °C, whereas the precipitation amounts rose by 10.6% on average with a high spatial and temporal variability [21]. Application of ensembles of climate scenarios shows increasing trends in floods for the Elbe basin in Germany [22] as well as in the Czech Republic [23], especially in wintertime.

Climate change will have direct and indirect impacts on river water quantity and quality [24–28]. With the rising temperature, water availability will decrease due to increased evapotranspiration, and a reduced discharge will facilitate algal growth and reduce dilution of point source pollutants. Higher temperatures and lower flow velocities would additionally stimulate turnover processes and increase system respiration rates, causing oxygen deficits in river reaches. All these processes lead to the degradation of water quality and ecological status of a waterbody connected with a higher probability of algal blooms [24,29,30] and increasing problems for water supply and treatment [25].

As phytoplankton growth is a key factor for water quality in lowland river ecosystems [31], the algal processes should be included in evaluating impacts of global change on water quality. Light and nutrients are generally deemed to be the most important external drivers of phytoplankton dynamics in rivers, along with temperature which also has a positive relationship with phytoplankton, and discharge which has a negative relationship [29–31].

Additionally, climate change would influence nutrient turnover and transport processes (denitrification, nitrification, volatilization and leaching) in the catchments, due to altered temperature and precipitation [32–34], and the terrestrial processes will also influence the final river water quality at the outlet of the basin. River systems are also affected by anthropogenic impacts (land use composition, population and industry) causing nutrient pollution from point (factories, sewage treatment plants) and diffuse sources (agricultural fields), which lead to eutrophication processes and a decrease in river water quality [35–38].

Due to the high pressure of rising population, growing industry and increasing transport demand on landscapes and natural vegetation, many changes in land use could be recognized in Europe in the past. The current tendencies include a decreasing trend in crop and pasture acreages in Spain, Czech Republic and Northern Germany, slowly growing forested areas in Northern Europe and Portugal, and notably growing urban areas in France and Western Germany [5,39]. It is expected that these trends will continue in the coming 10–20 years.

Population density and human activities are important underlying factors for point and diffuse nutrient pollution entering rivers [40]. During the last decades, many efforts to reduce nutrient inputs to the rivers by construction and improvement of sewage treatment plants and optimization of fertilizer application on cropland were undertaken in Europe. They resulted in a remarkable reduction of phosphorus emissions (mainly from point sources), but only a small decrease of nitrogen inputs (mainly from diffuse sources) due to the lag time of diffuse nutrients in soils [41–43]. It is widely recognized that the control of diffuse source emissions is much more difficult. So it is expected that the inputs of nutrients from households and industry will be further reduced in the future, and diffuse inputs from fertilizers and manure will become the main sources of nutrient pollution in Europe [42].

Climate as well as land use change impacts on river water quality superimpose each other and create a very complex system of interactions and feedbacks [27,44,45]. The nitrate loads in the rivers, for example, are climate-dependent, and were likely influenced by former climate variations, so it is difficult to define and interpret the pure effects of management changes in the past [41]. Furthermore, adaptation measures and policy responses to projected climate change, e.g., subvention for bio-fuels or control of greenhouse gas emissions, affect freshwater quality as well [24]. A combined land use and climate change impact assessment would be an important step facilitating an integrated river basin management. The system characteristics and variable boundary conditions should be taken into account by default in modern management strategies [31] to support the implementation of adaptation measures in river basins.

The process-based eco-hydrological watershed models driven by climate and land use parameters can be useful for assessing potential future developments in a changing world. Watershed models including both landscape and in-stream nutrient processes, which are able to simulate nutrient turnover processes in a catchment and river network, may represent effective tools for the evaluation of river water quality at the basin scale [46,47]. However, it should be kept in mind that current water quality modeling at the watershed scale still has more weaknesses and uncertainties compared to pure hydrological modeling due to the higher complexity of modeled processes and the requirement to include more input data and parameters.

In former applications of the semi-distributed eco-hydrological Soil and Water Integrated Model (SWIM) [48] for water quality modeling in river basins in Germany, it was observed that nutrient retention and decomposition solely in the soils of the watershed is not sufficient for tackling nutrients coming from different sources, especially in larger basins [49]. Therefore, SWIM was extended by a new module representing nutrient and oxygen turnover processes and algal growth in rivers, which was already tested for the mesoscale Saale river, a sub-catchment of the Elbe river with an area of about 25,000 km^2 [50]. The aim of this study is to apply the new SWIM version for a combined climate and land use change impact assessment on the entire Elbe basin including the upstream part in the Czech Republic and the lower part in Germany (total drainage area about 150,000 km^2). This can support the development of management strategies and adaptation measures to potential changes in the future in this large-scale river basin.

In particular, the following objectives were pursued in this study:

- Testing the in-stream SWIM module for the large scale by modeling water quality parameters (nitrate nitrogen (NO_3-N), ammonium nitrogen (NH_4-N), phosphate phosphorus (PO_4-P), dissolved oxygen (DOX), and chlorophyll *a* (Chla)) at the basin outlet and at confluences of the large Elbe tributaries in the historical period;
- Analysis of climate scenarios for the region provided by the European ENSEMBLES project [51], and climate change impact assessment on water quantity and quality for two future periods (2021–2050 and 2071–2100) with unchanged management;
- Simulation of selected land use change and management scenarios aiming at the reduction of point and diffuse nutrient loads in the basin; and
- Analysis of the combined climate and land use change impacts on water quantity and quality to derive ideas for suitable measures for adaptation to climate change.

The model-based assessments of climate and land use change impacts on water quality are rare in literature so far compared to impact assessments on the hydrological cycle, especially at the large scale. The recently implemented in-stream module enables a more realistic representation of all interrelated processes for the impact study. Therefore, this study is an important step forward to large-scale application of water quality models with distributed calibration for impact studies in general, as well as towards a fully integrated water resources assessment in the Elbe catchment in particular.

2. Study Area: The Elbe Catchment

The Elbe river (1094 km) originates at 1386 m a.s.l. in the Giant Mountains located between Poland and the Czech Republic, drains an area of 148,268 km^2 and flows into the North Sea [52]. The Elbe has the fourth largest catchment area among the European rivers [31]. 65.5% of its catchment belongs to Germany, 33.7% to the Czech Republic, 0.6% to Austria and 0.2% to Poland (see Figure 1). The discharge regime (861 m^3·s^{-1} on average) usually shows high water levels in winter and spring and low water levels in summer and autumn. In total, 24.5 million people live in the Elbe basin, which also includes the large cities Berlin, Hamburg and Prague [52].

Table 1 gives an overview of the main characteristics of the Elbe basin until the gauge Neu Darchau, and its six main tributaries, covering catchment areas above 5000 km^2. In this table some topography-specific differences can be detected between the tributary subbasins, namely in regard to climate parameters, soil conditions and, as a consequence, land use distribution, which also affect nutrient concentrations in the rivers. So, in the sub-catchments with dominating agricultural land use due to fertile loess soils (e.g., Saale and Mulde) the nitrate and nitrogen concentrations are higher (see Table 1), resulting from fertilizer application and leaching. In contrast, the catchment of the Havel river has the highest share of low fertile soils consisting of almost two-thirds of sandy grained particles (about 70% of the total area) and shows the lowest nitrogen pollution but the highest phosphorus level. The high phosphate concentrations of the Havel river can be mainly explained by desorption from historically polluted sediments [53].

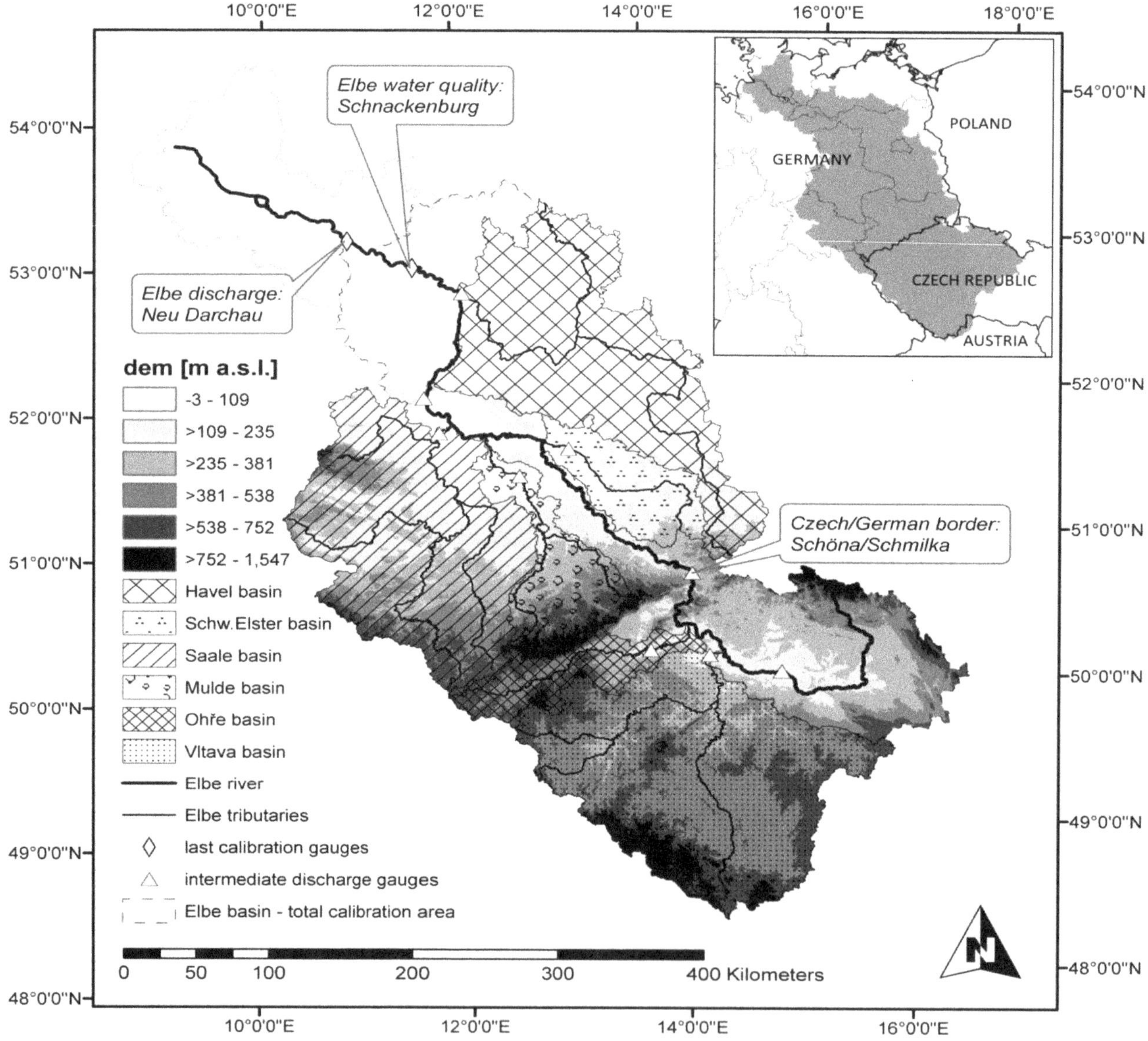

Figure 1. Location and digital elevation model of the Elbe river basin and six catchments of its main tributaries, as well as location of the observation gauges used for calibration.

According to the German classification of water quality [54], which uses the 90th percentile for nutrients and the 10th percentile for dissolved oxygen to compare with certain water quality thresholds, the highest nitrate level results in water quality class III (Mulde and Saale), the highest ammonium value belongs also to class III (Vltava), the maximum phosphate phosphorus level represents water quality class II–III (Vltava and Havel), and the lowest dissolved oxygen concentration results in water quality class II (Havel). There is quite high diversity between the rivers in this respect, and no river exists which has the worst or best status for all components.

Table 1. Characteristics of the Elbe river catchment and its main tributaries of second (classical) order for the time period 2001–2010.

River *gauges (discharge/water quality)*		**Elbe** *Neu Darchau/Schnackenburg*	**Vltava** *Vraňany/Zelčín*	**Ohře** *Louny/Terezín*	**Schwarze Elster** *Löben/Gorsdorf*	**Mulde Bad** *Düben/Dessau*	**Saale** *Calbe-Grizehne/Groß Rosenburg*	**Havel** *Havelberg/Toppel*	**Unit**
Length [1]		907	430	305	179*	314	434	334	km
Mean discharge [1]		711	145	38	21	67	117	114	$m^3 \cdot s^{-1}$
Catchment area [1]		131,950	28,090	5614	5705	7400	24,079	23,858	km^2
Average altitude		281	523	507	131	394	287	74	m a.s.l.
Average temperature [2]		8.9	7.8	7.6	9.7	8.9	9.2	9.6	°C
Average sum of precipitation [2]		698	713	771	652	822	680	616	$mm \cdot y^{-1}$
Land use [3]	*Agriculture*	51.3	49.7	42.2	48.1	53.3	63.0	38.6	%
	Forest	31.7	36.8	37.7	35.0	28.8	23.3	38.2	
	Grassland	8.4	7.8	13.6	7.2	6.9	4.6	11.1	
	Settlements	6.3	4.3	3.9	5.9	9.4	7.6	7.9	
Soil texture [4]	*clay*	16.2	20.3	22.1	8.5	17.1	20.0	9.0	%
	silt	38.2	37.4	39.6	29.8	47.9	54.9	26.3	
	sand	45.6	42.3	38.3	61.7	35.0	25.1	64.7	
Point sources [5]	*TN*	22318	4704	570	183	1673	3557	2768	$t \cdot y^{-1}$
	TP	1870	564	73	29	155	357	167	
Nutrients [6]	*NO_3-N av.*	3.17	3.73	2.38	2.31	4.35	4.68	0.82	$mg \cdot L^{-1}$
	NO_3-N 90-perc.	4.60	4.91	3.20	4.30	6.10	6.15	1.56	
	NH_4-N av.	0.16	0.31	0.08	0.20	0.16	0.21	0.10	
	NH_4-N 90-perc.	0.33	0.94	0.11	0.53	0.36	0.46	0.23	
	PO_4-P av.	0.07	0.12	0.03	0.02	0.06	0.09	0.13	
	PO_4-P 90-perc.	0.11	0.22	0.05	0.03	0.09	0.13	0.24	
	DOX av.	11.7	11.7	10.6	9.7	10.6	10.3	10.6	
	DOX 10-perc.	9.7	9.4	8.0	7.8	8.5	7.8	6.7	
Chlorophyll [6]	*CHLA av.*	77.1	36.7	8.0	9.3	10.7	21.8	37.6	$\mu g \cdot L^{-1}$
	CHLA 90-perc.	184.0	96.1	14.2	18.0	28.5	61.7	73.0	

Notes: * Wikipedia. Data sources: [1] [55], [2] DWD/PIK, [3] Corine2000, [4] Germany: BÜK1000, Czech Republic: [56], [5] Germany: [57], Czech Republic: [58], [6] German gauges: FIS, Czech gauges: IKSE.

The Elbe river is the most important river draining the eastern part of Germany. The natural flow regime is influenced by reservoirs and regulation of small rivers, drainage of wetlands and brown coal mining [59]. Due to former political and socio-economic conditions, the Elbe was one of the most polluted rivers in Europe with a low ecological potential. The water quality improved after the German reunification in 1990 due to closure or upgrading of sewage treatment plants and industrial enterprises in the basin, as well as to a substantial decrease in fertilization rates on agricultural land [58,60]. However, contamination problems still exist, especially looking at sediments, which are characterized by a high adsorption of heavy metals and other polluting substances [61]. There are also no significant improvements regarding chlorophyll *a* concentrations in the Elbe river [60].

In general, the Elbe river is characterized by a strong phytoplankton growth in the free-flowing section due to inputs from the reservoirs of the upper Elbe and Vltava and high nutrient loads from tributaries [62]. The high primary productivity leads to substantial differences in nutrient concentrations along the river with remarkable intra- and interannual variations [62], and the season of main biological activity is between March and October [31]. Low flow velocities in the lowland tributaries with many natural lakes in the river course (e.g., Havel) and in rivers influenced by weirs and barrages (e.g., Vltava, Saale) facilitate good conditions for algal growth and cause high concentrations of chlorophyll *a*.

The middle course of the Elbe river in Germany contains several protected natural areas with a high diversity of flora, fauna and landscape types. Large parts of the river in Germany are free-flowing and not influenced by barrages. However, the original floodplain areas have often been cut off by flood protection measures for settlements, agriculture and industry during the last centuries. Approximately 84% of the floodplain along the Elbe river course in Germany is protected by dikes [63]. The reduced flooding area not only causes problems in times of very high water levels (e.g., during the last decades when immense flood events and damage occurred), but also hinders the natural nutrient retention capacity of the river ecosystems. This induces an intensification of nutrient pollution problems in the river waters.

3. Materials and Methods

3.1. Soil and Water Integrated Model (SWIM)

The Soil and Water Integrated Model (SWIM) is an eco-hydrological model of intermediate complexity simulating the hydrological cycle and vegetation growth integrated with nutrient turnover processes within a river basin driven by climate parameters and taking soil conditions and land use management into account. SWIM was developed on the base of the Soil and Water Assessment Tool (SWAT) [64] and the MATSALU model [65] specifically as a tool for the analysis of climate and land use change impacts on hydrological processes, agricultural production and water quality at the regional scale. More details can be found in [48].

Being a spatially semi-distributed dynamic model working with a daily time step, SWIM calculates all hydrological, vegetation and nutrient processes on a hydrotope level (set of elementary units in a subbasin with the same land use class and soil type). Lateral fluxes (surface, subsurface and groundwater flow with associated nutrients) are summarized at the subbasin level and routed through the river network to the outlet of the catchment.

Hydrological processes on the hydrotope level are based on the water balance equation, taking precipitation, evapotranspiration, percolation, surface and subsurface runoff, capillary rise and ground water recharge into account.

The available water content in soil is influenced by crop and vegetation types, which are parameterized in a database connected to SWIM [48]. The crop database is the same as in SWAT [66], and only some parameters were adapted during calibration. The vegetation affects nutrient turnover as well, as plants are important nutrient consumers as well as sources (from plant residue).

The nitrogen module of the applied SWIM version (compare Hesse *et al.* [50]) calculates nutrient processes in the soil profile and includes several pools: nitrate and ammonium nitrogen, active and stable organic nitrogen, and organic nitrogen in plant residues. They are influenced by fertilization, mineralization, volatilization and (de-)nitrification processes, plant uptake, wet deposition, wash-off, leaching and erosion. Leaching is calculated differently for nitrate and ammonium nitrogen, as the latter has much higher bonding capacity to soil particles.

The soil phosphorus module includes labile phosphate phosphorus, active and stable mineral phosphorus, organic phosphorus and phosphorus in the plant residue. The phosphorus pools are influenced by fertilization, (de-)sorption, mineralization, plant uptake, erosion, and leaching. The equation applied to calculate leaching of phosphate phosphorus through the soil profile can be found in Hesse *et al.* [49].

Processes related to diffuse source nitrogen and phosphorus flows to the river network are surface runoff, subsurface runoff, groundwater flow, wash-off, leaching, erosion and retention of nutrients in the landscape. After simulating all nutrient-specific processes in the soil profile, nitrogen and phosphorus are transported with surface, subsurface and groundwater flows to the rivers. During their passage through the basin, nutrients are subject to retention and transformation processes in soils, wetlands and in the river system. These processes and model equations, as well as the testing of different retention methods, were described in detail in previous publications [50,67,68].

Additional information about the general SWIM model concept, necessary input and output data, calibration parameters, process equations as well as the GIS interface for model setup can be found in the User Manual [48].

3.2. Data Preparation and Model Setup for Calibration

SWIM model setup requires spatial and temporal data sets as well as major water and land use management information. The spatial maps include a digital elevation model (DEM), a soil map with soil parameterization, a land use map and a subbasin map. The temporal data sets include the daily historical observed or projected future climate parameters (temperature, precipitation, solar radiation and relative air humidity) as external drivers of the model. The observed river discharge and nutrient concentrations, at least close to the catchment's outlet, are necessary for the model calibration and validation. Additional monitoring data at intermediate gauges and tributaries allow a multi-site calibration, which is more reliable, especially for large-scale catchments. Necessary management data include water abstraction, storage or transfer, major crops with their planting and harvesting dates, as well as fertilization rates and dates and effluents from industrial sites or waste water treatment plants.

The model setup for the Elbe river was based on spatial maps with a 250 m resolution. The DEM map was resampled from the data provided by the NASA Shuttle Radar Topographic Mission (SRTM). The general German soil map "BÜK1000" delivered by the Federal Institute for Geosciences and Natural Resources (BGR) was combined with the soil map and soil parameterization of the Czech Republic [59] and the European Soil Database (ESDB) provided by the Joint Research Centre of the European Commission to cover the entire Elbe river catchment. The land use map was obtained from the CORINE land cover (CLC2000) data set of the European Environment Agency (EEA) and reclassified to the 15 SWIM land use classes required by the model. The subbasin map was combined from the standard maps of the Federal Environment Agency (UBA) for Germany and the T.G.M. Water Research Institute for the Czech Republic, and included 2268 subbasins.

The historical climate data of 348 climate observation stations located within and 20 km around the Elbe catchment were used to interpolate the climate parameters to the centroids of all subbasins by an inverse distance method for calibration and validation of the SWIM model, taking climate information of at least one to maximum four neighboring stations into account. The station density with available climate data was higher in the German than in the Czech part of Elbe river catchment.

The observed discharge and water quality data for selected gauges located at the Elbe river and its main tributaries in Germany originated from the Data Information System (FIS) of the River Basin Community Elbe (FGG-Elbe) and were downloaded in December 2012. The Czech monitoring data with a monthly time step were taken from the publications of the International Commission for the Protection of the Elbe river (IKSE). In case the observed nutrient concentrations were indicated to be below the detection limit, the minimum detectable concentration was halved and assumed for this day. Data on nutrient inputs from point sources at the German part of the basin were taken from FGG-Elbe [60]. For the Czech part, assumed data on nutrient emissions from point sources were derived from a report of the IKSE for the year 2000 [58]. As there were only temporally aggregated data available, the point source emissions were implemented in the model as daily constant values.

The standard SWIM does not consider crop rotation management on agricultural fields so that only one main crop type could be assumed on the entire arable land. According to data in the statistical yearbooks of the German federal states in Germany considerably overlaying with the Elbe basin (Thuringia, Saxony-Anhalt, Brandenburg, Saxony and Mecklenburg-West Pomerania), winter wheat was selected to be the main crop. Assuming some nutrient storage in the soils, 100 kg N/ha and 12 kg P/ha were assumed as an average fertilization level in accordance with recommendations of the federal agriculture agencies. However, fertilization is recommended to be increased with increasing yield expectations [69]. To implement this option, arable land was classified according to the expected yield as simulated by SWIM (as a function of soil quality, water availability and climate conditions under constant fertilization). Then the medium yield class received the average fertilization, and fertilization of the low/high yield classes was reduced/increased by 20%.

In order to better represent specific behavior of vegetation in lowland areas with its connection to groundwater and the increased evapotranspiration potential, the simpler of two approaches for wetland simulation as described in Hattermann *et al.* [70] was used in SWIM. In total, 22.6% of the entire Elbe river basin belongs to wetlands, with especially high shares in the Schwarze Elster catchment (41%), the lower Elbe reaches (40%), and the Havel river catchment (33%). In the catchments of the other large tributaries (Saale, Mulde, Ohře and Vltava), wetlands make up 10%–16% of their total areas.

The model calibration and validation for the whole basin was performed for five years, each within the period 2001–2010, considering observed data at the last gauges at Neu Darchau (discharge, Elbe, km 536.4) and Schnackenburg (water quality, Elbe, km 474.5), which are undisturbed by tidal influences. The nutrient loads at the gauge at Schnackenburg were calculated as products of concentration and discharge using the discharge of the gauge at Wittenberge (km 453.9) with the correction factor 1.001 [71].

The calibration of water discharge (Q) and nutrient loads was done by adjusting the main model calibration parameters described in the SWIM manual [48], and listed in former SWIM model applications [49,50,68,72,73]. During the model calibration it was realized that a global calibration parameter set was not sufficient to represent the basin- and river-specific water and nutrient processes for the several catchments of the Elbe tributaries, which can be highly variable due to different combinations of elevation, soil, land use and river characteristics. Therefore, it was decided to use the most important calibration parameters spatially distributed for the seven large river catchments, which were calibrated individually. Table 2 lists and describes those parameters for water quantity and quality calibration used in a distributed mode within the Elbe river basin.

Table 2. SWIM calibration parameters applied spatially distributed in the Elbe river basin.

Module	Parameter	Description	Unit
Hydrology	bff	baseflow factor used to calculate return flow travel time	-
	delay	time needed for water leaving root zone to reach shallow aquifer	day
	roc2/roc4	coefficients to correct the storage time constants for surface and subsurface flows	-
Soil nutrients	ret	retention times of nitrate nitrogen (NO_3-N), ammonium nitrogen (NH_4-N) and phosphate phosphorus (PO_4-P) in the lateral subsurface and groundwater flows (6 parameters)	day
	deg	degradation rates of NO_3-N, NH_4-N and PO_4-P in the lateral subsurface and groundwater flows (6 parameters)	day^{-1}
	deth	soil water content threshold for denitrification of NO_3-N	%
	dad/dkd	ratios of adsorbed NH_4-N and PO_4-P to that in soil water	-
In-stream processes	mumax	maximum specific algal growth rate	day^{-1}
	tempo	optimal temperature for algal growth	°C
	lio	optimal radiation for algal growth	ly
	pr20	predation rate in the reach at 20 °C	day^{-1}
	ai1/ai2	fractions of algal biomass that is nitrogen and phosphorus	$mg \cdot mg^{-1}$
	rs1	local algal settling rate in the reach at 20 °C	$m \cdot day^{-1}$
	rs2/rs3	benthic source rates for PO_4-P and NH_4-N in the reach at 20 °C	$mg \cdot (m^2 \cdot day)^{-1}$
	rs5	organic phosphorus settling rate in the reach at 20 °C	day^{-1}
	rk2	oxygen reaeration rate in the reach at 20 °C	day^{-1}
	bc3	rate constant for hydrolysis of organic nitrogen to NH_4-N at 20 °C	day^{-1}
	bc4	rate constant for mineralization of organic phosphorus to PO_4-P at 20 °C	day^{-1}

3.3. Evaluation of Model Results

The ability of SWIM to simulate water and nutrient processes in the Elbe catchment and to reproduce the observed monitoring values was evaluated in different ways for water quantity and quality.

The simulated daily and/or monthly discharges were assessed using the Nash-and-Sutcliffe efficiency (NSE, [74]) as well as the deviation in water balance (DB) (compare [49]). The non-dimensional NSE is a measure to analyze the squared differences between the observed and simulated values, and DB describes the long-term differences of the observed values against the simulated ones for the whole simulation period in percent.

The model's efficiency to represent the water quality parameters was evaluated at the long-term average monthly basis using three criteria, $\Delta\mu$, $\Delta\sigma$ and r, according to Gudmundsson *et al.* [75]. Here $\Delta\mu$ is a balance measure defined as the relative bias of the mean annual observed and simulated values. Criterion $\Delta\sigma$ evaluates the amplitude or the spread from the lowest to the highest monthly values of the mean annual cycle by comparing the relative difference in standard deviations of the observed and the simulated values. Also, the usual Pearson's correlation coefficient r, which is sensitive to differences in the shape as well as in the timing of the mean annual cycle, was applied.

Table 3 lists the possible ranges, optima and aspired results of the different performance criteria used in this study.

Table 3. Description of performance criteria used in this study to evaluate model results.

Criterium	Range	Optimum	Aim in This Study
NSE	$-\infty$ to 1	1	>0.65
DB	$-\infty$ to $+\infty$	0	>−20% to <20%
$\Delta\mu$	$-\infty$ to $+\infty$	0	>−0.2% to <0.2%
$\Delta\sigma$	$-\infty$ to $+\infty$	0	>−0.2% to <0.2%
r	−1 to 1	1	>0.5

3.4. Description, Evaluation and Processing of Climate Scenario Data

The ENSEMBLES project [51] delivered projections for a possible climatic future of Europe by running an ensemble of different Regional Climate Models (RCMs) using the boundary conditions produced by several General Circulation Models (GCMs). All models assumed the A1B emission scenario with a balanced use of fossil and non-fossil fuels in a world with a rapidly growing economy, population growth until 2050 and a decline afterwards, and fast development of new and effective technologies [76]. According to this scenario an average global temperature rise of 2.8 °C (with a range between 1.8 and 4.4 °C) is estimated [1] until the end of the 21st century.

The resulting ENSEMBLES climate scenarios differ in resolution (25 or 50 km grids) and simulation period (1951/1961–2050/2100). For our study, 19 scenarios covering the period until 2100 were chosen (Table 4). As climate data necessary for SWIM modeling were not available for all scenarios until 2100, only data until 2098 were considered in all cases. A scenario-specific number of grid cells with data were treated as virtual climate stations for climate interpolation to the centroids of the 2268 subbasins within the Elbe basin using an inverse distance method.

Table 4. Numbering of the chosen climate scenarios as combinations of General Circulation Models (GCMs) and Regional Climate Models (RCMs), the responsible institute, resolution [km], starting year, and number of grid cells used for interpolation of the projected climate in the Elbe catchment.

ID	Institute	GCM	RCM	Resolution	Start Year	Number of Grid Cells
1	SMHI	HadCM3Q3	RCA	25	1951	316
2	HC	HadCM3Q0	HadRM3Q0	25	1951	316
3	HC	HadCM3Q3	HadRM3Q3	25	1951	316
4	HC	HadCM3Q16	HadRM3Q16	25	1951	316
5	C4I	HadCM3Q16	RCA3	25	1951	316
6	ETHZ	HadCM3Q0	CLM	25	1951	316
7	KNMI	ECHAM5-r3	RACMO	25	1951	316
8	SMHI	BCM	RCA	25	1961	316
9	SMHI	ECHAM5-r3	RCA	25	1951	316
10	MPI	ECHAM5-r3	REMO	25	1951	316
11	CNRM	ARPEGE_RM5.1	Aladin	25	1951	300
12	DMI	ARPEGE	HIRHAM	25	1951	316
13	DMI	ECHAM5-r3	DMI-HIRHAM5	25	1951	316
14	DMI	BCM	DMI-HIRHAM5	25	1961	316
15	ICTP	ECHAM5-r3	RegCM	25	1951	282
16	KNMI	ECHAM5-r1	RACMO	50	1951	79
17	KNMI	ECHAM5-r2	RACMO	50	1951	79
18	KNMI	ECHAM5-r3	RACMO	50	1951	79
19	KNMI	MIROC	RACMO	50	1951	79

To analyze the projected trends of single climate scenarios, climate change signals were calculated for two future periods for temperature, precipitation and solar radiation. Climate change signals describe the differences between the mean climate parameter values in a future period and in the reference period of the same scenario. The signals were derived taking all scenario grid cells into account and were evaluated for the annual mean climate parameter values as well as for their seasonal dynamics.

The 19 climate scenarios were used to drive the calibrated SWIM model, each for the reference period 1971–2000 (p0) and the two future periods 2021–2050 (p1) and 2071–2098 (p2).

It is very important which downscaling approach is used to generate climate scenarios, and whether it is statistical or dynamical. The choice of a hydrological model is less important in terms of its contribution to uncertainty, especially when only the long-term mean annual changes are compared [77]. Often it was detected that results achieved with one hydrological model under two or more climate scenarios differ more than the results of different hydrological models achieved by using only one climate scenario [78,79]. Hence, many authors suggest using an ensemble of climate change scenarios to get the full range of uncertainty between the different scenario projections [22,80,81]. The last two authors also mentioned that there is no direct link between the climate model performance in the historical period and the robustness of trends in the future, and thus the application of a smaller number of best fitting scenarios could not be recommended. Therefore, in our study we did not try to find the most probable future climate scenarios by their comparison with the historical measurements.

The observed climate data are also often used for bias correction of climate scenarios before applying them for impact assessment in order to avoid unrealistic simulations of runoff or nutrient loads. However, there is no consistent opinion on the usefulness of bias correction for impact assessments. While Teutschbein and Seibert [82] recommend an application of bias correction, other authors complain about the lack of physical justifications of corrections damaging the physical consistency between the variables [77,83]. The latter do not appreciate this method as a "valid procedure", and complain that an additional uncertainty is added to the model chain. In our study it was decided not to use bias correction and to simply compare the simulations driven by 19 RCMs between periods to detect trends and the relative changes caused by climate change.

3.5. Processing of Socio-Economic Change Experiments

In addition to climate change simulations, five land use change experiments were applied for testing the effects of specific socio-economic measures aimed at reducing point or diffuse nutrient emissions.

The applied land use change scenarios are listed in Table 5 together with the description of the changes implemented in model input data. Two scenarios are dealing with the direct reduction of nutrient emissions ("Point sources" and "Fertilization") by 10% or 20%. The decrease of point source emissions was assumed with different percentages for the two nutrients, as it was supposed that phosphorus reduction potential in sewage treatment is higher. The third scenario ("Retention") indirectly tested the effects of a possible increase of the retention potential and decomposition rate in the soils of the Elbe catchment by 10%. This could be achieved by different measures, for example by extension of wetland areas around the watercourses or intensified cultivation of plant communities with a high nutrient intake rate (mycorrhiza, legumes). In addition, two such measures were tested directly in the model ("Buffer" and "Slope") by changing the land use composition in the catchment.

Due to the spatial resolution of the SWIM project with 250 × 250 m raster maps, water courses in agricultural areas were converted to 250 m raster cells in the "Buffer" experiment, containing extensive meadows without fertilization. In the "Slope" scenario, all agricultural areas with a slope >4% were converted to extensively used meadows to study the effects on water quantity and quality in the catchment (see e.g., [84] where hillsides with slopes >4% are considered as being a risk of facilitating erosion).

Table 5. Description of applied land use change experiments in the Elbe river catchment.

Scenario Name	Description
Point sources	Reduction of emissions from point sources (nitrogen −10%, phosphorus −20%)
Fertilization	Reduction of N and P fertilizers on agricultural land by 20%
Retention	Increase of nutrient retention time and decomposition rate in soils by 10% each
Buffer	Conversion of all agricultural lands around water courses to extensive meadow
Slope	Conversion of agricultural lands to extensive meadows on hillsides with slopes >4%

The socio-economic experiments were run under the 19 ENSEMBLES climate scenarios to allow evaluation of the combined climate and land use change impacts on water quantity and quality in addition to the land use change impacts only. As 19 climate scenarios were applied with specific climate conditions, the results were different, not only for the combined impacts, but also for the land use change impacts. To show the possible effects of scenarios, the 19 single percental changes of the model outcomes were analyzed regarding their medians and 25th/75th percentile values, representing the most probable 50% range of all scenario results.

4. Results

4.1. SWIM Model Calibration and Validation

A successful calibration of a model for water quality requires a well-calibrated hydrological model. During the hydrological and water quality calibration, the observed and simulated values at the most downstream Elbe gauges, at the gauges located close to the German-Czech border, as well as at the main tributaries were compared and statistically evaluated for the period of 2001–2010.

Figure 2 presents the observed and simulated daily discharges for the 10-year period (left), and the long-term daily averages (right) at the two main Elbe gauges Schöna and Neu Darchau. The discharge dynamic is well reproduced, reaching the good to very good performance ratings. The performance criteria for the daily model results are better at the downstream gauge Neu Darchau. The long-term seasonal dynamics are reproduced well at both gauges.

However, not all simulation results at the tributaries reach the same model performance (Table 6). The most difficult river to simulate was the Schwarze Elster, which is highly influenced by human activities and regulation (opencast lignite mining, discharge regulation and stream straightening), so that the hydrological processes are no longer natural. As these site-specific management measures were not implemented in the model, the model does not perform well enough at the Löben gauge. Similar problems apply to the lowland catchment of the Havel river, which is characterized by a high number of wetland areas and stream lakes, and is also highly affected by lignite mining in its upper course, all this leading to lower NSE values at gauge Havelberg.

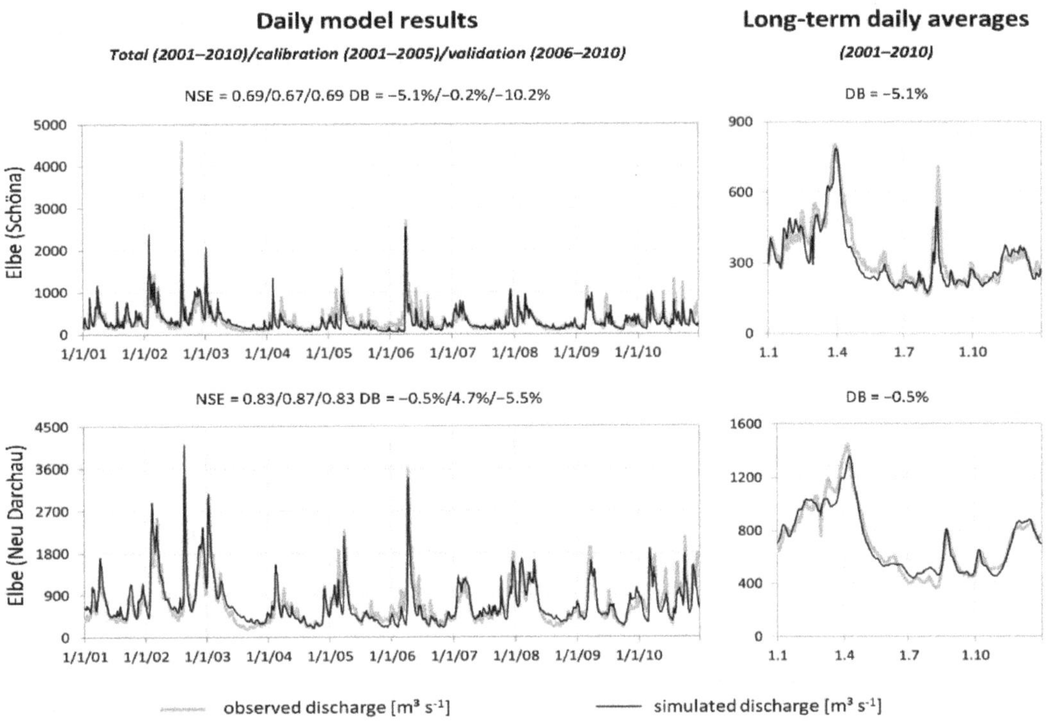

Figure 2. Calibration results for the Elbe river discharge at the most downstream gauge Neu Darchau and the intermediate Elbe gauge Schöna (Czech/German border) for the time period 2001–2010, separated into calibration and validation sub-periods.

Table 6. Model performances for four discharge gauges of the Elbe river and six gauges of its main tributaries from the upstream to downstream location of tributaries.

River	Gauge	Time Period	NSE (–)		DB (%)
			Daily	*Monthly*	
Elbe	Nymburk	11/2002–10/2010		0.75	−13.5
Vltava	Vranany	11/2002–10/2010		0.64	−10.5
Ohře	Louny	11/2002–10/2010		0.86	−0.3
Elbe	Schöna	2001–2010	0.69	0.77	−5.1
Schwarze Elster	Löben	2001–2008	0.25	0.60	13.4
Mulde	Bad Düben	2001–2010	0.74	0.88	1.7
Saale	Calbe-Griezehne	2001–2010	0.61	0.84	1.5
Elbe	Magdeburg	2001–2010	0.82	0.86	1.1
Havel	Havelberg	2001–2010	0.54	0.68	−1.5
Elbe	Neu Darchau	2001–2010	0.83	0.86	−0.5

Only monthly measurements for a shorter time period were available for the three gauges located in the Czech part of the Elbe basin. The best results here could be achieved for the smaller and mountainous river Ohře. The upper part of the Elbe river (gauge Nymburk), as well as the largest Elbe tributary, Vltava, show a slight underestimation of discharge. This could be explained by water regulation measures in the water course of these rivers, with a high number of barrages and dams to ensure water availability for shipping and for flood protection, which were not implemented in the model. However, the hydrological model performance in terms of NSE and DB for the Elbe and its tributaries mostly meet the aim (compare Table 3), so that it was used for the subsequent water quality calibration.

Figure 3 presents the results of water quality calibration for two main gauges in the Elbe river: Schmilka at the Czech-German border and the most downstream Elbe gauge Schnackenburg. The long-term average daily observed loads were calculated based on interpolated values between biweekly measurements and have some degree of uncertainty. The calibration was aimed at visually and statistically matching the inner-annual dynamics and minimizing the deviation in balance of the mean annual nutrient loads for the 10-year period of 2001–2010.

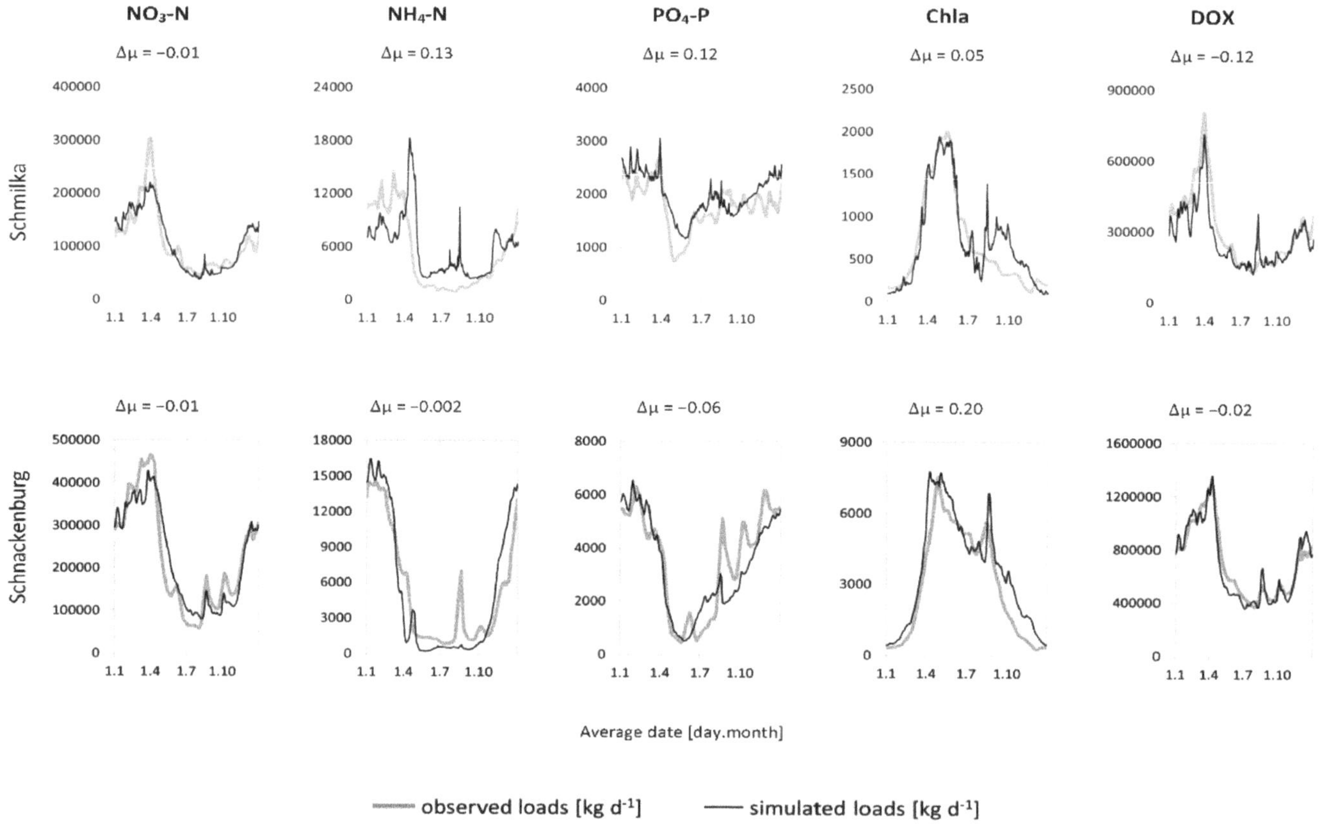

Figure 3. Long-term average daily observed and simulated loads of nitrate nitrogen (NO_3-N), ammonium nitrogen (NH_4-N), phosphate phosphorus (PO_4-P), chlorophyll *a* (Chla) and dissolved oxygen (DOX) at the two Elbe gauges Schmilka (corresponds to the total Czech loads) and Schnackenburg (most downstream gauge) for the time period 2001–2010.

In Figure 3, a specific annual cycle of the three nutrients can be observed, which is reproduced quite well by the SWIM model. The nitrate nitrogen loads (mainly coming from diffuse sources) generally follow the discharge curve with a spring peak and low values in summer. Ammonium nitrogen and phosphate phosphorus are more algae-influenced. The periods with high concentrations of chlorophyll *a* especially result in ammonium depletion in the river due to the high ammonium preference factor of the algae defined in the model. Algal influences on the phosphate loads are less significant, but also obvious, especially during the spring algal bloom. The dissolved oxygen loads are highly connected to the water amounts and are simulated very well. The balance measure $\Delta\mu$ is low in all cases and is located within the aimed range, also reflecting sufficiently good calibration results.

Figure 4 and Table 7 show results on water quality for the main tributaries of the Elbe river and for selected Elbe gauges. Figure 4 plots the simulated *versus* observed long-term average monthly values and illustrates the variation of the long-term seasonal cycle ratios around a diagonal of perfect fit, and Table 7 analyzes the model's performance statistically.

Table 7. Model ability to simulate the long-term monthly average loads of water quality variables for six main tributaries as well as three selected Elbe water quality observation gauges in the time period 2001–2010 after spatially distributed model calibration. The model performance variables were calculated according to [75] and are described in Section 3.3.

	Vltava *Zelčín*	**Ohře** *Terezín*	**Elbe** *Schmilka*	**Schwarze Elster** *Gorsdorf*	**Mulde** *Dessau*	**Saale** *Groß Rosenburg*	**Elbe** *Magdeburg*	**Havel** *Toppel*	**Elbe** *Schnackenburg*
Time period	**2001–2004**	**2007–2010**	**2001–2010**	**2004–2010**	**2001–2010**	**2001–2010**	**2001–2010**	**2001–2010**	**2001–2010**
					NO_3-N				
Δμ	−0.01	−0.09	−0.01	−0.13	0.04	0.10	0.09	−0.01	−0.01
Δσ	−0.11	−0.23	−0.04	−0.27	−0.20	0.18	0.03	−0.42	−0.07
r	0.91	0.93	0.93	0.96	0.93	0.94	0.93	0.97	0.95
					NH_4-N				
Δμ	0.02	0.12	0.13	0.17	−0.08	−0.01	0.17	0.05	0.01
Δσ	0.17	−0.25	−0.23	0.02	−0.31	−0.19	0.08	0.13	0.26
r	0.64	0.74	0.68	0.53	0.85	0.97	0.92	0.94	0.95
					PO_4-P				
Δμ	−0.01	0.11	0.12	0.12 *	0.03	−0.11	0.10	−0.10	−0.06
Δσ	−0.40	0.01	−0.02	0.39 *	−0.37	−0.03	0.00	−0.19	−0.03
r	0.73	0.70	0.87	0.13 *	0.87	0.93	0.92	0.24	0.94
					Chla				
Δμ	0.08	−0.02	0.05	0.03	−0.08	0.04	0.06	0.12	0.20
Δσ	0.06	0.12	−0.07	0.05	0.17	0.04	−0.07	0.18	0.00
r	0.95	0.82	0.94	0.82	0.80	0.82	0.91	0.78	0.97
					DOX				
Δμ	−0.06	−0.12	−0.12	0.02	−0.01	0.06	0.04	0.03	−0.02
Δσ	−0.28	−0.39	−0.30	−0.50	−0.37	0.08	−0.02	−0.25	0.02
r	0.85	0.95	0.98	0.91	0.95	0.97	0.96	0.98	0.98

Note: * 2008–2010.

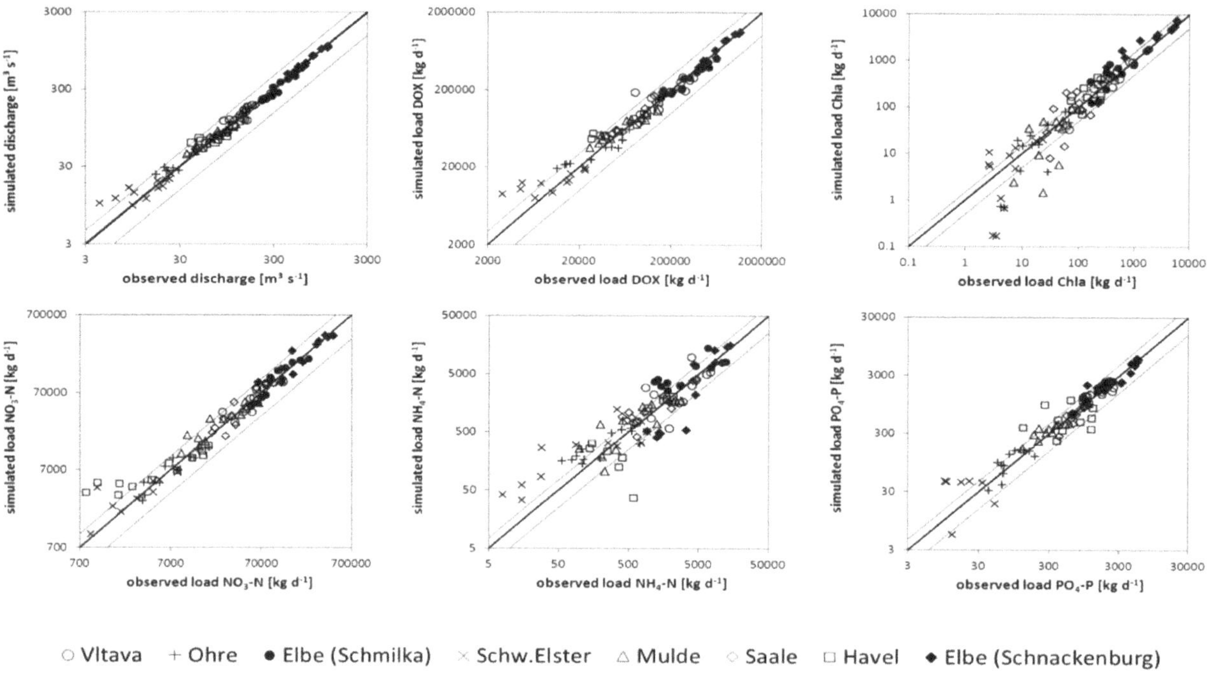

Figure 4. The long-term average monthly observed and simulated discharge and loads per tributary and at two selected Elbe gauges in the period 2001–2010 (diagonals: black—perfect fit, grey—± 50% intervals).

As already detected in the hydrological calibration, the largest discrepancies between the observed and simulated values can be seen for the Schwarze Elster river. The intensive human activities within this catchment (e.g., surface water management due to lignite mining) influence nutrient processes but are not fully implemented in the model, resulting in model outputs different from observations. Some problems can also be seen for the Havel and (partly) the Mulde tributaries. The largest dispersion around the diagonal of perfect fit is obvious for ammonium nitrogen, which is difficult to model as it is highly influenced by point source emissions involving input data uncertainty as well as by algal consumption processes (parameter uncertainty). The latter, due to their complex behavior influenced by many physical, chemical and biological interactions, are difficult to model, especially in large basins. The results in terms of statistical criteria (Table 7) with mostly high r and low $\Delta\mu$ and $\Delta\sigma$ confirm the visual impression.

Generally, the calibrated SWIM model for the large-scale Elbe river basin matches observations well, and can be used for climate and land use change impact assessment.

4.2. Climate Change Signals of the ENSEMBLES Scenarios

Before applying the 19 ENSEMBLES climate scenarios to the Elbe basin, they were analyzed for their trends in temperature, precipitation and solar radiation averaged over the whole basin by comparing two future scenario periods, p1 and p2, with the reference period p0. The comparison was done for the long-term average annual values as well as for the long-term average monthly values of all scenarios and periods.

The climate change signals per scenario can be found in Table 8. The results show an increase in temperature by 1.3 °C for the first period and by 3 °C for the second period on average, as well as an increase in precipitation by +41/+57 mm on average for all 19 climate scenarios. The increase in precipitation is accompanied by a decrease in solar radiation of −15/−27 J cm^{-2} on average, probably due to increased cloudiness with higher precipitation amounts. There is a wide spread in signals between the scenarios, which is increasing in the second period. Regarding temperature, all scenarios agree on increasing trend, but the increase in period p2 ranges between 2 and 5 °C depending on the scenario. The agreement of the single scenarios with the overall trends is lower for precipitation (15 of 19 scenarios agree with the trend) and solar radiation (14 scenarios agree). However, a majority of scenarios correspond to the average trends.

Table 8. Climate change signals for temperature, precipitation and radiation of 19 analyzed ENSEMBLES climate scenarios and on average for the two future periods 2021–2050 (p1) and 2071–2098 (p2) compared to the reference period 1971–2000 (p0) for the Elbe basin.

Scenario	Temperature (°C)		Precipitation (mm)		Radiation (J·cm^{-2})	
	p1-p0	*p2-p0*	*p1-p0*	*p2-p0*	*p1-p0*	*p2-p0*
S1	1.5	2.9	67	95	−26	−76
S2	2.1	4.0	−2	16	27	28
S3	1.7	3.4	34	17	8	7
S4	2.2	5.0	48	−49	1	43
S5	1.8	4.1	104	94	−57	−67
S6	1.7	3.5	24	13	−22	−12
S7	0.9	2.6	35	110	−12	−20
S8	0.7	1.9	63	86	−30	−48
S9	0.8	2.4	47	112	−29	−65
S10	0.9	2.6	14	47	−18	−42
S11	1.1	2.8	−4	−68	5	36
S12	0.9	2.0	14	−31	−16	−111
S13	0.6	2.0	57	157	−21	−73
S14	0.9	2.5	37	99	−29	−47
S15	0.9	2.6	29	87	1	7
S16	1.0	3.1	52	63	−12	−5
S17	1.4	3.3	65	54	−22	−11
S18	0.9	2.6	36	99	−16	−20
S19	1.8	3.9	50	76	−26	−40
$mean_{all}$	1.3	3.0	41	57	−15	−27
$stdev_{all}$	0.5	0.8	26	60	18	42

The seasonal climate change signals are visualized in Figure 5. Looking at the changes per month, it is obvious that the value as well as the spread of the climate change signals is higher in the second period. The increase in temperature is confirmed for the entire course of the year, and it is lowest in May and highest in winter months (December–February) and in August. The changes in precipitation and solar radiation vary around the zero-line and show an opposite behavior (probably due to connection of precipitation and cloudiness). In the first period precipitation is slightly decreasing in July and August, and in the second period negative changes in precipitation are projected from June to September. The changes in solar radiation show almost the opposite trends. In general, the 19 ENSEMBLES climate scenarios project a warmer and wetter climate with less sunshine hours from autumn to spring, but a warmer, dryer and sunnier climate in the summer months for the region.

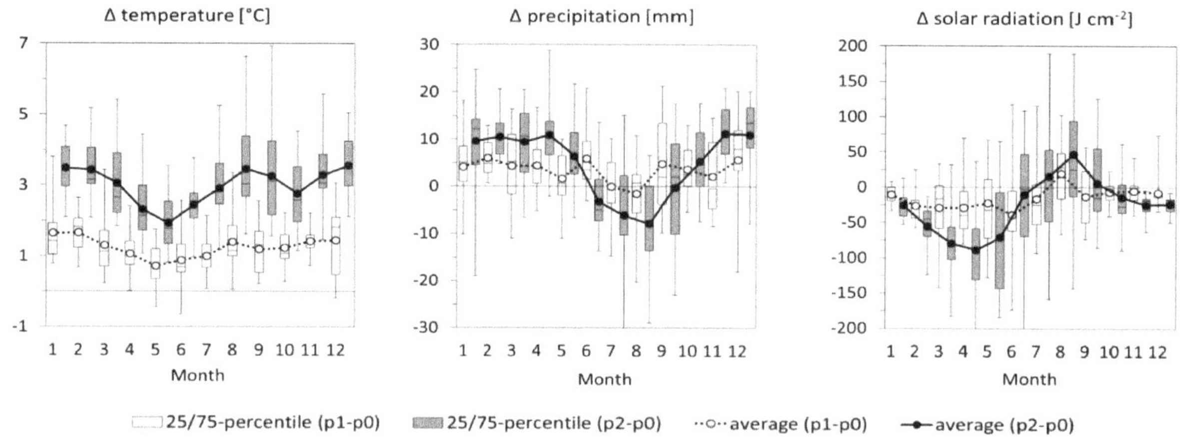

Figure 5. Ranges of seasonal climate change signals for temperature, precipitation and solar radiation of 19 ENSEMBLES climate scenarios for the two future periods compared to the reference period of the same scenario for the Elbe basin. The plots represent median (**line**), 25th/75th percentiles (**box**), min/max values (**whiskers**) and the average (**dots**) change of all 19 scenarios.

4.3. *Climate Change Impacts*

The projected changes in climate lead to changes in simulated water quantity and quality variables in the Elbe basin in future periods. The results are shown Figure 6 for the two Elbe river gauges Schöna and Neu Darchau. They present changes in the long-term average seasonal dynamics comparing the average and the 25th/75th percentile ranges of six variables from simulations driven by 19 climate scenarios in the future and the average of the reference period 1971–2000.

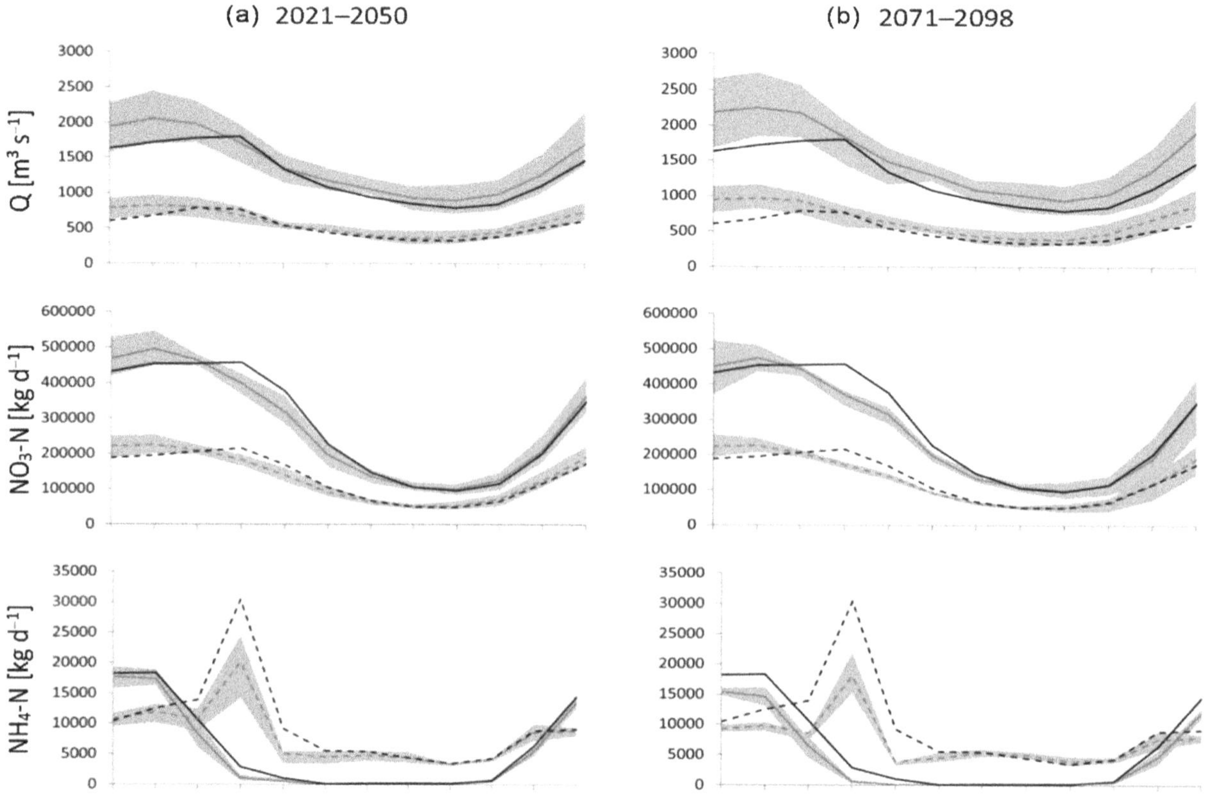

Figure 6. *Cont.*

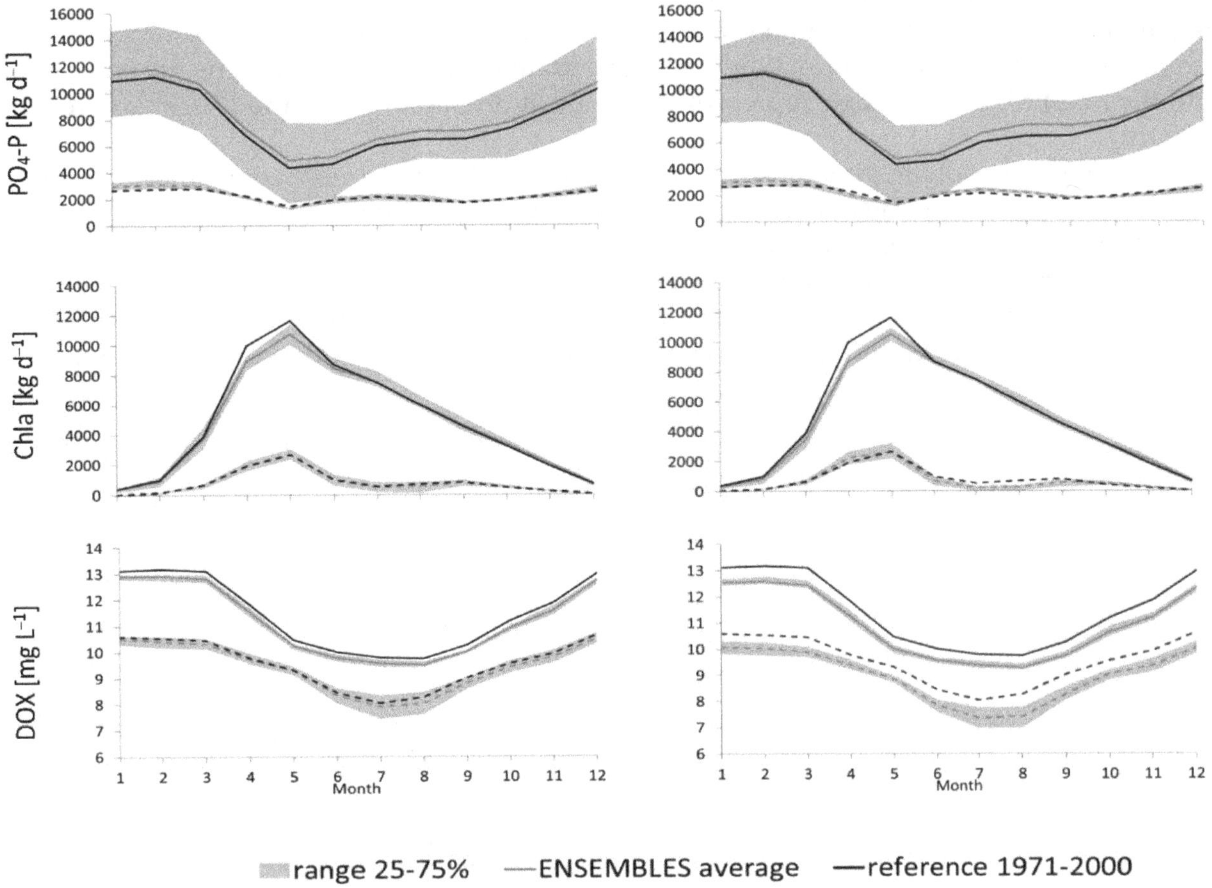

Figure 6. The long-term average monthly values of simulated discharge (Q), nutrient and chlorophyll *a* loads (NO_3-N, NH_4-N, PO_4-P, Chla) and dissolved oxygen concentrations (DOX) with uncertainty ranges (25th/75th percentiles corresponding to 19 simulations) at the two Elbe gauges Neu Darchau (**full lines**) and Schöna (**dashed lines**) for the future periods 2021–2050 (**p1, a**) and 2071–2078 (**p2, b**) in comparison to the corresponding average values of the reference period 1971–2000 (p0).

Following the increasing trend for precipitation in the Elbe basin, the discharge is projected to increase as well, both at the last Elbe gauge and at the gauge of the Czech-German border. The increase can be observed during almost the whole year, with the highest values in winter months (due to higher rainfall) and the lowest values, or even negative changes in the p1 period, in April (due to lower or missing snow melt peaks). Though a decrease in precipitation is projected in the summertime (compare Figure 5), the projected discharge in summer months is higher than in the reference period, probably due to the capability of soils to retain additional winter and spring water causing delayed subsurface and groundwater flows. However, the uncertainty ranges for the projected discharge are quite high, especially at the most downstream gauge.

The nitrate nitrogen load performs similarly to the discharge, as nitrate nitrogen comes to the river mainly dissolved in water from diffuse sources. A moderate increase can be observed in the first winter months, followed by some decrease in spring, whereas the second half of the season shows only minor changes on average compared to the reference period (due to higher retention time of nitrate nitrogen compared to water as well as impacts of vegetation).

The ammonium nitrogen loads are higher on average in the upstream part of the Elbe (gauge Schöna) than downstream (gauge Neu Darchau) due to higher loads in the Czech part of the catchment as well as to progressively increasing phytoplankton concentration downstream of the Elbe. The decrease in ammonium load caused by changes in climate conditions is obvious in the first half of the season (especially during spring flood). The decrease in NH_4-N loads is probably connected to the rising temperatures, as mineralization processes and the emergence of leachable ammonium in soils are temperature-related and occur mainly within a certain temperature range. The uncertainty ranges around the ENSEMBLES average, representing the most probable 50% of the 19 scenario results, are quite narrow.

The average phosphate phosphorus load shows a slight and almost constant increasing trend throughout the season, but the uncertainty ranges are the largest for this nutrient, caused by the high uncertainty and climate-dependence of phosphorus-related processes in the Havel catchment (compare with Figure 7). The increase in loads is probably connected to increasing erosion and leaching processes with higher precipitation in the future, washing more phosphorus from sandy and highly permeable soils. It could also be a result of less ingestion by a decreasing algae population in the future.

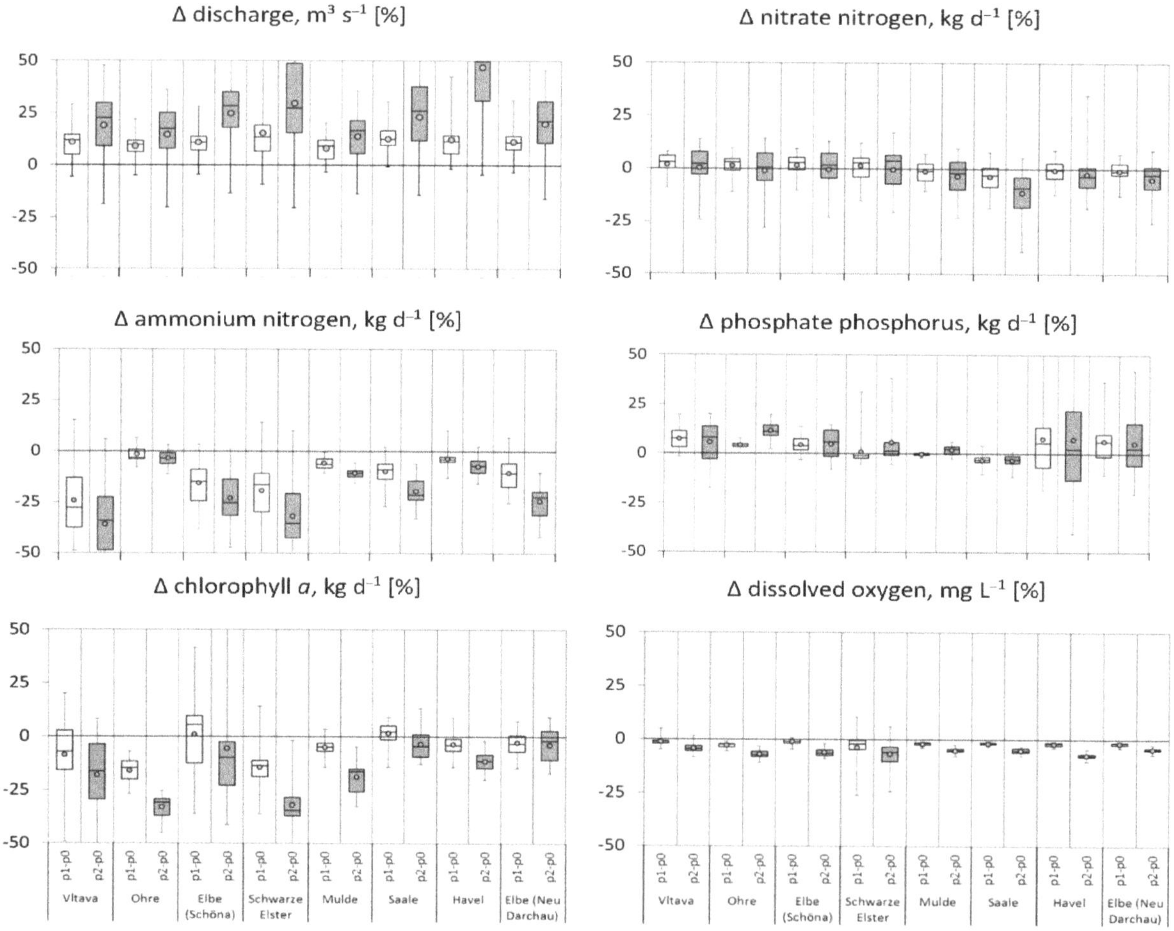

Figure 7. Ranges of the percental changes of 30-year-average river discharges, nutrients and chlorophyll *a* loads, as well as dissolved oxygen concentrations in the Elbe river and its main tributaries simulated with SWIM driven by 19 ENSEMBLES climate scenarios (in future periods p1 (**light**) and p2 (**dark**) compared to the reference period p0 of the same scenario). The plots visualize the following ranges: min/max (**whiskers**), 25th/75th percentiles (**boxes**), median (**line**) and average (**dots**) changes of all 19 scenarios.

The chlorophyll *a* load is projected to decrease in the spring blossom time, when warmer temperatures (temperature stress) and lower solar radiation (below the optimum value) may hamper phytoplankton growth and less ammonium is available for algae consumption.

The dissolved oxygen concentration in the Elbe river is projected to decrease, and the changes remain almost constant throughout the season. This is probably connected to the increasing water temperature, resulting in lower values of oxygen saturation in the water. The uncertainty ranges for future dissolved oxygen concentrations are higher upstream, probably due to the generally higher ammonium loads modeled in the upper river reaches, where oxygen is used for nitrification in the water column.

In addition to the temporal analysis of climate impacts, Figure 7 illustrates some spatially distributed results for the Elbe and its tributaries. For that, average percental changes were calculated for six main tributaries of the Elbe and two Elbe gauges (the same as in Figure 6).

The overall trend for the entire basin can be generally detected regarding different variables in Figure 7, though there are some outlying sub-catchments. For all gauges an increasing discharge is projected, which becomes higher in the second period. Also, the uncertainty ranges increase in p2. The differences between gauges are small.

The nitrate nitrogen load decreases on average for the entire Elbe river basin (Neu Darchau). The decrease is largest for the Saale catchment, which is characterized by the highest share of agricultural areas due to very fertile soils with a high nutrient retention capability. There are also some sub-catchments where a small increase (or no change) in nitrate load on average is simulated. This is probably connected to an increased diffuse pollution with increased precipitation in these sub-areas.

The impacts on ammonium nitrogen loads are almost all negative, and show a high diversity between the sub-catchments. The uncertainty ranges are highest in the Vltava and Schwarze Elster sub-catchments, where ammonium pollution is generally at its highest level, and have more space for variability due to climate change impacts.

Except for the Saale sub-catchment with its fertile soils and high nutrient retention potential, the climate change impact on phosphate phosphorus shows increasing loads due to increased leaching and erosion processes. The uncertainty ranges are extremely high in the Havel sub-catchment, where phosphorus contamination is the highest in the Elbe drainage area, and a high share of permeable and sandy soils causes a high phosphorus leaching potential with higher precipitation amounts.

Chlorophyll *a* demonstrates a decreasing trend on average almost everywhere. The uncertainty ranges, especially in the upper tributaries, are quite high, due to the high complexity of algae processes simulated in the model, which are influenced by many system-internal and external drivers.

Changes in the dissolved oxygen concentrations have a very small uncertainty range and show a decreasing trend on average for all gauges due to increased temperatures and lower oxygen saturation capacity. The highest range in average changes can be observed for the Schwarze Elster sub-catchment, which is quite heavily polluted with ammonium nitrogen. The latter is highly sensitive to climate change impacts and is connected to the oxygen processes in the river water.

4.4. Socio-Economic Change Impacts under Climate Change

In addition to the climate change impact assessment, five land use change experiments were run to test the model's reaction on certain management measures aimed at reducing nutrient inputs to the river network. The aim was to check whether such measures are able to be reversed, intensify or revoke climate change impacts. The land use change experiments were run 19 times, driven by the 19 ENSEMBLES climate scenarios for the near future period 2021–2050 (p1), and the results were compared with the results achieved under the reference management conditions for the period 1971–2000 (p0) of the same scenarios (combined impacts) as well as with the climate scenario-driven results with the reference management for period p1 (land use change impacts only).

The single and combined impacts were analyzed for the two Elbe gauges Schöna (Czech/German border) and Neu Darchau (Elbe outlet) as well as for the outlets of the two selected tributaries Saale and Havel (Figure 8). The results are shown as median values with a 25th/75th percentile range. In some cases, even the single land use change impact shows some range of relative changes caused by different behavior of temperature- and water-dependent nutrient processes under different climate conditions used as an external driver.

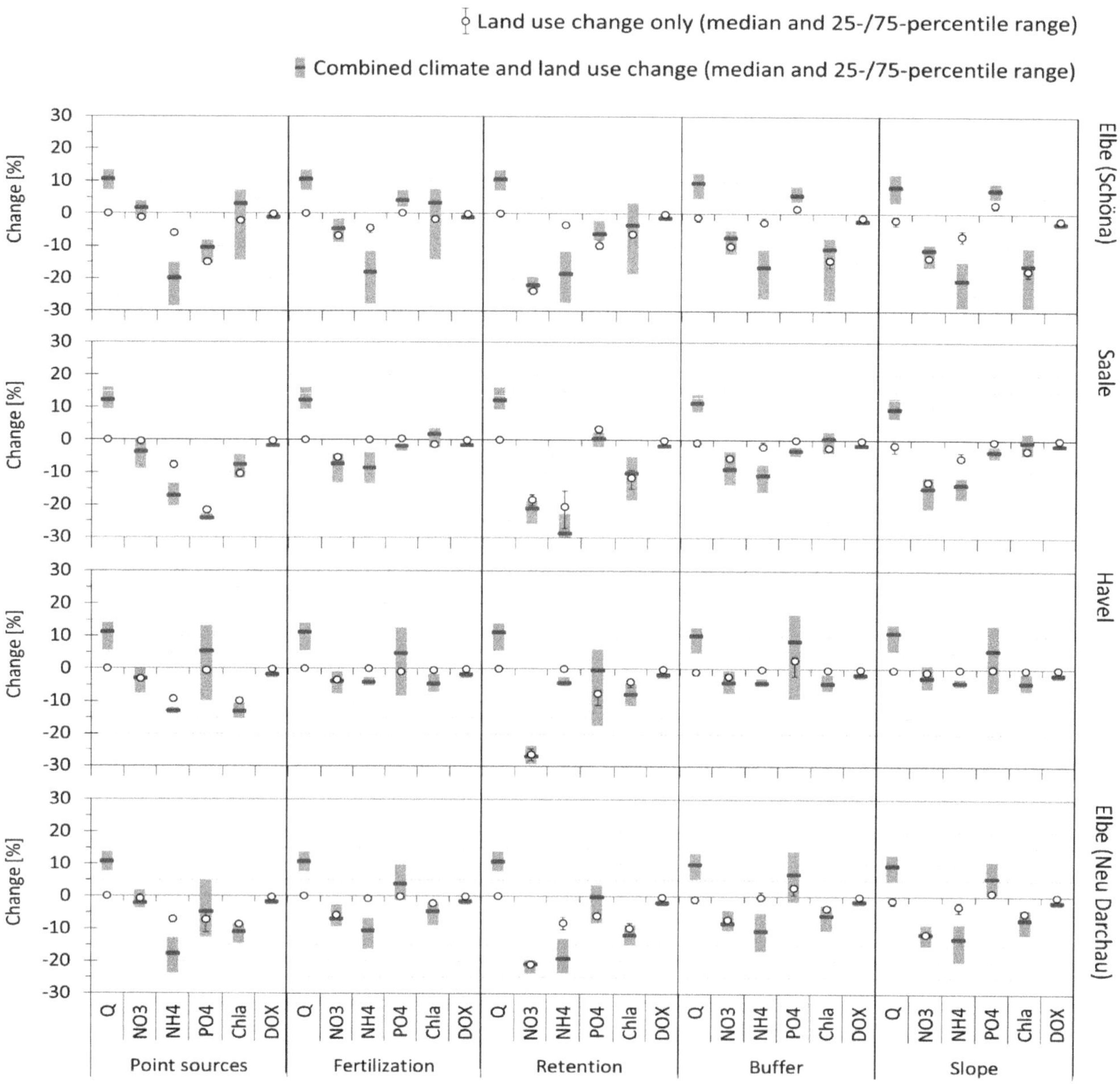

Figure 8. Impacts of socio-economic changes and combined climate and socio-economic changes on the average water discharge (Q), nutrient (NO_3-N, NH_4-N, PO_4-P) and chlorophyll *a* (Chla) loads and dissolved oxygen concentrations (DOX) of the Elbe river at two stations and at two main German tributaries. The dark grey bars and white dots show the median of 19 percental changes together with their 25th/75th percentile ranges (light grey ranges and black whiskers).

The socio-economic changes related to nutrient inputs to the river network (experiments "Point sources" and "Fertilization") and an increased nutrient retention potential in soils (experiment "Retention") have no influence on water discharge. Only the combined impacts show an increase in discharge of about 10% due to climate change. The solely socio-economic impacts of a changed land use composition ("Buffer" and "Slope") on river discharge show a decrease (due to increased evapotranspiration of the enlarged grassland areas), but it is quite low, and cannot compensate the increase in Q caused by the projected climate change, so that all combined impacts for these experiments have a positive direction.

The reduction of point source emissions has the highest influence on phosphate and ammonium loads, as these nutrients mainly originate from anthropogenic inputs of water treatment plants or industrial units. The projected climate change even intensifies the reduction of ammonium nitrogen loads in the rivers, whereas the decrease of phosphate phosphorus is reduced by climate change impacts (except for the Saale basin). The sole reduction of point source emissions predominantly results in a decrease of chlorophyll *a* loads in the rivers due to less available ammonium and phosphate as algal food.

The decrease in fertilizer application causes lower nitrate loads in all analyzed river parts, as this nutrient originates mainly from diffuse sources (predominantly from agricultural fields). The reduction is only marginally influenced by climate change. A decrease in fertilization affects NH_4-N only partly, and causes decreased ammonium loads, particularly in the upper part of the Elbe basin. As the changes in NH_4-N and PO_4-P loads are less distinct under the "Fertilization" experiment, chlorophyll *a* loads are only marginally influenced. The on-average-increasing chlorophyll *a* trend caused by climate change impacts in the upper Elbe and Saale catchments cannot be reversed by a simple reduction of fertilization in the combined experiments.

An increased nutrient retention and decomposition potential in the soils of the landscape ("Retention") has the highest impact on nutrient loads. Especially the diffuse nitrate nitrogen loads are affected, but also ammonium and phosphate show some reactions, though with different magnitudes for the four analyzed gauges. The diversity in the magnitude of changes for the river parts can be explained by the heterogeneity and distribution of land use patterns and point sources as well as by the diversity in projected climate change within the catchment. As NH_4-N and/or PO_4-P are remarkably reduced under the retention experiment, chlorophyll *a* shows a decreasing trend due to a lack of nutrients. This reduction is even able to reverse the increasing trend in chlorophyll *a* caused by climate changes in the upper Elbe and Saale sub-catchments.

Two experiments dealing with a changed land use composition ("Buffer" and "Slope") result in more meadows and less agricultural areas in the sub-catchments and show similar results in the different river parts. Nitrate nitrogen is reduced most in the majority of cases due to less agricultural area with fertilizer application and hence lower total fertilizer loading under the experiments. The highest diversity of changes can be seen under the "Slope" experiment in the upper Elbe and Saale sub-catchments, which are characterized by a high share of mountainous areas, where the share of transformed land use areas is higher than in the lowland sub-catchment of the Havel river. For the latter, the "Slope" experiment has nearly no impact on the model outputs, and the combined changes result only from the climate scenario impacts.

The concentrations of dissolved oxygen are not visibly influenced by the changes in land use or management. The decreasing trend due to increased water temperature is more obvious in the upper part of the Elbe basin (gauge Schöna), probably due to less oxygen production with decreasing chlorophyll *a* loads in the river.

In general, the shares of cropland and distribution of point sources, as well as the distribution of soils with their specific nutrient retention potentials, are very important factors influencing the nutrient loads coming with the rivers. However, in the model application presented here, it is often difficult to distinguish between the single impacts on nutrient loads caused by certain land use or management changes and the secondary impacts due to altered chlorophyll *a* concentrations and a resulting change in nutrient uptake in the water body. The in-stream processes include a complex behavior of nutrients with a high number of interactions and feedbacks with the algae population. Chlorophyll *a*, for example, increases with decreasing NH_4-N availability and *vice versa*, causing increase (or decrease) of PO_4-P due to less (or more) algal uptake. Therefore, the resulting impacts are not only directly caused by land use changes, but are also indirectly caused by the subsequently changed conditions in the river water.

5. Discussion

A comparison of obtained results with the results of previous studies dealing with global change impacts in the Elbe river catchment is sometimes difficult, as different scenarios and downscaling methods were used by different authors. The whole range of model outputs illustrates a high uncertainty in climate change impact assessment.

The majority of published studies for the Elbe basin deal with climate change impacts on the hydrological cycle. The simulated effects of climate change on water cycle and river discharge presented in literature are diverse and differ in intensity and even in direction of change, resulting mainly from the diversity of precipitation change signals projected by different climate scenarios. Studies using 100 realizations of the Statistical Analogue Resampling Scheme (STARS), with a distinct decrease in summer precipitation and a moderate increase of precipitation in winter, project lower river discharge in the Elbe basin [85–88]. However, we have to note that recently the STARS model was critically discussed [89,90] regarding its suitability for producing climate scenarios. Model runs using the Inter-Sectoral Impact Model Intercomparison Project (ISI-MIP) climate scenarios [91] for the Elbe basin also project a decreased discharge on average, but the magnitude of changes is less pronounced [85]. Huang *et al.* [92] report diverse results depending on the driving climate model: the projections driven by the empirical-statistical model WETTREG (WETTerlagen-basierte REGionalisierungsmethode) produce negative trends in flood occurrence, whereas the projections forced by the two dynamical regional climate models REMO (REgional MOdelling) and CCLM (COSMO-Climate Limited-area Modelling) show various results with the prevalence of increasing trends in flood occurrence for this region. Applications of the ENSEMBLES scenarios for assessing future risks of floods and droughts in Germany showed an increasing trend of floods but no significant increase in droughts for the Elbe basin [22]. This is also reflected in our study, where under the same ENSEMBLES climate scenarios' higher discharges are projected on average (compare Figure 7).

There are some studies on the management change impacts on river water quality for often only small parts of the Elbe region (e.g., [93–95]), but only a few publications exist covering water quality issues under climate change. So, Quiel *et al.* [62] used the outputs of a model chain driven by selected realizations of the statistical model STARS as boundary conditions to run the river model QSim for a 700 km reach of the Elbe river. Soluble phosphorus concentrations decrease in all tributaries under all scenarios compared with the reference period for the same scenario. This results in an increased phytoplankton growth along the studied river reach and a shift of the chlorophyll *a* maximum under the dry and medium scenarios, but in a decrease in chlorophyll *a* concentration under the wet scenario conditions [62]. The latter is in accordance with the results for chlorophyll *a* presented in our study (compare Figure 7), as the ENSEMBLES climate projections produce a wetter climate in the future as well.

It seems that the projected nutrient loads presented in literature often correspond to the precipitation change signals, especially when eco-hydrological models using a simple routing of nutrients through the river network are applied. The increased precipitation causes higher nitrogen leaching through soils as well as higher phosphorus erosion rates with surface flow to the river network. Both processes increase nutrient loading to the river waters. Therefore, statistically downscaled climate scenarios with a negative trend in precipitation (e.g., STARS) mostly project decreasing nutrient loads in the Elbe catchment, whereas dynamic climate scenarios (e.g., REMO or the wet years of ISI-MIP scenarios) mostly result in increasing nutrient loads in the river due to positive precipitation change signals (e.g., [96,97]). This simple relationship between precipitation change signals and final nutrient load projections can be disturbed by including in-stream and algal processes in the eco-hydrological models, due to included transformation and ingestion of nutrients in the river network.

Nevertheless, the same conclusion as for river discharge is valid for the water quality impact assessment: a wide range in projections can be found in literature (as well as in our study). However, the diversity in discharge and water quality projections may not necessarily be the result of the application of different model approaches or climate scenario sets. Even with one scenario and one model, a high spatial variability can be observed, and some sub-regional trends can actually be opposite to the overall average trend of a large-scale basin, or local effects can be masked by large-scale aggregation [78,79]. This could also be seen in our study, where changes of model outputs due to climate impact differ in magnitude and intensity or even in the direction of change when comparing several tributaries of the Elbe (Figure 7).

In climate and socio-economic impact assessments in addition to the general (and often large) uncertainty associated with climate scenarios as drivers, there is also uncertainty connected to applied watershed models. The so-called structural and parameterization uncertainty is related to the ability of eco-hydrological models to represent the interrelated processes in landscape, vegetation and river network. The parameterization uncertainty can be especially large in a model with a high number of calibration parameters influencing each other, as in the SWIM model with implemented in-stream processes. Often several calibration parameter combinations exist, delivering the same or very similar model performance, so that it could happen to be "right for the wrong reasons" [98,99]. In general, such uncertainty rises with the rising model complexity, and goes along with a rising need in calibration efforts [68], and this should be taken into account when the model is extended by adding new processes. To overcome the limitations and weaknesses of a single eco-hydrological or climate model approach, it is useful to apply several models with the same input parameter sets (model intercomparison) and ensembles of climate scenarios for a more comprehensive assessment of uncertainties and elicitation of robust outputs [91,100].

The land use change experiments applied in this study do not represent the "full" set of potential future land use scenarios in the Elbe region, which could be elaborated considering options of future socio-economic development. For example, changes in urbanization or forest patterns could also have effects on the environment and the water resources [101,102]. In our study only the effects of single measures connected to nutrient sources and agricultural practices (which are currently considered in the planning of land/water management) on water quantity and quality were tested, also in combination with climate change. This could be regarded as a first step to finding suitable methods for adaptation to climate change impacts. However, for further studies it is recommended to apply a combination of different measures under consideration of the future socio-economic development for a more realistic land use change impact assessment. As climate change can strengthen, revoke or even inverse the land use change impacts, this aspect should always be included in such studies.

6. Summary and Conclusions

The SWIM model supplemented by an in-stream module was successfully calibrated and validated for the entire Elbe river basin, and applied for climate and land use change impact assessment in the region. For that, the commonly used technique was applied, using 19 climate scenario data sets provided by the ENSEMBLES project to drive an eco-hydrological model for 30-year periods in order to evaluate changes in water quantity and quality for the two future periods of 2021–2050 and 2071–2098 in comparison to the reference period of 1971–2000.

The calibration and validation of the extended SWIM for the Elbe region was complicated due to the high number of calibration parameters and the spatial variability within the catchment. Satisfactory model results could be still achieved by applying spatially distributed calibration parameter sets to capture variability in soil type distribution, land use pattern and economic development in the sub-catchments.

The analysis of a potential future climate, as projected by 19 scenarios for the Elbe catchment, indicates increasing trends in temperature and precipitation, but a decreasing trend in solar radiation on average. However, looking at the climate change signals of the 19 scenarios separately, differences can be seen in the intensity and—for precipitation and solar radiation—also in the direction of change signals. The standard deviation of the whole set of climate change signals increases in the second future period.

The results of the climate change impact assessment on water quantity and quality show a high spatial variability within the catchment according to the individual characteristics of the tributaries within the basin. For the entire Elbe catchment, river discharge is projected to increase by 11% and 20% on average for the two future periods. Dissolved oxygen concentration is projected to decrease by 2% and 5%, mainly due to the increased water temperature. The projected changes in nutrient loads do not show the same change direction. While NO_3-N loads slightly decrease on average (−1% and −5%), and NH_4-N shows a distinct decreasing trend (−11% and −24%), PO_4-P loads are expected to increase by 6% and 5% on average. The simulated reaction of nutrient loads to climate change is always influenced by the phytoplankton population, and *vice versa*. The chlorophyll *a* concentrations decrease slightly under the future conditions, by 3% and 4% on average, at the last downstream Elbe gauge.

Five simulation experiments dealing with possible changes in nutrient emissions were applied in the study, also in combination with climate change scenarios. Water discharge was mainly influenced by climate change impacts, and land use change measures had little or no influence on runoff. A reduction of agricultural area or fertilizer application mainly influenced the resulting nitrate nitrogen loads in the Elbe, whereas the reduction of point source emissions had the highest impacts on ammonium nitrogen and phosphate phosphorus loads. The chlorophyll *a* concentrations reacted to a changed food supply in the river, and would be reduced with a reduced nitrogen and phosphorus availability. An increase in nutrient retention and decomposition potential within the catchments would certainly be beneficial to reduce all types of nutrient loads in the river waters.

Nevertheless, the model application in the Elbe basin comes along with a certain degree of structural, parameterization and scenario uncertainty. Due to the lack of more detailed information on the case-specific observations and processes, not all possible methods to reduce uncertainties could be applied in this study, and the climate scenario–related uncertainty is unavoidable. The climate change impact assessment and land use change simulation experiments presented here deliver the first results and rough estimation on probable future developments in the Elbe river basin under climate change. For future research, in order to diminish and better assess (but not to eliminate) uncertainty, it could be recommended to apply two to three eco-hydrological models, as well as a "full" set of socio-economic scenarios, for a more reliable combined climate and land use change impact assessment. It could be also advantageous to additionally include management measures neglected in this model application so far (e.g., reservoirs or different crop types and rotations). These methods would help to identify future risks and threats more realistically, and to virtually test possible adaptation measures, as efforts to cope with the future climate conditions and their impacts are generally needed. Watershed models offer a suitable tool to guide decision-making on water quantity and quality for a sustainable management of water resources to match the requirements of the European Water Framework Directive (WFD).

Acknowledgments: The ENSEMBLES data used in the study were produced in the EU FP6 Integrated Project ENSEMBLES (contract No. 505539). The authors thank Shaochun Huang for delivering the prepared spatial SWIM input data of the entire Elbe river basin. The authors additionally thank Vanessa Wörner for download and preparation of observed water quality data, and Lasse Scheele for climate scenario data download, preparation and evaluation.

Author Contributions: The two authors planned and designed the methods to study climate and management change impacts on water quantity and quality in the Elbe river catchment. Cornelia Hesse set up, adjusted and calibrated the extended SWIM model with included in-stream processes for the Elbe river catchment, ran the climate and land use change scenarios and analyzed the temporal and spatial scenario output data. Cornelia Hesse also prepared the tables and figures for the publication, wrote the text and formatted the paper. Valentina Krysanova guided and supervised the whole process, discussed results during the modeling study, and edited the manuscript.

References

1. Intergovernmental Panel on Climate Change (IPCC). *Climate Change 2007: The Physical Change Basis. Contribution of Working Group I to the Fourth Assessment Report of the Intergovernmental Panel on Climate Change*; Solomon, S., Qin, D., Manning, M., Chen, Z., Marquis, M., Averyt, K.B., Tignor, M., Miller, H.L., Eds.; Cambridge University Press: Cambridge, UK; New York, NY, USA, 2007; p. 996.
2. Intergovernmental Panel on Climate Change (IPCC). Summary for Policymakers. In *Climate Change 2013: The Physical Science Basis. Contribution of Working Group I to the Fifth Assessment Report of the Intergovernmental Panel on Climate Change*; Stocker, T.F., Qin, D., Plattner, G.-K., Tignor, M., Allen, S.K., Boschung, J., Nauels, A., Xia, Y., Bex, V., Midgley, P.M., Eds.; Cambridge University Press: Cambridge, UK, 2013; pp. 3–29.
3. European Environment Agency (EEA). *The European Environment. State and Outlook 2010. Land Use*; EEA: Copenhagen, Denmark, 2010.
4. Cassardo, C.; Jones, J.A.A. Managing water in a Changing World. *Water* **2011**, *3*, 618–628. [CrossRef]
5. Research Institute for Knowledge Systems (RIKS). *Exploration of Land Use Trends under SOER 2010*; RIKS: Maastricht, The Netherlands, 2010.
6. Alcamo, J.; Flörke, M.; Märker, M. Future long-term changes in global water resources driven by socio-economic and climatic changes. *Hydrolog. Sci. J.* **2007**, *52*, 247–275. [CrossRef]
7. Christensen, O.B.; Christensen, J.H. Intensification of extreme European summer precipitation in a warmer climate. *Glob. Planet. Chang.* **2004**, *44*, 107–117. [CrossRef]
8. Kløve, B.; Ala-Aho, P.; Bertrand, G.; Gurdak, J.J.; Kupfersberger, H.; Kværner, J.; Muotka, T.; Mykrä, H.; Preda, E.; Rossi, P.; *et al.* Climate change impacts on groundwater and dependent ecosystems. *J. Hydrol.* **2014**, *518*, 250–266.
9. Kundzewicz, Z.W. Climate change impacts on the hydrological cycle. *Ecohydrol. Hydrobiol.* **2008**, *8*, 195–203. [CrossRef]
10. Wang, X.; Siegert, F.; Zhou, A.; Franke, J. Glacier and glacial lake changes and their relationship in the context of climate change, Central Tibetan Plateau 1972–2010. *Glob. Planet. Chang.* **2013**, *111*, 246–257. [CrossRef]
11. Linderholm, H.W. Growing season changes in the last century. *Agric. For. Meteorol.* **2006**, *137*, 1–14. [CrossRef]
12. Vittoz, P.; Cherix, D.; Gonseth, Y.; Lubini, V.; Maggini, R.; Zbinden, N.; Zumbach, S. Climate change impacts on biodiversity in Switzerland: A review. *J. Nat. Conserv.* **2013**, *21*, 154–162. [CrossRef]
13. Kittel, T.G.F. The Vulnerability of Biodiversity to Rapid Climate Change. In *Climate Vulnerability: Understanding and Addressing Threats to Essential Resources*; Pielke, R.A., Ed.; Elsevier Inc., Academic Press: Oxford, UK, 2013; Volumn 4, Chapter 4; pp. 185–201.
14. Lindner, M.; Fitzgerald, J.B.; Zimmermann, N.E.; Reyer, C.; Delzon, S.; van der Maaten, E.; Schelhaas, M.-J.; Lasch, P.; Eggers, J.; van der Maaten-Theunissen, M.; *et al.* Climate change and European forests: What do we know, what are the uncertainties, and what are the implications for forest management? *J. Environ. Manag.* **2014**, *146*, 69–83. [CrossRef] [PubMed]
15. Gabriel, K.M.A.; Endlicher, W.R. Urban and rural mortality rates during heat waves in Berlin and Brandenburg, Germany. *Environ. Pollut.* **2011**, *159*, 2044–2050. [CrossRef] [PubMed]
16. Haines, A.; Kovats, R.S.; Campbell-Lendrum, D.; Corvalan, C. Climate change and human health: Impacts, vulnerability and public health. *Public Health* **2006**, *120*, 585–596. [CrossRef] [PubMed]
17. Koch, H.; Vögele, S.; Hattermann, F.F.; Huang, S. Sensitivity of electricity generation in Germany to climate change and variability. *Meteorol. Z.* **2015**. in print.
18. Liersch, S.; Cools, J.; Kone, B.; Koch, H.; Diallo, M.; Reinhardt, J.; Fournet, S.; Aich, V.; Hattermann, F.F. Vulnerability of rice production in the Inner Niger Delta to water resources management under climate variability and change. *Environ. Sci. Policy* **2013**, *34*, 18–33. [CrossRef]
19. Alcamo, J.; Moreno, J.M.; Nováky, B.; Bindi, M.; Corobov, R.; Devoy, R.J.N.; Giannakopoulos, C.; Martin, E.; Olesen, J.E.; Shvidenko, A. Europe. In *Climate Change 2007: Impacts, Adaptation and Vulnerability; Contribution of Working Group II to the Fourth Assessment Report of the Intergovernmental Panel on Climate Change*; Parry, M.L., Canziani, O.F., Palutikof, J.P., van der Linden, P.J., Hanson, C.E., Eds.; Cambridge University Press: Cambridge, UK, 2007; pp. 541–580.
20. European Environment Agency (EEA). *Climate Change, Impacts and Vulnerability in Europe 2012*; EEA-Report No 12/2012; EEA: Copenhagen, Denmark, 2012; p. 300.

21. Federal Environmental Agency (UBA). *Monitoringbericht 2015 zur Deutschen Anpassungstrategie an den Klimawandel—Bericht der Interministeriellen Arbeitsgruppe Anpassungsstrategie der Bundesregierung*; UBA: Dessau-Roßlau, Germany, 2015.
22. Huang, S.; Krysanova, V.; Hattermann, F.F. Projections of climate change impacts on floods and droughts in Germany using an ensemble of climate change scenarios. *Reg. Environ. Chang.* **2015**, *15*, 461–473. [CrossRef]
23. Kyselý, J.; Beranová, R. Climate-change effects on extreme precipitation in central Europe: Uncertainties of scenarios based on regional climate models. *Theor. Appl. Climatol.* **2009**, *95*. [CrossRef]
24. Whitehead, P.G.; Wilby, R.L.; Battarbee, R.W.; Kernan, M.; Wade, A.J. A review of the potential impacts of climate change on surface water quality. *Hydrolog. Sci. J.* **2009**, *54*, 101–123. [CrossRef]
25. Kundzewicz, Z.W.; Mata, L.J.; Arnell, N.W.; Döll, P.; Kabat, P.; Jiménez, B.; Miller, K.A.; Oki, T.; Sen, Z.; Shiklomanov, I.A. Freshwater resources and their management. In *Climate Change 2007: Impacts, Adaptation and Vulnerability. Contribution of Working Group II to the Fourth Assessment Report of the Intergovernmental Panel on Climate Change*; Parry, M.L., Canziani, O.F., Palutikof, J.P., van der Linden, P.J., Hanson, C.E., Eds.; Cambridge University Press: Cambridge, UK, 2007; pp. 173–210.
26. Crossmann, J.; Futter, M.N.; Oni, S.K.; Whitehead, P.G.; Jin, L.; Butterfield, D.; Baulch, H.M.; Dillon, P.J. Impacts of climate change on hydrology and water quality: Future proofing management strategies in the Lake Simcoe watershed, Canada. *J. Great Lakes Res.* **2013**, *39*, 19–32. [CrossRef]
27. Dunn, S.M.; Brown, I.; Sample, J.; Post, H. Relationship between climate, water resources, land use and diffuse pollution and the significance of uncertainty in climate change. *J. Hydrol.* **2012**, *434–435*. [CrossRef]
28. Mimikou, M.A.; Baltas, E.; Varanou, E.; Pantazis, K. Regional impacts of climate change on water resources quantity and quality indicators. *J. Hydrol.* **2000**, *234*, 95–109. [CrossRef]
29. Desortová, B.; Punčochář, P. Variability of phytoplankton biomass in a lowland river: Response to climate conditions. *Limnol. Ecol. Manag. Inland Waters* **2011**, *41*, 160–166. [CrossRef]
30. Hardenbicker, P.; Rolinski, S.; Weitere, M.; Fischer, H. Contrasting long-term trends and shifts in phytoplankton dynamics in two large rivers. *Int. Rev. Hydrobiol.* **2014**, *99*. [CrossRef]
31. Scharfe, M.; Callies, U.; Blöcker, G.; Petersen, W.; Schroeder, F. A simple Lagrangian model to simulate temporal variability of algae in the Elbe River. *Ecol. Model.* **2009**, *220*, 2173–2189. [CrossRef]
32. Barclay, J.R.; Walter, M.T. Modeling denitrification in a changing climate. *Sustain. Water Qual. Ecol.* **2015**, *5*, 64–76. [CrossRef]
33. Whitehead, P.G.; Wilby, R.L.; Butterfield, D.; Wade, A.J. Impacts of climate change on in-stream nitrogen in a lowland chalk stream: An appraisal of adaptation strategies. *Sci. Total Environ.* **2006**, *365*, 260–273. [CrossRef] [PubMed]
34. Macleod, C.J.A.; Falloon, P.D.; Evans, R.; Haygarth, P.M. The effects of climate change on the mobilization of diffuse substances from agricultural systems. *Adv. Agron.* **2012**, *115*, 41–77.
35. Vollenweider, R. *Scientific Fundamentals of the Eutrophication of Lakes and Flowing Waters, with Particular Reference to Nitrogen and Phosphorous as Factors in Eutrophication*; OECD Tech Rep. DAS/CSI/68.27; Organisation for Economic Co-operation and Development (OECD): Paris, France, 1968; p. 159.
36. Anderson, D.M.; Glibert, P.M.; Burkholder, J.M. Harmful algal blooms and eutrophication: Nutrient sources, composition, and consequences. *Estuaries* **2002**, *25*, 704–726. [CrossRef]
37. Pieterse, N.M.; Bleuten, W.; Jørgensen, S.E. Contribution of point sources and diffuse sources to nitrogen and phosphorus loads in lowland river tributaries. *J. Hydrol.* **2003**, *271*, 213–225. [CrossRef]
38. Schindler, D.W. Recent advances in the understanding and management of eutrophication. *Limnol. Oceanogr.* **2006**, *51*, 356–363. [CrossRef]
39. European Environment Agency (EEA). *Land Use Scenarios for Europe: Qualitative and Quantitative Analysis on a European Scale (PRELUDE)*; EEA Technical Report No 9/2007; EEA: Copenhagen, Denmark, 2007.
40. Seitzinger, S.P.; Mayorga, E.; Bouwman, A.F.; Kroeze, C.; Beusen, A.H.W.; Billen, G.; van Drecht, G.; Dumon, E.; Fekete, B.M.; Garnier, J.; *et al.* Global river nutrient export: A scenario analysis of past and future trends. *Glob. Biogeochem. Cy.* **2010**, *24*. [CrossRef]
41. Bouraoui, F.; Grizzetti, B. Long term change of nutrient concentrations of rivers discharging in European seas. *Sci. Total Environ.* **2011**, *409*, 4899–4916. [CrossRef] [PubMed]
42. De Wit, M.; Behrendt, H.; Bendoricchio, G.; Bleuten, W.; van Gaans, P. *The Contribution of Agriculture to Nutrient Pollution in Three European Rivers, with Reference to the European Nitrates Directive*; European Water Management Online, Official Publication of the European Water Association (EWA): Hennef, Germany, 2002.

43. Grizzetti, B.; Bouraoui, F.; Aloe, A. Changes of nitrogen and phosphorus loads to European seas. *Glob. Chang. Biol.* **2012**, *18*, 769–782. [CrossRef]
44. Mehdi, B.; Ludwig, R.; Lehner, B. Evaluating the impacts of climate change and crop land use change on streamflow, nitrates and phosphorus: A modelling study in Bavaria. *J. Hydrol. Reg. Stud.* **2015**, *4*. [CrossRef]
45. Huttunen, I.; Lehtonen, H.; Huttunen, M.; Piirainen, V.; Korppoo, M.; Veijalainen, N.; Viitasalo, M.; Vehviläinen, B. Effects of climate change and agricultural adaptation on nutrient loading from Finnish catchments to the Baltic Sea. *Sci. Total Environ.* **2015**, *529*, 168–181. [CrossRef] [PubMed]
46. Horn, A.L.; Rueda, F.J.; Hörmann, G.; Fohrer, N. Implementing river water quality modelling issues in mesoscale watershed models for water policy demands—An overview on current concepts, deficits, and future tasks. *Phys. Chem. Earth* **2004**, *29*, 725–737. [CrossRef]
47. Daniel, E.B.; Camp, J.V.; LeBoeuf, E.J.; Penrod, J.R.; Dobbins, J.P.; Abkowitz, M.D. Watershed Modeling and its Applications: A State-of-the-Art Review. *Open Hydrol. J.* **2011**, *5*, 26–50. [CrossRef]
48. Krysanova, V.; Wechsung, F.; Arnold, J.; Srinivasan, R.; Williams, J. *SWIM (Soil and Water Integrated Model.): User Manual*; PIK Report No. 69; Potsdam Institute for Climate Impact Research (PIK): Potsdam, Germany, 2000; p. 239.
49. Hesse, C.; Krysanova, V.; Päzolt, J.; Hattermann, F.F. Eco-hydrological modelling in a highly regulated lowland catchment to find measures for improving water quality. *Ecol. Model.* **2008**, *218*, 135–148. [CrossRef]
50. Hesse, C.; Krysanova, V.; Voß, A. Implementing In-Stream Nutrient Processes in Large-Scale Landscape Modeling for the Impact Assessment on Water Quality. *Environ. Model. Assess.* **2012**, *17*, 589–611. [CrossRef]
51. Van der Linden, P., Mitchell, J.F., Eds.; *ENSEMBLES: Climate Change and Its Impacts: Summary of Research and Results from the ENSEMBLES Project*; Met Office Hadley Centre: Exeter, UK, 2009; p. 160.
52. International Commission for the Protection of the Elbe River (IKSE). *Die Elbe und ihr Einzugsgebiet—Ein Geographisch-Hydrologischer und Wasserwirtschaftlicher Überblick*; Schlüter GmbH & Co. KG: Schönebeck (Elbe), Germany, 2005; p. 258.
53. Bronstert, A.; Itzerott, S. Bewirtschaftungsmöglichkeiten im Einzugsgebiet der Havel: Abschlussbericht zum BMBF-Forschungsprojekt. 2006, p. 212. Available online: http://www.havelmanagement.net/Havel-ger/Publikationen/Endberichte/Endbericht_Verbund.pdf (accessed on 28 November 2014).
54. German Working Group on water issues (LAWA). *Beurteilung der Wasserbeschaffenheit von Fließgewässern in der Bundesrepublik Deutschland—Chemische Gewässergüteklassifikation*; Kulturbuchverlag: Berlin, Germany, 1998; pp. 1–35.
55. Koskova, R.; Nemecková, S.; Hesse, C. Using of the soil parametrization based on soil samples databases in rainfall-runoff modelling. In Proceedings of the Adolf Patera Workshop "Extreme Hydrological Events in Catchments"; Jakubíková, A., Broza, V., Szolgay, J., Eds.; Czech Technical University: Prague, Czech Republic, 2007; pp. 241–249. (In Czech).
56. River Basin Community Elbe (FGG-Elbe). *Zusammenfassender Bericht der Flussgebietsgemeinschaft Elbe über die Analysen nach Artikel 5 der Richtlinie 2000/60/EG (A-Bericht)*; FGG-Elbe: Magdeburg, Germany, 2004.
57. International Commission for the Protection of the Elbe River (IKSE). *Bestandsaufnahme von bedeutenden punktuellen kommunalen und industriellen Einleitungen von prioritären Stoffen im Einzugsgebiet der Elbe Appendix 3: Bestandsaufnahme der kommunalen Abwassereinleitungen größer 20000 EGW im Einzugsgebiet der Elbe in der Tschechischen Republik (Stand 1995)*; IKSE: Magdeburg, Germany, 1995.
58. Hussian, M.; Grimvall, A.; Petersen, W. Estimation of the human impact on nutrient loads carried by the Elbe river. *Environ. Monit. Assess.* **2004**, *96*, 15–33. [CrossRef] [PubMed]
59. Klöcking, B.; Haberlandt, U. Impact of land use changes on water dynamics—A case study in temperate meso- and macroscale river basins. *Phys. Chem. Earth* **2002**, *27*, 619–629. [CrossRef]
60. Lehmann, A.; Rode, M. Long-term behaviour and cross-correlation water quality analysis of the river Elbe, Germany. *Water Res.* **2001**, *35*, 2153–2160. [CrossRef]
61. Schneider, P.; Reincke, H. Contaminated Sediments in the Elbe Basin and its Tributary Mulde. In *Uran. Environ.*; Springer: Berlin, Germany, 2006; pp. 655–662.
62. Quiel, K.; Becker, A.; Kirchesch, V.; Schöl, A.; Fischer, H. Influence of global change on Phytoplankton and nutrient cycling in the Elbe river. *Reg. Environ. Chang.* **2011**, *11*, 405–421. [CrossRef]
63. Grossmann, M. Economic value of the nutrient retention function of restored floodplain wetlands in the Elbe River basin. *Ecol. Econ.* **2012**, *83*, 108–117. [CrossRef]

64. Arnold, J.; Allan, P.; Bernhardt, G. A comprehensive surface-groundwater flow model. *J. Hydrol.* **1993**, *142*, 47–69. [CrossRef]
65. Krysanova, V.; Meiner, A.; Roosaare, J.; Vasilyev, A. Simulation modelling of the coastal waters pollution from agricultural watershed. *Ecol. Model.* **1989**, *49*, 7–29. [CrossRef]
66. Neitsch, S.L.; Arnold, J.G.; Kiniry, J.R.; Srinivasan, R.; Williams, J.R. *Soil and Water Assessment Tool User's Manual: Version 2000*; TWRI Report TR-192; Texas Water Resources Institute (TWRI): Temple, TX, USA, 2002; p. 412.
67. Hattermann, F.F.; Krysanova, V.; Habeck, A.; Bronstert, A. Integrating wetlands and riparian zones in river basin modelling. *Ecol. Model.* **2006**, *199*, 379–392. [CrossRef]
68. Hesse, C.; Krysanova, V.; Vetter, T.; Reinhardt, J. Comparison of several approaches representing terrestrial and in-stream nutrient retention and decomposition in watershed modelling. *Ecol. Model.* **2013**, *269*. [CrossRef]
69. Thuringian Regional Office for Agriculture (TLL). *Leitlinie zur effizienten und umwelt-verträglichen Erzeugung von Winterweizen*; 2011; Available online: http://www.tll.de/ainfo/archiv/weiz0508.pdf (accessed on 12 December 2013).
70. Hattermann, F.F.; Krysanova, V.; Hesse, C. Modelling wetland processes in regional applications. *Hydrol. Sci. J.* **2008**, *53*, 1001–1012. [CrossRef]
71. Internatiopnal Kommission zum Schutz der Elbe (IKSE). *Informationsdokumente zum Internationalen Messprogramm Elbe 2007, Übersicht der Messstationen und Messstellen*, 2007. Available online: http://www.ikse-mkol.org/index.php?id=212 (accessed on 21 November 2012).
72. Hattermann, F.F.; Wattenbach, M.; Krysanova, V.; Wechsung, F. Runoff simulations on the macroscale with the ecohydrological model SWIM in the Elbe catchment—validation and uncertainty analysis. *Hydrol. Proc.* **2005**, *19*. [CrossRef]
73. Huang, S.; Hesse, C.; Krysanova, V.; Hattermann, F. From meso- to macro-scale dynamic water quality modelling for the assessment of land use change scenarios. *Ecol. Model.* **2009**, *220*, 2543–2558. [CrossRef]
74. Nash, J.E.; Sutcliffe, J.V. River flow forecasting through conceptual models part I: A discussion of principles. *J. Hydrol.* **1970**, *10*, 282–290. [CrossRef]
75. Gudmundsson, L.; Wagener, T.; Tallaksen, L.M.; Engeland, K. Evaluation of nine large-scale hydrological models with respect to the seasonal runoff climatology in Europe. *Water Resour. Res.* **2012**, *48*. [CrossRef]
76. Nakicenovic, N., Swart, R., Eds.; *Special Report on Emissions Scenarios. Report of Working Group III of the Intergovernmental Panel on Climate Change*; Cambridge University Press: Cambridge, UK, 2000; p. 570.
77. Gädeke, A.; Hölzel, H.; Koch, H.; Pohle, I.; Grünewald, U. Analysis of uncertainties in the hydrological response of a model-based climate change impact assessment in a subcatchment of the Spree River, Germany. *Hydrol. Process.* **2014**, *28*. [CrossRef]
78. Arheimer, B.; Dahné, J.; Donnelly, C. Climate change impact on riverine nutrient load and land-based remedial measures of the Baltic action plan. *Ambio* **2012**, *41*, 600–612. [CrossRef] [PubMed]
79. Piniewski, M.; Kardel, I.; Giełczewski, M.; Marcinkowski, P.; Okruszko, T. Climate change and agricultural development: adapting Polish agriculture to reduce future nutrient loads in a coastal watershed. *Ambio* **2014**, *43*. [CrossRef] [PubMed]
80. Tebaldi, C.; Knutti, R. The use of the multi-model ensemble in probabilistic climate projections. *Philos. Trans. Roy. Soc.* **2007**, *365*. [CrossRef] [PubMed]
81. Kling, H.; Fuchs, M.; Paulin, M. Runoff conditions in the upper Danube basin under an ensemble of climate change scenarios. *J. Hydrol.* **2012**, *424–425*, 264–277. [CrossRef]
82. Teutschbein, C.; Seibert, J. Regional Climate Models for Hydrological Impact Studies at the Catchment Scale: A Review of Recent Modeling Strategies. *Geogr. Compass.* **2010**, *4*, 834–860. [CrossRef]
83. Ehret, U.; Zehe, E.; Wulfmeyer, V.; Warrach-Sagi, K.; Liebert, J. HESS Opinions "Should we apply bias correction to global and regional climate model data?". *Hydrol. Earth Syst. Sci.* **2012**, *16*. [CrossRef]
84. Bundesministerium für Verbraucherschutz, Ernährung und Landwirtschaft (BMVEL). *Gute Fachliche Praxis zur Vorsorge gegen Bodenschadverdichtungen und Bodenerosion*; BMVEL: Bonn, Germany, 2002.
85. Roers, M.; Wechsung, F. Neubewertung der Auswirkungen des Klimawandels auf den Wasserhaushalt im Elbegebiet. *Hydrol. Wasserbewirts* **2015**, *59*, 109–119.
86. Huang, S.; Krysanova, V.; Österle, H.; Hattermann, F.F. Simulation of spatiotemporal dynamics of water fluxes in Germany under climate change. *Hydrol. Process.* **2010**, *24*. [CrossRef]

87. Hattermann, F.F.; Post, J.; Krysanova, V.; Conradt, T.; Wechsung, F. Assessment of water availability in a Central-European River Basin (Elbe) under climate change. *Adv. Clim. Chang. Res.* **2008**, *4*, 42–50.
88. Conradt, T.; Koch, H.; Hattermann, F.F.; Wechsung, F. Spatially differentiated management-revised discharge scenarios for an integrated analysis of multi-realisation climate and land use scenarios. *Reg. Environ. Chang.* **2012**, *12*. [CrossRef]
89. Wechsung, F.; Wechsung, M. Dryer years and brighter sky—The predictable simulation outcomes for Germany's warmer climate from the weather resampling model STARS. *Int. J. Climatol.* **2014**. [CrossRef]
90. Wechsung, F.; Wechsung, M. A methodological critique on using temperature-conditioned resampling for climate projections as in the paper of Gerstengarbe *et al.* (2013) winter storm- and summer thunderstorm-related loss events in Theoretical and Applied Climatology (TAC). *Theor. Appl. Climatol.* **2015**, *1–5*. [CrossRef]
91. Warszawski, L.; Frieler, K.; Huber, V.; Piontek, F.; Serdeczny, O.; Schewe, J. The Inter-Sectoral Impact Model Intercomparison Project (ISI-MIP). *Proc. Natl. Acad. Sci. USA* **2013**, *111*. [CrossRef]
92. Huang, S.; Hattermann, F.F.; Krysanova, V.; Bronstert, A. Projections of climate change impacts on river flood conditions in Germany by combining three different RCMs with a regional eco-hydrological model. *Clim. Chang.* **2013**, *116*. [CrossRef]
93. Kersebaum, K.C.; Steidl, J.; Bauer, O.; Piorr, H.-P. Modelling scenarios to assess the effects of different agricultural management and land use options to reduce diffuse nitrogen pollution into the river Elbe. *Phys. Chem. Earth* **2003**, *28*, 537–545. [CrossRef]
94. Meissner, R.; Seeger, J.; Rupp, H. Effects of agricultural land use changes on diffuse pollution of water resources. *Irrig. Drain.* **2002**, *51*. [CrossRef]
95. Ullrich, A.; Volk, M. Application of the Soil and Water Assessment Tool (SWAT) to predict the impact of alternative management practices on water quality and quantity. *Agric. Water Manag.* **2009**, *96*. [CrossRef]
96. Martinkova, M.; Hesse, C.; Krysanova, V.; Vetter, T.; Hanel, M. Potential impact of climate change on nitrate load from the Jizera catchment (Czech Republic). *Phys. Chem. Earth* **2011**, *36*, 673–683. [CrossRef]
97. Roers, M.; Venohr, M.; Wechsung, F.; Müller, E.N. Nährstoffeinträge im Elbegebiet—Modellstudien zu deren zukünftiger Änderung und zur Wirkung von Reduktionsmaßnahmen unter Klimawandel. *Hydrol. Wasserbewirts* Unpublished work. **2015**.
98. Dayton, P.K. Two Cases of Resource Partitioning in an Intertidal Community: Making the Right Prediction for the Wrong Reason. *Am. Nat.* **1973**, *107*, 662–670. [CrossRef]
99. Van der Laan, M.; Annandale, J.G.; Bristow, K.L.; Stirzaker, R.J.; du Preez, C.C.; Thorburn, P.J. Modelling nitrogen leaching: Are we getting the right answer for the right reason? *Agric. Water Manag.* **2014**, *133*. [CrossRef]
100. Schewe, J.; Heinke, J.; Gerten, D.; Haddeland, I.; Arnell, N.W.; Clark, D.B.; Dankers, R.; Eisner, S.; Fekete, B.; Colón-González, F.J.; *et al.* Multi-model assessment of water scarcity under climate change. *Proc. Natl. Acad. Sci. USA* **2013**, *111*. [CrossRef]
101. O'Driscoll, M.; Clinton, S.; Jefferson, A.; Manda, A.; McMillan, S. Urbanization Effects on Watershed Hydrology and In-Stream Processes in the Southern United States. *Water* **2010**, *2*. [CrossRef]
102. Wei, X.; Liu, W.; Zhou, P. Quantifying the Relative Contributions of Forest Change and Climatic Variability to Hydrology in Large Watersheds: A Critical Review of Research Methods. *Water* **2013**, *5*. [CrossRef]

PERMISSIONS

All chapters in this book were first published in MDPI; hereby published with permission under the Creative Commons Attribution License or equivalent. Every chapter published in this book has been scrutinized by our experts. Their significance has been extensively debated. The topics covered herein carry significant findings which will fuel the growth of the discipline. They may even be implemented as practical applications or may be referred to as a beginning point for another development.

The contributors of this book come from diverse backgrounds, making this book a truly international effort. This book will bring forth new frontiers with its revolutionizing research information and detailed analysis of the nascent developments around the world.

We would like to thank all the contributing authors for lending their expertise to make the book truly unique. They have played a crucial role in the development of this book. Without their invaluable contributions this book wouldn't have been possible. They have made vital efforts to compile up to date information on the varied aspects of this subject to make this book a valuable addition to the collection of many professionals and students.

This book was conceptualized with the vision of imparting up-to-date information and advanced data in this field. To ensure the same, a matchless editorial board was set up. Every individual on the board went through rigorous rounds of assessment to prove their worth. After which they invested a large part of their time researching and compiling the most relevant data for our readers.

The editorial board has been involved in producing this book since its inception. They have spent rigorous hours researching and exploring the diverse topics which have resulted in the successful publishing of this book. They have passed on their knowledge of decades through this book. To expedite this challenging task, the publisher supported the team at every step. A small team of assistant editors was also appointed to further simplify the editing procedure and attain best results for the readers.

Apart from the editorial board, the designing team has also invested a significant amount of their time in understanding the subject and creating the most relevant covers. They scrutinized every image to scout for the most suitable representation of the subject and create an appropriate cover for the book.

The publishing team has been an ardent support to the editorial, designing and production team. Their endless efforts to recruit the best for this project, has resulted in the accomplishment of this book. They are a veteran in the field of academics and their pool of knowledge is as vast as their experience in printing. Their expertise and guidance has proved useful at every step. Their uncompromising quality standards have made this book an exceptional effort. Their encouragement from time to time has been an inspiration for everyone.

The publisher and the editorial board hope that this book will prove to be a valuable piece of knowledge for researchers, students, practitioners and scholars across the globe.

LIST OF CONTRIBUTORS

Marjolein H. J. Van Huijgevoort and Kelly H. S. Nijhuis
KWR Water Research Institute, 3430BB Nieuwegein, The Netherlands

Bernard R. Voortman
KWR Water Research Institute, 3430BB Nieuwegein, The Netherlands
Moisture Matters, 3533ED Utrecht, The Netherlands

Sjoerd Rijpkema
Vitens, 6802CA Arnhem, The Netherlands

Jan-Philip M. Witte
KWR Water Research Institute, 3430BB Nieuwegein, The Netherlands
Flip Witte Ecohydrologie, 6862DC Oosterbeek, The Netherlands

Darius Jakimavičius, Gintaras Adžgauskas, Diana Šarauskienė and Jūratė Kriauči ūnienė
Laboratory of Hydrology, Lithuanian Energy Institute, Breslaujos st. 3, LT–444003 Kaunas, Lithuania

Rossana Escanilla-Minchel, Hernán Alcayaga and Marco Soto-Alvarez
Escuela de Ingeniería en Obras Civiles, Universidad Diego Portales, Santiago 8370109, Chile

Christophe Kinnard
Department of Environmental Sciences, University of Québec at Trois-Rivières, Trois-Rivières, QC G8Z 4M3, Canada

Roberto Urrutia
Facultad de Ciencias Ambientales/Centro EULA-Chile de la Universidad de Concepción, Concepción 409541, Chile

Rashid Mahmood and Shaofeng Jia
Institute of Geographic Science and Natural Resources Research/Key Laboratory of Water Cycle and Related Land Surface Processes, Chinese Academy of Sciences, Beijing 100101, China

Mukand S. Babel
Water Engineering and Management, Asian Institute of Technology, Pathumthani 12120, Thailand

Vivek Shandas and Rosa Lehman
Toulan School of Urban Studies and Planning, Portland State University, Portland, OR 97201-0751, USA

Kelli L. Larson
Schools of Geographical Sciences and Urban Planning and Sustainability, Arizona State University, AZ 85287-5302, USA

Jeremy Bunn
Herrera Inc., Seattle, WA 98121, USA

Heejun Chang
Department of Geography, Portland State University, Portland, OR 97201-0751, USA

Jungsool Park and Jaewon Kwak
Forecast and Control Division, Nakdong River Flood Control Office, Busan 49300, Korea

Soojun Kim
Columbia Water Center, Columbia University, New York, NY 10027, USA

Gilho Kim
Department of Hydro Science and Engineering, Korea Institute of Civil Engineering and Building Technology, Goyang-si, Gyeonggi-do 10223, Korea

Vijay P. Singh
Department of Biological and Agricultural Engineering and Zachry Department of Civil Environmental Engineering, Texas A & M University, College Station, TX 77843-2117, USA

Hung Soo Kim
Department of Civil Engineering, Inha University, Incheon 22212, Korea

Yingkui Li
Department of Geography, University of Tennessee, Knoxville, TN 37996, USA

Michael A. Urban
Department of Geography, University of Missouri, Columbia, MO 65211, USA

Hsiao-Ping Wei
National Science and Technology Center for Disaster Reduction, New Taipei City 23143, Taiwan

Keh-Chia Yeh
Department of Civil Engineering in National Chiao Tung University, Hsinchu 300, Taiwan

Jun-Jih Liou, Yung-Ming Chen and Chao-Tzuen Cheng
National Science and Technology Center for Disaster Reduction, New Taipei City 23143, Taiwan

Ilias G. Pechlivanidis, Jonas Olsson and Thomas Bosshard
Swedish Meteorological and Hydrological Institute, Norrköping 60176, Sweden

Devesh Sharma and K.C. Sharma
Department of Environmental Science, Central University of Rajasthan, Kishangarh, Dist-Ajmer, Rajasthan 305817, India

Willibroad Gabila Buma, Sang-Il Lee and Jae Young Seo
Department of Civil and Environmental Engineering, Dongguk University, Seoul 04620, Korea

Wen-Ting Chan
Department of Civil Disaster Prevention Engineering, National United University, Miaoli 36063, Taiwan

Wen-Cheng Liu
Department of Civil Disaster Prevention Engineering, National United University, Miaoli 36063, Taiwan
Taiwan Typhoon and Flood Research Institute, National Applied Research Laboratories, Taipei 10093, Taiwan

Hejia Wang
Department of Hydraulic Engineering, Tsinghua University, Haidian District, Beijing 100084, China
State Key Laboratory of Simulation and Regulation of Water Cycle in River Basin, China Institute of Water Resources and Hydropower Research, 1-A Fuxing Road, Haidian District, Beijing 100038, China

Weihua Xiao, Jianhua Wang, Yicheng Wang, Baodeng Hou and Chuiyu Lu
State Key Laboratory of Simulation and Regulation of Water Cycle in River Basin, China Institute of Water Resources and Hydropower Research, 1-A Fuxing Road, Haidian District, Beijing 100038, China

Ya Huang
State Key Laboratory of Simulation and Regulation of Water Cycle in River Basin, China Institute of Water Resources and Hydropower Research, 1-A Fuxing Road, Haidian District, Beijing 100038, China
College of Civil Engineering and Architecture, Guangxi University, Nanning 530004, China

Cornelia Hesse and Valentina Krysanova
Potsdam-Institute for Climate Impact Research, Potsdam 14412, Germany

Index

www.ingramcontent.com/pod-product-compliance
Lightning Source LLC
LaVergne TN
LVHW060527210926
848485LV00005B/1161
* 9 7 8 1 6 4 7 4 0 4 3 7 6 *